Reverend William Henry Hall

THE GENIUS OF
JOHN HENRY NEWMAN

THE GENIUS OF
John Henry Newman

Selections from his Writings

Edited with an introduction by
IAN KER

CLARENDON PRESS · OXFORD
1989

Oxford University Press, Walton Street, Oxford OX2 6DP
Oxford New York Toronto
Delhi Bombay Calcutta Madras Karachi
Petaling Jaya Singapore Hong Kong Tokyo
Nairobi Dar es Salaam Cape Town
Melbourne Auckland
and associated companies in
Berlin Ibadan

Oxford is a trade mark of Oxford University Press

Published in the United States
by Oxford University Press, New York

British Library Cataloguing in Publication Data
Newman, John Henry, 1801–1890
The genius of John Henry Newman.
1. Catholic Church. Christian Doctrine. Theories of
Newman, John Henry, 1801–1890
I. Title   II. Ker, Ian
230'.2'0924
ISBN 0–19–826682–0

Library of Congress Cataloging in Publication Data
Newman, John Henry, 1801–1890
The genius of John Henry Newman: selections from his writings/
edited with an introduction by Ian Ker.
Bibliography: p.   Includes index.
1. Theology.   2. Education, Higher.   I. Ker, Ian.   II. Title.
BX890.N43   1989   230'.2—dc20   89–15946
ISBN 0–19–826682–0

Typeset by Cotswold Typesetting Limited, Gloucester

Printed in Great Britain by
Courier International Ltd, Tiptree, Essex

For
James Reidy

ACKNOWLEDGEMENTS

The idea of a new selection of Newman's writings came to me in the summer of 1984 during a Newman conference at University College, Dublin. I discussed it then with Father James Reidy, whose enthusiasm and interest encouraged me to proceed further with the project, the completion of which owes much to his advice and practical help. I am also grateful to Dr Don Briel for his helpful comments.

I wish to thank Mrs Virginia Lyons for her careful help in preparing this volume for publication, which has gone beyond the bounds of normal secretarial services. I am grateful to the College of St Thomas for the generous facilities I have enjoyed as the holder of the Endowed Chair in Theology and Philosophy. I am particularly indebted to my student assistant, K. L. Dvorak, Jun., and also to L. K. Wilcox for their help with reading the proofs.

Selections from *The Letters and Diaries of John Henry Newman* are published by kind permission of the Birmingham Oratory. Thanks are due to the National Portrait Gallery, London, for permitting, and to Jane Stuart-Smith of Oxford University Press for arranging, the reproduction of the George Richmond drawing on the cover.

I.T.K.

St Paul, Minnesota, USA
June 1989

CONTENTS

INTRODUCTION

In spite of John Henry Newman's lifelong, single-minded dedication to the cause of revealed religion, there is nothing narrow or restricted about his achievement as a writer. Although his genius was devoted so exclusively to one object and to one purpose, it was by no means confined to one kind of expression. A list of his half dozen or so most famous writings would certainly include: *Apologia pro Vita Sua, An Essay on the Development of Christian Doctrine, An Essay in Aid of a Grammar of Assent, The Idea of a University*, and *Parochial and Plain Sermons*. These five works are all established classics in their respective fields of literature, theology, philosophy of religion, education, and preaching or spirituality. However, while it is true that the *Apologia, The Idea of a University*, and his *Sermons* have never lacked admirers since his death, the appreciation of Newman's other books has greatly increased during the last few decades, and for two quite separate reasons.

First, the past thirty or so years have witnessed an enormous growth of interest in the Victorians, from which Newman's reputation has obviously benefited. His importance as one of the leading 'sages' of the period, to be ranked alongside Matthew Arnold, Carlyle, George Eliot, Mill, and Ruskin, is now taken for granted. Moreover, it has become evident even from the most secular and unfriendly point of view that Newman's writings exerted too pervasive an influence on Victorian culture and thought to be discounted as simply reflecting an outmoded religious outlook. But while there has never been any lack of admiration of his eloquent prose style and his rhetorical art, and while the standing of the *Apologia* has only been enhanced by the contemporary fascination with biography as a literary genre, nevertheless Newman has other significant literary claims, which to this day have not had justice done to them. Thus he is still virtually unrecognized not only as one of the great English satirical writers, but also as one of the most remarkable letter-writers in the language. It seems fair to say that the revaluation of the 'literary' Newman lags behind the reassessment of the thinker. It is hoped that this volume of selections will help to correct this imbalance.

The second reason for the increased interest in Newman is clearly of a religious nature and is closely connected with the Second Vatican Council (1962–5), which has been called 'Newman's Council' and which inaugurated not only extensive changes and reforms in the

Roman Catholic Church itself but opened in effect a whole new era in Christian history. The almost revolutionary return by Catholic theology to its scriptural and patristic roots has invested Newman's own theology, which was so deeply, even exclusively, rooted in the Bible and the Fathers, with a wholly new importance and significance. Those ideas and insights of Newman, which were so often sharply rejected in his own time, have come not simply to be accepted but to be taken for granted in the Church of his adoption. Moreover, the fact that Newman left the Church of England has not in the long run proved so much to be an ecumenical embarrassment as a fruitful source of reconciliation. Thus his *Lectures on Justification* and his *Letter to Pusey*, to take two obvious examples, one from his Anglican and the other from his Catholic period, have recently emerged as early landmarks in the theology of ecumenism. It is also noteworthy how much attention his philosophy (or phenomenology) of religious belief, which was the object of suspicion to scholastic philosophers in his own time, is attracting from contemporary philosophers of religion.

Newman draws different kinds of readers with different interests. To some he is first and foremost a historic religious leader, to others a writer principally, to some an educator, to others a theologian or philosopher. Nevertheless it is as hard to read Newman from a simply literary point of view, without regard for the religious and theological content, as it is difficult to dissociate his theology and philosophy from his imagination and its artistic expression. The champion and defender of a dogmatic religion is also at one and the same time the supremely imaginative writer, so that it is impossible to separate the two. There are not some writings which are 'theological' and others which are 'literary', any more than *The Idea of a University* is a purely educational work.

One obvious consequence of this difficulty of dividing Newman neatly into the educator, the philosopher, the preacher, the theologian, and the writer, is the problem this creates for the potential reader, who is confronted with some 40 published (including posthumous) works, as well as a vast correspondence which will eventually fill another 31 volumes. Since only four or so of the most famous titles are regularly reprinted, the need for a comprehensive selection is as urgent as it is in the case of such prolific Victorian writers as Carlyle and Ruskin. It is true that *The Idea of a University* contains most of Newman's educational thought and writings, but where does one go to comprehend the nature and range of Newman's theological ideas? There is no single work where Newman summarizes or synthesizes his religious thought, nor is there any book which can claim to be at all fully representative of it. For the fact is that Newman was not a 'systematic' thinker. He was above all an 'occasional' writer who wrote for particular, usually controversial, occasions, and in that sense he may be called as 'occasional' thinker, so long as that description is not taken to imply any kind of incoherence in a body of thought which is as notable for its lack of 'system' as it is impressive for its consistency and unity.

There have been a number of anthologies of Newman's writings, but this is the first to attempt to provide representative selections intended specifically to introduce the reader to the five main areas of his achievement. The reason for this division is not any desire to schematize or to impose artificial boundaries which do not exist in fact. Indeed, the selections themselves show only too clearly that such lines of demarcation do not apply in reality to Newman's multifaceted genius. But rather, the ordering of material is meant to help readers to discover for themselves something of the fullness and variety of a body of work which defies the attempt to select but which also threatens to overwhelm the explorer by its own richness and profusion.

A final word should be added about the principle on which these selections have been made. In spite of what has been said above, my primary purpose has not been to portray Newman as the 'Eminent Victorian' and 'the Father of Vatican II'. I believe that Newman's significance transcends both these considerations, important as they are. It follows that there may be passages which have not been included here which are of particular interest to the Victorian specialist or to the historian of modern theology. For my overriding concern throughout has been to paint as full a portrait as possible of the genius who was at the same time one of the great masters of English prose (arguably, I suggest, the very greatest writer of non-fiction prose in the language) and also one of those very few Christian thinkers who may be mentioned in the same breath as the Fathers of the Church.

NOTE ON THE TEXT

Newman collected his works in a uniform edition (36 vols., 1868–81), but he continued to make minor alterations and changes in subsequent editions or reprints. From 1886, until the stock was destroyed in the Second World War, all the volumes were published by Longmans, Green and Co. of London. The text of these selections from the published works is in every case taken from the volumes in this standard Longmans edition which were published after Newman's death.

The text of the letters is from *The Letters and Diaries of John Henry Newman*, edited by Charles Stephen Dessain *et al.*, vols. i–vi (Oxford, 1978–84), xi–xxii (London, 1961–72), xxiii–xxxi (Oxford, 1973–7).

Except where an extract begins or ends with a broken-off sentence, omission prints are only supplied before and after selections from individual essays, letters, and sermons which might otherwise purport to be self-contained writings. Selections from books are presumed to be of a fragmentary character with omissions of text normally preceding and following the extracts.

Extracts from books are followed by chapter or equivalent references in square brackets; selections from essays and sermons are followed by references in square brackets to the volumes in which they are collected.

A CHRONOLOGY OF JOHN HENRY NEWMAN

1801	21 February: born in London.
1808	Enters Ealing School.
1816	Converted to a dogmatic Christianity under the influence of the Revd Walter Mayers, an Evangelical schoolmaster.
1817	8 June: enters Trinity College, Oxford.
1818	Wins college scholarship.
1820	Obtains poor BA degree.
1822	Elected fellow of Oriel College, Oxford.
1824	Ordained deacon and appointed curate at St Clement's, Oxford.
1825	Appointed Vice-Principal of Alban Hall. Ordained priest.
1826	Appointed tutor of Oriel.
1828	Appointed Vicar of St Mary's, the University Church.
1832	Completes his first book, *The Arians of the Fourth Century*. December: sails from Falmouth for Mediterranean with Hurrell Froude.
1833	May: illness in Sicily. 8 July: returns to England. 14 July: Keble's Assize Sermon on 'National Apostasy' marks beginning of Oxford Movement. September: begins *Tracts for the Times*.
1834	Publishes first volume of *Parochial Sermons*.
1837	*Lectures on the Prophetical Office of the Church*.
1838	*Lectures on the Doctrine of Justification*.
1841	*The Tamworth Reading Room*. *Tract 90*.
1842	Moves to Littlemore.
1843	*Oxford University Sermons*. September: resigns the living of St Mary's. *Sermons on Subjects of the Day*.
1845	3 October: resigns Oriel fellowship. 9 October: received into the Roman Catholic Church. *Essay on the Development of Christian Doctrine*.
1846	23 February: leaves Oxford for Maryvale, near Birmingham.
1847	30 May: ordained priest in Rome. Writes *Loss and Gain*.
1848	1 February: founds the Oratory of St Philip Neri at Birmingham.
1849	*Discourses Addressed to Mixed Congregations*.

1850   *Lectures on Certain Difficulties felt by Anglicans in submitting to the Catholic Church.*

1851   Restoration of Catholic hierarchy to England.
Lectures on the Present Position of Catholics in England.
Appointed Rector of the Catholic University of Ireland (resigns 1858).

1852   *Discourses on the Scope and Nature of University Education.*

1856   *Callista: A Sketch of the Third Century.*

1857   *Sermons Preached on Various Occasions.*

1859   *Lectures and Essays on University Subjects.*
Publishes *On Consulting the Faithful in Matters of Doctrine* as article in *Rambler* magazine.

1864   *Apologia pro Vita Sua.*

1865   *The Dream of Gerontius.*

1866   *A Letter to the Rev. E. B. Pusey.*

1868   Collected poetry published as *Verses on Various Occasions.*

1870   *An Essay in Aid of a Grammar of Assent.*
Papal infallibility defined by Vatican Council.

1873   *The Idea of a University.*

1875   *A Letter to the Duke of Norfolk.*

1877   *Via Media.*
Elected honorary fellow of Trinity College, Oxford.

1879   Created Cardinal.

1890   11 August: died.

I
THE EDUCATOR

In 1863 Newman noted in his journal that his chief interest had always lain in education. This may seem a surprising claim, even though he was admittedly using the word in a wide sense. But he observed that his constant concern since his conversion to the Roman Catholic Church had been to try and raise the intellectual level among English Catholics. As an Anglican, his public career at Oxford had begun with his election to a coveted fellowship at Oriel College. And, as he once wryly pointed out, the Oxford Movement itself might never have started but for an internal college dispute about the tutorial office, which eventually led to his being dismissed as a college tutor. Newman then in effect ceased to be a teaching fellow and became what we should now call a research fellow of Oriel, with time and leisure for that intensive study of the Church Fathers which would finally lead him out of the Church of England.

On 15 April 1851 Newman received a letter from Archbishop Paul Cullen requesting his help in founding the new Catholic University of Ireland. Cullen suggested that Newman might be prepared to give a few lectures on education. In the end, Newman agreed to become Rector of the University. It had been a dream of his ever since becoming a Catholic that there should be a university for English-speaking Catholics. The first of what came to be called the *Discourses on the Scope and Nature of University Education* was delivered in Dublin on 10 May 1852. In all, five lectures were delivered and another five written but never delivered publicly. They were published in one volume, dated 1852, but not in fact published till 1853. At the time he thought it was one of his two most 'perfect' works from a literary point of view, the other being *Lectures on the Present Position of Catholics in England*.

These *Discourses* (with one omission) constitute the first half of Newman's classic work on education, *The Idea of a University*, which was published in 1873. The second, slightly longer half consists of ten essays and lectures, which were first published in book-form in 1859 under the title *Lectures and Essays on University Subjects*. They have often been neglected, even ignored, in discussions of Newman's educational ideas, but in fact, far from being a mere appendage to *The Idea of a University*, they complement and illuminate the first half by providing concrete applications of the theory of the *Discourses*. For Walter Pater, the final achievement was a triumph—'the perfect handling of a theory'.

The power of Newman's rhetoric is inseparable from his remarkable ability to hold two (or more) different points of view in sharp equipoise. In the *Idea of a University* the subtlety of his rhetorical art depends very largely on the careful balance he maintains between his unconditional insistence on the absolute value of knowledge and education on the one hand, and on the other hand his no less firm conviction that in themselves they are emphatically less important than religion and religious faith. But there is no irreconcilable opposition between the intellectual and the religious, because it is ultimately a religious conviction which supports Newman's supremely confident vision of the wholeness of knowledge and truth: underlying the belief in the 'imperial intellect' is the prior belief in God as Creator. Indeed, far from being opposed to each other, religion and knowledge are indivisibly connected, if only because religion forms part of the subject-matter of knowledge. However, it remains true that if there has to be a choice between being an educated unbeliever and an uneducated believer, then Newman opts without hesitation for the latter.

The rare balance then, which Newman contrives to maintain between the claims of revealed religion and the value of intellectual culture produces what the historian G. M. Young has hailed as the definitive statement of Christian humanism: '. . . he employs all his magic to enlarge and refine and exalt this conception of intellectual cultivation as a good in itself, worth while for itself, to be prized and esteemed for itself beyond all knowledge and all professional skill; while, all the time, so earnestly does he affirm its inadequacy, its shortcomings on the moral side, its need to be steadied and purified by religion, that at the end we feel that what we have heard is the final utterance, never to be repeated or needing to be supplemented, of Christian Humanism: as if the spirit evoked by Erasmus had found its voice at last.'

THE IDEA OF A UNIVERSITY

Bearing of Theology on Other Knowledge

. . . all knowledge forms one whole, because its subject-matter is one; for the universe in its length and breadth is so intimately knit together, that we cannot separate off portion from portion, and operation from operation, except by a mental abstraction; and then again, as to its Creator, though He of course in His own Being is infinitely separate from it, and Theology has its departments towards which human knowledge has no relations, yet He has so implicated Himself with it, and taken it into His very bosom, by His presence in it, His providence over it, His impressions upon it, and His influences through it, that we cannot truly or fully contemplate it without in some main aspects contemplating Him. Next, sciences are the results of that mental abstraction, which I have spoken of, being the logical record of this or that aspect of the whole subject-matter of knowledge. As they all belong to one and the same circle of objects, they are one and all connected together; as they are but aspects of things, they are severally incomplete in their relation to the things themselves, though complete in their own idea and for their own respective purposes; on both accounts they at once need and subserve each other. And further, the comprehension of the bearings of one science on another, and the use of each to each, and the location and limitation and adjustment and due appreciation of them all, one with another, this belongs, I conceive, to a sort of science distinct from all of them, and in some sense a science of sciences, which is my own conception of what is meant by Philosophy, in the true sense of the word, and of a philosophical habit of mind, and which in these Discourses I shall call by that name. This is what I have to say about knowledge and philosophical knowledge generally; and now I proceed to apply it to the particular science, which has led me to draw it out.

I say, then, that the systematic omission of any one science from the catalogue prejudices the accuracy and completeness of our knowledge altogether, and that, in proportion to its importance. Not even Theology itself, though it comes from heaven, though its truths were given once for all at the first, though they are more certain on account of the Giver than those of mathematics, not even Theology, so far as it is relative to us, or is the Science of Religion, do I exclude from the law to which every mental exercise is subject, viz., from that imperfection, which ever must attend the abstract, when it would determine the concrete. Nor do I speak only of National Religion; for even the teaching of the Catholic Church, in certain of its aspects, that is, its religious teaching, is variously influenced by the other sciences. Not to insist on the introduction of the Aristotelic philosophy into its phraseology, its explanation of dogmas is influenced by ecclesiastical acts or events; its interpretations of prophecy are directly affected by the issues of history; its comments upon Scripture by the conclusions of the astronomer and the geologist; and its casuistical decisions by the various experience, political, social, and psychological, with which times and places are ever supplying it.

[Discourse 3]

Bearing of Other Knowledge on Theology

. . . I observe, then, that, if you drop any science out of the circle of knowledge, you cannot keep its place vacant for it; that science is forgotten; the other sciences close up, or, in other words, they exceed their proper bounds, and intrude where they have no right. For instance, I suppose, if ethics were sent into banishment, its territory would soon disappear, under a treaty of partition, as it may be called, between law, political economy, and physiology; what, again, would become of the province of experimental science, if made over to the Antiquarian Society; or of history, if surrendered out and out to Metaphysicians? The case is the same with the subject-matter of Theology; it would be the prey of a dozen various sciences, if Theology were put out of possession; and not only so, but those sciences would be plainly exceeding their rights and their capacities in seizing upon it. They would be sure to teach wrongly, where they had no mission to teach at all. The enemies of Catholicism ought to be the last to deny this:—for they have never been blind to a like usurpation, as they have called it, on the part of theologians; those who accuse us of wishing, in accordance with Scripture language, to make the sun go round the earth, are not the men to deny that a science which exceeds its limits falls into error.

[Discourse 4]

Knowledge Its Own End

I have said that all branches of knowledge are connected together, because the subject-matter of knowledge is intimately united in itself, as being the acts and the work of the Creator. Hence it is that the Sciences, into which our knowledge may be said to be cast, have multiplied bearings one on another, and an internal sympathy, and admit, or rather demand, comparison and adjustment. They complete, correct, balance each other. This consideration, if well-founded, must be taken into account, not only as regards the attainment of truth, which is their common end, but as regards the influence which they exercise upon those whose education consists in the study of them. I have said already, that to give undue prominence to one is to be unjust to another; to neglect or supersede these is to divert those from their proper object. It is to unsettle the boundary lines between science and science, to disturb their action, to destroy the harmony which binds them together. Such a proceeding will have a corresponding effect when introduced into a place of education. There is no science but tells a different tale, when viewed as a portion of a whole, from what it is likely to suggest when taken by itself, without the safeguard, as I may call it, of others.

Let me make use of an illustration. In the combination of colours, very different effects are produced by a difference in their selection and juxtaposition; red, green, and white, change their shades, according to the contrast to which they are submitted. And, in like manner, the drift and meaning of a branch of knowledge varies with the company in which it is introduced to the student. If his reading is confined simply to one subject, however such division of labour may favour the advancement of a particular pursuit, a point into which I do not here enter, certainly it has a tendency to contract his mind. If it is incorporated with others, it depends on those others as to the kind of influence which it exerts upon him. . . .

It is a great point then to enlarge the range of studies which a University professes, even for the sake of the students; and, though they cannot pursue every subject which is open to them, they will be the gainers by living among those and under those who represent the whole circle. This I conceive to be the advantage of a seat of universal learning, considered as a place of education. An assemblage of learned men, zealous for their own sciences, and rivals of each other, are brought, by familiar intercourse and for the sake of intellectual peace, to adjust together the claims and relations of their respective subjects of investigation. They learn to respect, to consult, to aid each other. Thus is created a pure and clear atmosphere of thought, which the student also breathes, though in his own case he only pursues a few sciences out of the multitude. He profits by an intellectual tradition, which is independent of particular teachers, which guides him in his choice of subjects, and duly interprets for him those which he chooses. He apprehends the great outlines of knowledge, the principles on which it rests, the scale of its parts, its lights and its shades, its great points and its little, as he otherwise cannot apprehend them. Hence it is that his education is called "Liberal." A habit of mind is formed which lasts through life, of which the attributes are, freedom, equitableness, calmness, moderation, and wisdom; or what in a former Discourse I have ventured to call a philosophical habit. This then I would assign as the special fruit of the education furnished at a University, as contrasted with other places of teaching or modes of teaching. This is the main purpose of a University in its treatment of its students.

And now the question is asked me, What is the *use* of it? . . .

Cautious and practical thinkers, I say, will ask of me, what, after all, is the gain of this Philosophy, of which I make such account, and from which I promise so much. Even supposing it to enable us to exercise the degree of trust exactly due to every science respectively, and to estimate precisely the value of every truth which is anywhere to be found, how are we better for this master view of things, which I have been extolling? Does it not reverse the principle of the division of labour? will practical objects be obtained better or worse by its cultivation? to what then does it lead? where does it end? what does it do? how does it profit? what does it promise? Particular sciences are respectively the basis of definite arts, which carry on to results tangible and beneficial the truths which are the subjects of the knowledge attained; what is the Art of this science of sciences? what is the fruit of such a Philosophy? . . .

I am asked what is the end of University Education, and of the Liberal or Philosophical Knowledge which I conceive it to impart: I answer, that . . . it has a very tangible, real, and sufficient end, though the end cannot be divided from that knowledge itself. Knowledge is capable of being its own end. Such is the constitution of the human mind, that any kind of knowledge, if it be really such, is its own reward. And if this is true of all knowledge, it is true also of that special Philosophy, which I have made to consist in a comprehensive view of truth in all its branches, of the relations of science to science, of their mutual bearings, and their respective values. What the worth of such an acquirement is, compared with other objects which we seek,—wealth or power or honour or the conveniences and comforts of life, I do not profess here to discuss; but I would maintain, and mean to show, that it is an object, in its own nature so really and undeniably good, as to be the compensation of a great deal of thought in the compassing, and a great deal of trouble in the attaining.

Now, when I say that Knowledge is, not merely a means to something beyond it, or the preliminary of certain arts into which it naturally resolves, but an end sufficient to rest in and to pursue for its own sake, surely I am uttering no paradox, for I am stating what is both intelligible in itself, and has ever been the common judgment of philosophers and the ordinary feeling of mankind. . . . That further advantages accrue to us and redound to others by its possession, over and above what it is in itself, I am very far indeed from denying; but, independent of these, we are satisfying a direct need of our nature in its very acquisition; and, whereas our nature, unlike that of the inferior creation, does not at once reach its perfection, but depends, in order to it, on a number of external aids and appliances, Knowledge, as one of the principal of these, is valuable for what its very presence in us does for us after the manner of a habit, even though it be turned to no further account, nor subserve any direct end. . . .

Things, which can bear to be cut off from every thing else and yet persist in living, must have life in themselves; pursuits, which issue in nothing, and still maintain their ground for ages, which are regarded as admirable, though they have not as yet proved themselves to be useful, must have their sufficient end in themselves, whatever it turn out to be. And we are brought to the same conclusion by considering the force of the epithet, by which the knowledge under consideration is popularly designated. It is common to speak of "*liberal* knowledge," of the "*liberal* arts and studies," and of a "*liberal* education," as the especial characteristic or property of a University and of a gentleman; what is really meant by the word? Now, first, in its grammatical sense it is opposed to *servile*; and by "servile work" is understood . . . bodily labour, mechanical employment, and the like, in which the mind has little or no part. Parallel to such servile works are those arts, if they deserve the name . . . which owe their origin and their method to hazard, not to skill; as, for instance, the practice and operations of an empiric. As far as this contrast may be considered as a guide into the meaning of the word, liberal education and liberal pursuits are exercises of mind, of reason, of reflection.

But we want something more for its explanation, for there are bodily exercises which are liberal, and mental exercises which are not so. For instance, in ancient times the practitioners in medicine were commonly slaves; yet it was an art as intellectual in its nature, in spite of the pretence, fraud, and quackery with which it might then, as now, be debased, as it was heavenly in its aim. And so in like manner, we contrast a liberal education with a commercial education or a professional; yet no one can deny that commerce and the professions afford scope for the highest and most diversified powers of mind. There is then a great variety of intellectual exercises, which are not technically called "liberal;" on the other hand, I say, there are exercises of the body which do receive that appellation. Such, for instance, was the palæstra, in ancient times; such the Olympic games, in which strength and dexterity of body as well as of mind gained the prize. . . . War, too, however rough a profession, has ever been accounted liberal, unless in cases when it becomes heroic, which would introduce us to another subject.

Now comparing these instances together, we shall have no difficulty in determining the principle of this apparent variation in the application of the term which I am examining. Manly games, or games of skill, or military prowess, though bodily, are, it seems, accounted liberal; on the other hand, what is merely professional, though highly intellectual, nay, though liberal in comparison of trade and manual labour, is not simply called liberal, and mercantile occupations are not liberal at all. Why this distinction? because that alone is liberal knowledge, which stands on its own pretensions, which is independent of sequel, expects no complement, refuses to be *informed* (as it is called) by any end, or absorbed into any art, in order duly to present itself to our contemplation. The most ordinary pursuits have this specific character, if they are self-sufficient and complete; the highest lose it, when they minister to something beyond them. It is absurd to balance, in point of worth and importance, a treatise on reducing fractures with a game of cricket or a fox-chase; yet of the two the bodily exercise has that quality which we call "liberal," and the intellectual has it not. And so of the learned professions altogether, considered merely as professions; although one of them be the most popularly beneficial, and another the most politically important, and the third the most intimately divine of all human pursuits, yet the very greatness of their end, the health of the body, or of the commonwealth, or of the soul, diminishes, not increases, their claim to the appellation "liberal," and that still more, if they are cut down to the strict exigencies of that end. If, for instance, Theology, instead of being cultivated as a contemplation, be limited to the purposes of the pulpit or be represented by the catechism, it loses,—not its usefulness, not its divine character, not its meritoriousness (rather it gains a claim upon these titles by such charitable condescension),—but it does lose the particular attribute which I am illustrating; just as a face worn by tears and fasting loses its beauty, or a labourer's hand loses its delicateness;— for Theology thus exercised is not simple knowledge, but rather is an art or a business making use of Theology. And thus it appears that even what is supernatural need not be liberal, nor need a hero be a gentleman, for the plain reason that one idea is not another idea. And in like manner the Baconian Philosophy, by using its physical sciences in the service of man, does thereby transfer them from the order of Liberal Pursuits to, I do not say the inferior, but the distinct class of the Useful. And, to take a different instance, hence again, as is evident, whenever personal gain is the motive, still more distinctive an effect has it upon the character of a given pursuit; thus racing, which was a liberal exercise in Greece, forfeits its rank in times like these, so far as it is made the occasion of gambling. . . .

. . . the word "liberal" as applied to Knowledge and Education, expresses a specific idea, which ever has been, and ever will be, while the nature of man is the same, just as the idea of the Beautiful is specific, or of the Sublime, or of the Ridiculous, or of the Sordid. It is in the world now, it was in the world then; and, as in the case of the dogmas of faith, it is illustrated by a continuous historical tradition, and never was out of the world, from the time it came into it. There have indeed been differences of opinion from time to time, as to what pursuits and what arts came under that idea, but such differences are but an additional evidence of its reality. That idea must have a substance in it, which has maintained its ground amid these conflicts and changes, which has ever served as a standard to measure things withal, which has passed from mind to mind unchanged, when there was so much to colour, so much to influence any notion or thought whatever, which was not founded in our very nature. Were it a mere generalization, it would have varied with the subjects from which it was generalized; but though its subjects vary with the age, it varies not itself. . . .

I consider, then, that I am chargeable with no paradox, when I speak of a Knowledge which is its own end, when I call it liberal knowledge, or a gentleman's knowledge, when I educate for it, and make it the scope of a University. And still less am I incurring such a charge, when I make this acquisition consist, not in Knowledge in a vague and ordinary sense, but in that Knowledge which I have especially called Philosophy or, in an extended sense of the word, Science; for whatever claims Knowledge has to be considered as a good, these it has in a higher degree when it is viewed not vaguely, not popularly, but precisely and transcendently as Philosophy. Knowledge, I say, is then especially liberal, or sufficient for itself, apart from every external and ulterior object, when and so far as it is philosophical, and this I proceed to show.

. . . Philosophy, then, or Science, is related to Knowledge in this way:— Knowledge is called by the name of Science or Philosophy, when it is acted upon, informed, or if I may use a strong figure, impregnated by Reason. Reason is the principle of that intrinsic fecundity of Knowledge, which, to those who possess it, is its especial value, and which dispenses with the necessity of their looking abroad for any end to rest upon external to itself. Knowledge, indeed, when thus exalted into a scientific form, is also power; not only is it excellent in itself, but whatever such excellence may be, it is something more, it has a result beyond itself. Doubtless; but that is a further consideration, with which I am not concerned. I only say that, prior to its being a power, it is a good; that it is, not only an instrument, but an end. I know well it may resolve itself into an art, and terminate in a mechanical process, and in tangible fruit; but it also may fall back upon that Reason which informs it, and resolve itself into Philosophy. In one case it is called Useful Knowledge, in the other Liberal. The same person may cultivate it in both ways at once; but this again is a matter foreign to my subject; here I do but say that there are two ways of using Knowledge, and in matter of fact those who use it in one way are not likely to use it in the other, or at least in a very limited measure. You see, then, here are two methods of Education; the end of the one is to be philosophical, of the other to be mechanical; the one rises towards general ideas, the other is exhausted upon what is particular and external. Let me not be thought to deny the necessity, or to decry the benefit, of such attention to what is particular and practical, as belongs to the useful or mechanical arts; life could not go on without them; we owe our daily welfare to them; their exercise is the duty of the many, and we owe to the many a debt of gratitude for fulfilling that duty. I only say that Knowledge, in proportion as it tends more and more to be particular, ceases to be Knowledge. It is a question whether Knowledge can in any proper sense be predicated of the brute creation; without pretending to metaphysical exactness of phraseology, which would be unsuitable to an occasion like this, I say, it seems to me improper to call that passive sensation, or perception of things, which brutes seem to possess, by the name of Knowledge. When I speak of Knowledge, I mean something intellectual, something which grasps what it perceives through the senses; something which takes a view of things; which sees more than the senses convey; which reasons upon what it sees, and while it sees; which invests it with an idea. It expresses itself, not in a mere enunciation, but by an enthymeme: it is of the nature of science from the first, and in this consists its dignity. The principle of real dignity in Knowledge, its worth, its desirableness, considered irrespectively of its results, is this germ within it of a scientific or a philosophical process. This is how it comes to be an end in itself; this is why it admits of being called Liberal. Not to know the relative disposition of things is the state of slaves or children; to have mapped out the Universe is the boast, or at least the ambition, of Philosophy.

Moreover, such knowledge is not a mere extrinsic or accidental advantage, which is ours to-day and another's to-morrow, which may be got up from a book and easily forgotten again, which we can command or communicate at our pleasure, which we can borrow for the occasion, carry about in our hand, and take into the market; it is an acquired illumination, it is a habit, a personal possession, and an inward endowment. And this is the reason, why it is more correct, as well as more usual, to speak of a University as a place of education, than of instruction, though, when knowledge is concerned, instruction would at first sight have seemed the more appropriate word. We are instructed, for instances, to manual exercises, in the fine and useful arts, in trades, and in ways of business; for these methods, which have little or no effect upon the mind itself, are contained in rules committed to memory, to tradition, or to use, and bear upon an end external to themselves. But education is a higher word; it implies an action upon our mental nature, and the formation of a character; it is something individual and permanent, and is commonly spoken of in connexion with religion and virtue. When, then, we speak of the communication of Knowledge as being Education, we thereby really imply that that Knowledge is a state or condition of mind; and since cultivation of mind is surely worth seeking for its own sake, we are thus brought once more to the conclusion, which the word "Liberal" and the word "Philosophy" have already suggested, that there is a Knowledge, which is desirable, though nothing come of it, as being of itself a treasure, and a sufficient remuneration of years of labour. . . .

Useful Knowledge . . . has done its work; and Liberal Knowledge as certainly has not done its work,—that is, supposing, as the objectors assume, its direct end, like Religious Knowledge, is to make men better; but this I will not for an instant allow, and, unless I allow it, those objectors have said nothing to the purpose. I admit, rather I maintain, what they have been urging, for I consider Knowledge to have its end in itself. For all its friends, or its enemies, may say, I insist upon it, that it is as real a mistake to burden it with virtue or religion as with the mechanical arts. Its direct business is not to steel the soul against temptation or to console it in affliction, any more than to set the loom in motion, or to direct the steam carriage; be it ever so much the means or the condition of both material and moral advancement, still, taken by and in itself, it as little mends our hearts as it improves our temporal circumstances. And if its eulogists claim for it such a power, they commit the very same kind of encroachment on a province not their own as the political economist who should maintain that his science educated him for casuistry or diplomacy. Knowledge is one thing, virtue is another; good sense is not conscience, refinement is not humility, nor is largeness and justness of view faith. Philosophy, however enlightened, however profound, gives no command over the passions, no influential motives, no vivifying principles. Liberal Education makes not the Christian, not the Catholic, but the gentleman. It is well to be a gentlemen, it is well to have a cultivated intellect, a delicate taste, a candid, equitable, dispassionate mind, a noble and courteous bearing in the conduct of life;—these are the connatural qualities of a large knowledge; they are the objects of a University; I am advocating, I shall illustrate and insist upon them; but still, I repeat, they are no guarantee for sanctity or even for conscientiousness, they may attach to the man of the world, to the profligate, to the heartless,—pleasant, alas, and attractive as he shows when decked out in them. Taken by themselves, they do but seem to be what they are not; they look like virtue at a distance, but they are detected by close observers, and on the long run; and hence it is that they are popularly accused of pretence and hypocrisy, not, I repeat, from their own fault, but because their professors and their admirers persist in taking them for what they are not, and are officious in arrogating for them a praise to which they have no claim. Quarry the granite rock with razors, or moor the vessel with a thread of silk; then may you hope with such keen and delicate instruments as human knowledge and human reason to contend against those giants, the passion and the pride of man.

Surely we are not driven to theories of this kind, in order to vindicate the value and dignity of Liberal Knowledge. Surely the real grounds on which its pretensions rest are not so very subtle or abstruse, so very strange or improbable. Surely it is very intelligible to say, and that is what I say here, that Liberal Education, viewed in itself, is simply the cultivation of the intellect, as such, and its object is nothing more or less than intellectual excellence. Every thing has its own perfection, be it higher or lower on the scale of things; and the perfection of one is not the perfection of another. Things animate, inanimate, visible, invisible, all are good in their kind, and have a *best* of themselves, which is an object of pursuit. Why do you take such pains with your garden or your park? You see to your walks and turf and shrubberies; to your trees and drives; not as if you meant to make an orchard of the one, or corn or pasture land of the other, but because there is a special beauty in all that is goodly in wood, water, plain, and slope, brought all together by art into one shape, and grouped into one whole. Your cities are beautiful, your palaces, your public buildings, your territorial mansions, your churches; and their beauty leads to nothing beyond itself. There is a physical beauty and a moral: there is a beauty of person, there is a beauty of our moral being, which is natural virtue; and in like manner there is a beauty, there is a perfection, of the intellect. There is an ideal perfection in these various subject-matters, towards which individual instances are seen to rise, and which are the standards for all instances whatever. The Greek divinities and demigods, as the statuary has moulded them, with their symmetry of figure, and their high forehead and their regular features, are the perfection of physical beauty. The heroes, of whom history tells, Alexander, or Cæsar, or Scipio, or Saladin, are the representatives of that magnanimity or self-mastery which is the greatness of human nature. Christianity too has its heroes, and in the supernatural order, and we call them Saints. The artist puts before him beauty of feature and form; the poet, beauty of mind; the preacher, the beauty of grace: then intellect too, I repeat, has its beauty, and it has those who aim at it. To open the mind, to correct it, to refine it, to enable it to know, and to digest, master, rule, and use its knowledge, to give it power over its own faculties, application, flexibility, method, critical exactness, sagacity, resource, address, eloquent expression, is an object as intelligible (for here we are inquiring, not what the object of a Liberal

Education is worth, nor what use the Church makes of it, but what it is in itself), I say, an object as intelligible as the cultivation of virtue, while, at the same time, it is absolutely distinct from it.

This indeed is but a temporal object, and a transitory possession: but so are other things in themselves which we make much of and pursue. The moralist will tell us that man, in all his functions, is but a flower which blossoms and fades, except so far as a higher principle breathes upon him, and makes him and what he is immortal. Body and mind are carried on into an eternal state of being by the gifts of Divine Munificence; but at first they do but fail in a failing world; and if the powers of intellect decay, the powers of the body have decayed before them, and, as an Hospital or an Almshouse, though its end be ephemeral, may be sanctified to the service of religion, so surely may a University, even were it nothing more than I have as yet described it. We attain to heaven by using this world well, though it is to pass away; we perfect our nature, not by undoing it, but by adding to it what is more than nature, and directing it towards aims higher than its own.

[Discourse 5]

Knowledge Viewed in Relation to Learning

It were well if the English, like the Greek language, possessed some definite word to express, simply and generally, intellectual proficiency or perfection, such as "health," as used with reference to the animal frame, and "virtue," with reference to our moral nature. I am not able to find such a term;—talent, ability, genius, belong distinctly to the raw material, which is the subject-matter, not to that excellence which is the result of exercise and training. When we turn, indeed, to the particular kinds of intellectual perfection, words are forthcoming for our purpose, as, for instance, judgment, taste, and skill; yet even these belong, for the most part, to powers or habits bearing upon practice or upon art, and not to any perfect condition of the intellect, considered in itself. Wisdom, again, is certainly a more comprehensive word than any other, but it has a direct relation to conduct, and to human life. Knowledge, indeed, and Science express purely intellectual ideas, but still not a state or quality of the intellect; for knowledge, in its ordinary sense, is but one of its circumstances, denoting a possession or a habit; and science has been appropriated to the subject-matter of the intellect, instead of belonging in English, as it ought to do, to the intellect itself. The consequence is that, on an occasion like this, many words are necessary, in order, first, to bring out and convey what surely is no difficult idea in itself,—that of the cultivation of the intellect as an end; next, in order to recommend what surely is no unreasonable object; and lastly, to describe and make the mind realize the particular perfection in which that object consists. Every one knows practically what are the constituents of health or of virtue; and every one recognizes health and virtue as ends to be pursued; it is otherwise with intellectual excellence, and this must be my excuse, if I seem to any one to be bestowing a good deal of labour on a preliminary matter.

In default of a recognized term, I have called the perfection or virtue of the intellect by the name of philosophy, philosophical knowledge, enlargement of mind, or illumination; terms which are not uncommonly given to it by writers of this day: but, whatever name we bestow on it, it is, I believe, as a matter of history, the business of a University to make this intellectual culture its direct scope, or to employ itself in the education of the intellect,—just as the work of a Hospital lies in healing the sick or wounded, of a Riding or Fencing School, or of a Gymnasium, in exercising the limbs, of an Almshouse, in aiding and solacing the old, of an Orphanage, in protecting innocence, of a Penitentiary, in restoring the guilty. I say, a University, taken in its bare idea, and before we view it as an instrument of the Church, has this object and this mission; it contemplates neither moral impression nor mechanical production; it professes to exercise the mind neither in art nor in duty; its function is intellectual culture; here it may leave its scholars, and it has done its work when it has done as much as this. It educates the intellect to reason well in all matters, to reach out towards truth, and to grasp it. . . .

Knowledge then is the indispensable condition of expansion of mind, and the instrument of attaining to it; this cannot be denied, it is ever to be insisted on; I begin with it as a first principle; however, the very truth of it carries men too far, and confirms to them the notion that it is the whole of the matter. A narrow mind is thought to be that which contains little knowledge; and an enlarged mind, that which holds a great deal . . .

. . . The enlargement consists, not merely in the passive reception into the mind of a number of ideas hitherto unknown to it, but in the mind's energetic and simultaneous action upon and towards and among those new ideas, which are rushing in upon it. It is the action of a formative power, reducing to order and meaning the matter of our acquirements; it is a making the objects of our knowledge subjectively our own, or, to use a familiar word, it is a digestion of what we receive, into the substance of our previous state of thought; and without this no enlargement is said to follow. There is no enlargement, unless there be a comparison of ideas one with another, as they come before the mind, and a systematizing of them. We feel our minds to be growing and expanding *then*, when we not only learn, but refer what we learn to what we know already. It is not the mere addition to our knowledge that is the illumination; but the locomotion, the movement onwards, of that mental centre, to which both what we know, and what we are learning, the accumulating mass of our acquirements, gravitates. And therefore a truly great intellect, and recognized to be such by the common opinion of mankind, such as the intellect of Aristotle, or of St. Thomas, or of Newton, or of Goethe, (I purposely take instances within and without the Catholic pale, when I would speak of the intellect as such,) is one which takes a connected view of old and new, past and present, far and near, and which has an insight into the influence of all these one on another; without which there is no whole, and no centre. It possesses the knowledge, not only of things, but also of their mutual and true relations; knowledge, not merely considered as acquirement, but as philosophy.

Accordingly, when this analytical, distributive, harmonizing process is away, the mind experiences no enlargement, and is not reckoned as enlightened or comprehensive, whatever it may add to its knowledge. For instance, a great memory, as I have already said, does not make a philosopher, any more than a dictionary can be called a grammar. There are men who embrace in their minds a vast multitude of ideas, but with little sensibility about their real relations towards each other. These may be antiquarians, annalists, naturalists; they may be learned in law; they may be versed in statistics; they are most useful in their own place; I should shrink from speaking disrespectfully of them; still, there is nothing in such attainments to guarantee the absence of narrowness of mind. If they are nothing more than well-read men, or men of information, they have not what specially deserves the name of culture of mind, or fulfils the type of Liberal Education.

In like manner, we sometimes fall in with persons who have seen much of the world, and of the men who, in their day, have played a conspicuous part in it, but who generalize nothing, and have no observation, in the true sense of the word. They abound in information in detail, curious and entertaining, about men and things; and, having lived under the influence of no very clear or settled principles, religious or political, they speak of every one and every thing, only as so many phenomena, which are complete in themselves, and lead to nothing, not discussing them, or teaching any truth, or instructing the hearer, but simply talking. No one would say that these persons, well informed as they are, had attained to any great culture of intellect or to philosophy.

The case is the same still more strikingly where the persons in question are beyond dispute men of inferior powers and deficient education. Perhaps they have been much in foreign countries, and they receive, in a passive, otiose, unfruitful way, the various facts which are forced upon them there. Seafaring men, for example, range from one end of the earth to the other; but the multiplicity of external objects, which they have encountered, forms no symmetrical and consistent picture upon their imagination; they see the tapestry of human life, as it were on the wrong side, and it tells no story. They sleep, and they rise up, and they find themselves, now in Europe, now in Asia; they see visions of great cities and wild regions; they are in the marts of commerce, or amid the islands of the South; they gaze on Pompey's Pillar, or on the Andes; and nothing which meets them carries them forward or backward, to any idea beyond itself. Nothing has a drift or relation; nothing has a history or a promise. Every thing stands by itself, and comes and goes in its turn, like the shifting scenes of a show, which leave the spectator where he was. Perhaps you are near such a man on a particular occasion, and expect him to be shocked or perplexed at something which occurs; but one thing is much the same to him as another, or, if he is perplexed, it is as not knowing what to say, whether it is right to admire, or to ridicule, or to disapprove, while conscious that some expression of opinion is expected from him; for in fact he has no standard of judgment at all, and no landmarks to guide him to a conclusion. Such is mere acquisition, and, I repeat, no one would dream of calling it philosophy.

Instances, such as these, confirm, by the contrast, the conclusion I have already drawn from those which preceded them. That only is true enlargement of mind which is the power of viewing many things at once as one whole, of referring them severally to their true place in the universal system, of understanding their respective values, and determining their mutual dependence. Thus is that form of Universal Knowledge, of which I have on a former occasion spoken, set up in the individual intellect, and constitutes its perfection. Possessed of this real illumination, the mind never views any part of the extended subject-matter of Knowledge without recollecting that it is but a part, or without the associations which spring from this recollection. It makes every thing in some sort lead to every thing else; it would communicate the image of the whole to every separate portion, till that whole becomes in imagination like a spirit, every where pervading and penetrating its component parts, and giving them one definite meaning. Just as our bodily organs, when mentioned, recall their function in the body, as the word "creation" suggests the Creator, and "subjects" a sovereign, so, in the mind of the Philosopher, as we are abstractedly conceiving of him, the elements of the physical and moral world, sciences, arts, pursuits, ranks, offices, events, opinions, individualities, are all viewed as one, with correlative functions, and as gradually by successive combinations converging, one and all, to the true centre.

To have even a portion of this illuminative reason and true philosophy is the highest state to which nature can aspire, in the way of intellect; it puts the mind above the influences of chance and necessity, above anxiety, suspense, unsettlement, and superstition, which is the lot of the many. Men, whose minds are possessed with some one object, take exaggerated views of its importance, are feverish in the pursuit of it, make it the measure of things which are utterly foreign to it, and are startled and despond if it happens to fail them. They are ever in alarm or in transport. Those on the other hand who have no object or principle whatever to hold by, lose their way, every step they take. They are thrown out, and do not know what to think or say, at every fresh juncture; they have no view of persons, or occurrences, or facts, which come suddenly upon them, and they hang upon the opinion of others, for want of internal resources. But the intellect, which has been disciplined to the perfection of its powers, which knows, and thinks while it knows, which has learned to leaven the dense mass of facts and events with the elastic force of reason, such an intellect cannot be partial, cannot be exclusive, cannot be impetuous, cannot be at a loss, cannot but be patient, collected, and majestically calm, because it discerns the end in every beginning, the origin in every end, the law in every interruption, the limit in each delay; because it ever knows where it stands, and how its path lies from one point to another . . . . There are men who, when in difficulties, originate at the moment vast ideas or dazzling projects; who, under the influence of excitement, are able to cast a light, almost as if from inspiration, on a subject or course of action which comes before them; who have a sudden presence of mind equal to any emergency, rising with the occasion, and an undaunted magnanimous bearing, and an energy and keenness which is but made intense by opposition. This is genius, this is heroism; it is the exhibition of a natural gift, which no culture can teach, at which no Institution can aim; here, on the contrary, we are concerned, not with mere nature, but with training and teaching. That perfection of the Intellect, which is the result of Education, and its *beau ideal*, to be imparted to individuals in their respective measures, is the clear, calm, accurate vision and comprehension of all things, as far as the finite mind can embrace them, each in its place, and with its own characteristics upon it. It is almost prophetic from its knowledge of history; it is almost heart-searching from its knowledge of human nature; it has almost supernatural charity from its freedom from littleness and prejudice; it has almost the repose of faith, because nothing can startle it; it has almost the beauty and harmony of heavenly contemplation, so intimate is it with the eternal order of things and the music of the spheres. . . .

I say then, if we would improve the intellect, first of all, we must ascend; we cannot gain real knowledge on a level; we must generalize, we must reduce to method, we must have a grasp of principles, and group and shape our acquisitions by means of them. It matters not whether our field of operation be wide or limited; in every case, to command it, is to mount above it. Who has not felt the irritation of mind and impatience created by a deep, rich country, visited for the first time, with winding lanes, and high hedges, and green steeps, and tangled woods, and every thing smiling indeed, but in a maze? The same feeling comes upon us in a strange city, when we have no map of its streets. Hence you hear of practised travellers, when they first come into a place, mounting some high hill or church tower, by way of reconnoitring its neighbourhood. In like manner, you must be above your knowledge, not under it, or it will oppress you; and the more you have of it, the greater will be the load. . . .

Instances abound; there are authors who are as pointless as they are inexhaustible in their literary resources. They measure knowledge by bulk, as it lies in the rude block, without symmetry, without design. How many commentators are there on the Classics, how many on Holy Scripture, from whom we rise up, wondering at the learning which has passed before us, and wondering why it passed! . . . Recollect, the Memory can tyrannize, as well as the Imagination. Derangement, I believe, has been considered as a loss of control over the sequence of ideas. The mind, once set in motion, is henceforth deprived of the power of initiation, and becomes the victim of a train of associations, one thought suggesting another, in the way of cause and effect, as if by a mechanical process, or some physical necessity. No one, who has had experience of men of studious habits, but must recognize the existence of a parallel phenomenon in the case of those who have over-stimulated the Memory. In such persons Reason acts almost as feebly and as impotently as in the madman; once fairly started on any subject whatever, they have no power of self-control; they passively endure the succession of impulses which are evolved out of the original exciting cause; they are passed on from one idea to another and go steadily forward, plodding along one line of thought in spite of the amplest concessions of the hearer, or wandering from it in endless digression in spite of his remonstrances. Now, if, as is very certain, no one would envy the madman the glow and originality of his conceptions, why must we extol the cultivation of that intellect, which is the prey, not indeed of barren fancies but of barren facts, of random intrusions from without, though not of morbid imaginations from within? And in thus speaking, I am not denying that a strong and ready memory is in itself a real treasure; I am not disparaging a well-stored mind, though it be nothing besides, provided it be sober, any more than I would despise a bookseller's shop:—it is of great value to others, even when not so to the owner. Nor am I banishing, far from it, the possessors of deep and multifarious learning from my ideal University; they adorn it in the eyes of men; I do but say that they constitute no type of the results at which it aims; that it is no great gain to the intellect to have enlarged the memory at the expense of faculties which are indisputably higher.

Nor indeed am I supposing that there is any great danger, at least in this day, of over-education; the danger is on the other side. I will tell you, Gentlemen, what has been the practical error of the last twenty years,—not to load the memory of the student with a mass of undigested knowledge, but to force upon him so much that he has rejected all. It has been the error of distracting and enfeebling the mind by an unmeaning profusion of subjects; of implying that a smattering in a dozen branches of study is not shallowness, which it really is, but enlargement, which it is not; of considering an acquaintance with the learned names of things and persons, and the possession of clever duodecimos, and attendance on eloquent lecturers, and membership with scientific institutions, and the sight of the experiments of a platform and the specimens of a museum, that all this was not dissipation of mind, but progress. All things now are to be learned at once, not first one thing, then another, not one well, but many badly. Learning is to be without exertion, without attention, without toil; without grounding, without advance, without finishing. There is to be nothing individual in it; and this, forsooth, is the wonder of the age. What the steam engine does with matter, the printing press is to do with mind; it is to act mechanically, and the population is to be passively, almost unconsciously enlightened, by the mere multiplication and dissemination of volumes. Whether it be the school boy, or the school girl, or the youth at college, or the mechanic in the town, or the politician in the senate, all have been the victims in one way or other of this most preposterous and pernicious of delusions. Wise men have lifted up their voices in vain; and at length, lest their own institutions should be outshone and should disappear in the folly of the hour, they have been obliged, as far as they could with a good conscience, to humour a spirit which they could not withstand, and make temporizing concessions at which they could not but inwardly smile.

It must not be supposed that, because I so speak, therefore I have some sort of fear of the education of the people: on the contrary, the more education they have, the better, so that it is really education. Nor am I an enemy to the cheap publication of scientific and literary works, which is now in vogue: on the contrary, I consider it a great advantage, convenience, and gain; that is, to those to whom education has given a capacity for using them. Further, I consider such innocent recreations as science and literature are able to furnish will be a very fit occupation of the thoughts and the leisure of young persons, and may be made the means of keeping them from bad employments and bad companions. Moreover, as to that superficial acquaintance with chemistry, and geology, and astronomy, and political economy, and modern history, and biography, and other branches of knowledge, which periodical literature and occasional lectures and scientific institutions diffuse through the community, I think it a graceful accomplishment, and a suitable, nay, in this day a necessary accomplishment, in the case of educated men. Nor, lastly, am I disparaging or discouraging the thorough acquisition of any one of these studies, or denying that, as far as it goes, such thorough acquisition is a real education of the mind. All I say is, call things by their right names, and do not confuse together ideas which are essentially different. A thorough knowledge of one science and a superficial acquaintance with many, are not the same thing; a smattering of a hundred things or a memory for detail, is not a philosophical or comprehensive view. Recreations are not education; accomplishments are not education. Do not say, the people must be educated, when, after all, you only mean, amused, refreshed, soothed, put into good spirits and good humour, or kept from vicious excesses. I do not say that such amusements, such occupations of mind, are not a great gain; but they are not education. You may as well call drawing and fencing education, as a general knowledge of botany or conchology. Stuffing birds or playing stringed instruments is an elegant pastime, and a resource to the idle, but it is not education; it does not form or cultivate the intellect. Education is a high word; it is the preparation for knowledge, and it is the imparting of knowledge in proportion to that preparation. We require intellectual eyes to know withal, as bodily eyes for sight. We need both objects and organs intellectual; we cannot gain them without setting about it; we cannot gain them in our sleep, or by hap-hazard. The best telescope does not dispense with eyes; the printing press or the lecture room will assist us greatly, but we must be true to ourselves, we must be parties in the work. A University is, according to the usual designation, an Alma Mater, knowing her children one by one, not a foundry, or a mint, or a treadmill.

I protest to you, Gentlemen, that if I had to choose between a so-called University, which dispensed with residence and tutorial superintendence, and gave its degrees to any person who passed an examination in a wide range of subjects, and a University which had no professors or examinations at all, but merely brought a number of young men together for three or four years, and then sent them away as the University of Oxford is said to have done some sixty years since, if I were asked which of these two methods was the better discipline of the intellect,—mind, I do not say which is *morally* the better, for it is plain that compulsory study must be a good and idleness an intolerable mischief,—but if I must determine which of the two courses was the more successful in training, moulding, enlarging the mind, which sent out men the more fitted for their secular duties, which produced better public men, men of the world, men whose names would descend to posterity, I have no hesitation in giving the preference to that University which did nothing, over that which exacted of its members an acquaintance with every science under the sun. And, paradox as this may seem, still if results be the test of systems, the influence of the public schools and colleges of England, in the course of the last century, at least will bear out one side of the contrast as I have drawn it. What would come, on the other hand, of the ideal systems of education which have fascinated the imagination of this age, could they ever take effect, and whether they would not produce a generation frivolous, narrow-minded, and resourceless, intellectually considered, is a fair subject for debate; but so far is certain, that the Universities and scholastic establishments, to which I refer, and which did little more than bring together first boys and then youths in large numbers, these institutions, with miserable deformities on the side of morals, with a hollow profession of Christianity, and a heathen code of ethics,—I say, at least they can boast of a succession of heroes and statesmen, of literary men and philosophers, of men conspicuous for great natural virtues, for habits of business, for knowledge of life, for practical judgment, for cultivated tastes, for accomplishments, who have made England what it is,—able to subdue the earth, able to domineer over Catholics.

How is this to be explained? I suppose as follows: When a multitude of young men, keen, open-hearted, sympathetic, and observant, as young men are, come together and freely mix with each other, they are sure to learn one from another, even if there be no one to teach them; the conversation of all is a series of lectures to each, and they gain for themselves new ideas and views, fresh matter of thought, and distinct principles for judging and acting, day by day. An infant has to learn the meaning of the information which its senses convey to it, and this seems to be its employment. It fancies all that the eye presents to it to be close to it, till it actually learns the contrary, and thus by practice does it ascertain the relations and uses of those first elements of knowledge which are necessary for its animal existence. A parallel teaching is necessary for our social being, and it is secured by a large school or a college; and this effect may be fairly called in its own department an enlargement of mind. It is seeing the world on a small field with little trouble; for the pupils or students come from very different places, and with widely different notions, and there is much to generalize, much to adjust, much to eliminate, there are inter-relations to be defined, and conventional rules to be established, in the process, by which the whole assemblage is moulded together, and gains one tone and one character.

Let it be clearly understood, I repeat it, that I am not taking into account moral or religious considerations; I am but saying that that youthful community will constitute a whole, it will embody a specific idea, it will represent a doctrine, it will administer a code of conduct, and it will furnish principles of thought and action. It will give birth to a living teaching, which in course of time will take the shape of a self-perpetuating tradition, or a *genius loci*, as it is sometimes called; which haunts the home where it has been born, and which imbues and forms, more or less, and one by one, every individual who is successively brought under its shadow. Thus it is that, independent of direct instruction on the part of Superiors, there is a sort of self-education in the academic institutions of Protestant England; a characteristic tone of thought, a recognized standard of judgment is found in them, which, as developed in the individual who is submitted to it, becomes a twofold source of strength to him, both from the distinct stamp it impresses on his mind, and from the bond of union which it creates between him and others,—effects which are shared by the authorities of the place, for they themselves have been educated in it, and at all times are exposed to the influence of its ethical atmosphere. Here then is a real teaching, whatever be its standards and principles, true or false; and it at least tends towards cultivation of the intellect; it at least recognizes that knowledge is something more than a sort of passive reception of scraps and details; it is a something, and it does a something, which never will issue from the most strenuous efforts of a set of teachers, with no mutual sympathies and no intercommunion, of a set of examiners with no opinions which they dare profess, and with no common principles, who are teaching or questioning a set of youths who do not know them, and do not know each other, on a large number of subjects, different in kind, and connected by no wide philosophy, three times a week, or three times a year, or once in three years, in chill lecture-rooms or on a pompous anniversary.

Nay, self-education in any shape, in the most restricted sense, is preferable to a system of teaching which, professing so much, really does so little for the mind. Shut your College gates against the votary of knowledge, throw him back upon the searchings and the efforts of his own mind; he will gain by being spared an entrance into your Babel. Few indeed there are who can dispense with the stimulus and support of instructors, or will do any thing at all, if left to themselves. And fewer still (though such great minds are to be found), who will not, from such unassisted attempts, contract a self-reliance and a self-esteem, which are not only moral evils, but serious hindrances to the attainment of truth.

And next to none, perhaps, or none, who will not be reminded from time to time of the disadvantage under which they lie, by their imperfect grounding, by the breaks, deficiencies, and irregularities of their knowledge, by the eccentricity of opinion and the confusion of principle which they exhibit. They will be too often ignorant of what every one knows and takes for granted, of that multitude of small truths which fall upon the mind like dust, impalpable and ever accumulating; they may be unable to converse, they may argue perversely, they may pride themselves on their worst paradoxes or their grossest truisms, they may be full of their own mode of viewing things, unwilling to be put out of their way, slow to enter into the minds of others;—but, with these and whatever other liabilities upon their heads, they are likely to have more thought, more mind, more philosophy, more true enlargement, than those earnest but ill-used persons, who are forced to load their minds with a score of subjects against an examination, who have too much on their hands to indulge themselves in thinking or investigation, who devour premiss and conclusion together with indiscriminate greediness, who hold these sciences on faith, and commit demonstrations to memory, and who too often, as might be expected, when their period of education is passed, throw up all they have learned in disgust, having gained nothing really by their anxious labours, except perhaps the habit of application.

Yet such is the better specimen of the fruit of that ambitious system which has of late years been making way among us: for its result on ordinary minds, and on the common run of students, is less satisfactory still; they leave their place of education simply dissipated and relaxed by the multiplicity of subjects, which they have never really mastered, and so shallow as not even to know their shallowness. How much better, I say, is it for the active and thoughtful intellect, where such is to be found, to eschew the College and the University altogether, than to submit to a drudgery so ignoble, a mockery so contumelious! How much more profitable for the independent mind, after the mere rudiments of education, to range through a library at random, taking down books as they meet him, and pursuing the trains of thought which his mother wit suggests!

[Discourse 6]

Knowledge and Professional Skill

As the body may be sacrificed to some manual or other toil, whether moderate or oppressive, so may the intellect be devoted to some specific profession; and I do not call *this* the culture of the intellect. Again, as some member or organ of the body may be inordinately used and developed, so may memory, or imagination, or the reasoning faculty; and *this* again is not intellectual culture. On the other hand, as the body may be tended, cherished, and exercised with a simple view to its general health, so may the intellect also be generally exercised in order to its perfect state; and this *is* its cultivation.

Again, as health ought to precede labour of the body, and as a man in health can do what an unhealthy man cannot do, and as of this health the properties are strength, energy, agility, graceful carriage and action, manual dexterity, and endurance of fatigue, so in like manner general culture of mind is the best aid to professional and scientific study, and educated men can do what illiterate cannot; and the man who has learned to think and to reason and to compare and to discriminate and to analyze, who has refined his taste, and formed his judgment, and sharpened his mental vision, will not indeed at once be a lawyer, or a pleader, or an orator, or a statesman, or a physician, or a good landlord, or a man of business, or a soldier, or an engineer, or a chemist, or a geologist, or an antiquarian, but he will be placed in that state of intellect in which he can take up any one of the sciences or callings I have referred to, or any other for which he has a taste or special talent, with an ease, a grace, a versatility, and a success, to which another is a stranger. In this sense then, and as yet I have said but a very few words on a large subject, mental culture is emphatically *useful*.

If then I am arguing, and shall argue, against Professional or Scientific knowledge as the sufficient end of a University Education, let me not be supposed, Gentlemen, to be disrespectful towards particular studies, or arts, or vocations, and those who are engaged in them. In saying that Law or Medicine is not the end of a University course, I do not mean to imply that the University does not teach Law or Medicine. What indeed can it teach at all, if it does not teach something particular? It teaches *all* knowledge by teaching all *branches* of knowledge, and in no other way. I do but say that there will be this distinction as regards a Professor of Law, or of Medicine, or of Geology, or of Political Economy, in a University and out of it, that out of a University he is in danger of being absorbed and narrowed by his pursuit, and of giving Lectures which are the Lectures of nothing more than a lawyer, physician, geologist, or political economist; whereas in a University he will just know where he and his science stand, he has come to it, as it were, from a height, he has taken a survey of all knowledge, he is kept from extravagance by the very rivalry of other studies, he has gained from them a special illumination and largeness of mind and freedom and self-possession, and he treats his own in consequence with a philosophy and a resource, which belongs not to the study itself, but to his liberal education.

This then is how I should solve the fallacy, for so I must call it, by which Locke and his disciples would frighten us from cultivating the intellect, under the notion that no education is useful which does not teach us some temporal calling, or some mechanical art, or some physical secret. I say that a cultivated intellect, because it is a good in itself, brings with it a power and a grace to every work and occupation which it undertakes, and enables us to be more useful, and to a greater number. There is a duty we owe to human society as such, to the state to which we belong, to the sphere in which we move, to the individuals towards whom we are variously related, and whom we successively encounter in life; and that philosophical or liberal education, as I have called it, which is the proper function of a University, if it refuses the foremost place to professional interests, does but postpone them to the formation of the citizen, and, while it subserves the larger interests of philanthropy, prepares also for the successful prosecution of those merely personal objects, which at first sight it seems to disparage. . . .

. . . If then a practical end must be assigned to a University course, I say it is that of training good members of society. Its art is the art of social life, and its end is fitness for the world. It neither confines its views to particular professions on the one hand, nor creates heroes or inspires genius on the other. Works indeed of genius fall under no art; heroic minds come under no rule; a University is not a birthplace of poets or of immortal authors, of founders of schools, leaders of colonies, or conquerors of nations. It does not promise a generation of Aristotles or Newtons, of Napoleons or Washingtons, of Raphaels or Shakespeares, though such miracles of nature it has before now contained within its precincts. Nor is it content on the other hand with forming the critic or the experimentalist, the economist or the engineer, though such too it includes within its scope. But a University training is the great ordinary means to a great but ordinary end; it aims at raising the intellectual tone of society, at cultivating the public mind, at purifying the national taste, at supplying true principles to popular enthusiasm and fixed aims to popular aspiration, at giving enlargement and sobriety to the ideas of the age, at facilitating the exercise of political power, and refining the intercourse of private life. It is the education which gives a man a clear conscious view of his own opinions and judgments, a truth in developing them, an eloquence in expressing them, and a force in urging them. It teaches him to see things as they are, to go right to the point, to disentangle a skein of thought, to detect what is sophistical, and to discard what is irrelevant. It prepares him to fill any post with credit, and to master any subject with facility. It shows him how to accommodate himself to others, how to throw himself into their state of mind, how to bring before them his own, how to influence them, how to come to an understanding with them, how to bear with them. He is at home in any society, he has common ground with every class; he knows when to speak and when to be silent; he is able to converse, he is able to listen; he can ask a question pertinently, and gain a lesson seasonably, when he has nothing to impart himself; he is ever ready, yet never in the way; he is a pleasant companion, and a comrade you can depend upon; he knows when to be serious and when to trifle, and he has a sure tact which enables him to trifle with gracefulness and to be serious with effect. He has the repose of a mind which lives in itself, while it lives in the world, and which has resources for its happiness at home when it cannot go abroad. He has a gift which serves him in public, and supports him in retirement, without which good fortune is but vulgar, and with which failure and disappointment have a charm. The art which tends to make a man all this, is in the object which it pursues as useful as the art of wealth or the art of health, though it is less susceptible of method, and less tangible, less certain, less complete in its result.

[Discourse 7]

Knowledge and Religious Duty

Right Reason, that is, Reason rightly exercised, leads the mind to the Catholic Faith, and plants it there, and teaches it in all its religious speculations to act under its guidance. But Reason, considered as a real agent in the world, and as an operative principle in man's nature, with an historical course and with definite results, is far from taking so straight and satisfactory a direction. It considers itself from first to last independent and supreme; it requires no external authority; it makes a religion for itself. Even though it accepts Catholicism, it does not go to sleep; it has an action and development of its own, as the passions have, or the moral sentiments, or the principle of self-interest. Divine grace, to use the language of Theology, does not by its presence supersede nature; nor is nature at once brought into simple concurrence and coalition with grace. Nature pursues its course, now coincident with that of grace, now parallel to it, now across, now divergent, now counter, in proportion to its own imperfection and to the attraction and influence which grace exerts over it. And what takes place as regards other principles of our nature and their developments is found also as regards the Reason. There is, we know, a Religion of enthusiasm, of superstitious ignorance of statecraft; and each has that in it which resembles Catholicism, and that again which contradicts Catholicism. There is the Religion of a warlike people, and of a pastoral people; there is a Religion of rude times, and in like manner there is a Religion of civilized times, of the cultivated intellect, of the philosopher, scholar, and gentleman. This is that Religion of Reason, of which I speak. Viewed in itself, however near it comes to Catholicism, it is of course simply distinct from it; for Catholicism is one whole, and admits of no compromise or modification. Yet this is to view it in the abstract; in matter of fact, and in reference to individuals, we can have no difficulty in conceiving this philosophical Religion present in a Catholic country, as a spirit influencing men to a certain extent, for good or for bad or for both,—a spirit of the age, which again may be found, as among Catholics, so with still greater sway and success in a country not Catholic, yet specifically the same in such a country as it exists in a Catholic community. The problem then before us to-day, is to set down some portions of the outline, if we can ascertain them, of the Religion of Civilization, and to determine how they lie relatively to those principles, doctrines, and rules, which Heaven has given us in the Catholic Church.

And here again, when I speak of Revealed Truth, it is scarcely necessary to say that I am not referring to the main articles and prominent points of faith, as contained in the Creed. Had I undertaken to delineate a philosophy, which directly interfered with the Creed, I could not have spoken of it as compatible with the profession of Catholicism. The philosophy I speak of, whether it be viewed within or outside the Church, does not necessarily take cognizance of the Creed. Where the country is Catholic, the educated mind takes its articles for granted, by a sort of implicit faith; where it is not, it simply ignores them and the whole subject-matter to which they relate, as not affecting social and political interests. Truths about God's Nature, about His dealings towards the human race, about the Economy of Redemption,—in the one case it humbly accepts them, and passes on; in the other it passes them over, as matters of simple opinion, which never can be decided, and which can have no power over us to make us morally better or worse. I am not speaking then of belief in the great objects of faith, when I speak of Catholicism, but I am contemplating Catholicism chiefly as a system of pastoral instruction and moral duty; and I have to do with its doctrines mainly as they are subservient to its direction of the conscience and the conduct. I speak of it, for instance, as teaching the ruined state of man; his utter inability to gain Heaven by any thing he can do himself; the moral certainty of his losing his soul if left to himself; the simple absence of all rights and claims on the part of the creature in the presence of the Creator; the illimitable claims of the Creator on the service of the creature; the imperative and obligatory force of the voice of conscience; and the inconceivable evil of sensuality. I speak of it as teaching, that no one gains Heaven except by the free grace of God, or without a regeneration of nature; that no one can please Him without faith; that the heart is the seat both of sin and of obedience; that charity is the fulfilling of the Law; and that incorporation into the Catholic Church is the ordinary instrument of salvation. These are the lessons which distinguish Catholicism as a popular religion, and these are the subjects to which the cultivated intellect will practically be turned:—I have to compare and contrast, not the doctrinal, but the moral and social teaching of Philosophy on the one hand, and Catholicism on the other.

Now, on opening the subject, we see at once a momentous benefit which the philosopher is likely to confer on the pastors of the Church. It is obvious that the first step which they have to effect in the conversion of man and the renovation of his nature, is his rescue from that fearful subjection to sense which is his ordinary state. To be able to break through the meshes of that thraldom, and to disentangle and to disengage its ten thousand holds upon the heart, is to bring it, I might almost say, half way to Heaven. Here, even divine grace, to speak of things according to their appearances, is ordinarily baffled, and retires, without expedient or resource, before this giant fascination. Religion seems too high and unearthly to be able to exert a continued influence upon us: its effort to rouse the soul, and the soul's effort to co-operate, are too violent to last. It is like holding out the arm at full length, or supporting some great weight, which we manage to do for a time, but soon are exhausted and succumb. Nothing can act beyond its own nature; when then we are called to what is supernatural, though those extraordinary aids from Heaven are given us, with which obedience becomes possible, yet even with them it is of transcendent difficulty. We are drawn down to earth every moment with the ease and certainty of natural gravitation, and it is only by sudden impulses and, as it were, forcible plunges that we attempt to mount upwards. Religion indeed enlightens, terrifies, subdues; it gives faith, it inflicts remorse, it inspires resolutions, it draws tears, it inflames devotion, but only for the occasion. I repeat, it imparts an inward power which ought to effect more than this; I am not forgetting either the real sufficiency of its aids, nor the responsibility of those in whom they fail. I am not discussing theological questions at all, I am looking at phenomena as they lie before me, and I say that, in matter of fact, the sinful spirit repents, and protests it will never sin again, and for a while is protected by disgust and abhorrence from the malice of its foe. But that foe knows too well that such seasons of reprentance are wont to have their end: he patiently waits, till nature faints with the effort of resistance, and lies passive and hopeless under the next access of temptation. What we need then is some expedient or instrument, which at least will obstruct and stave off the approach of our spiritual enemy, and which is sufficiently congenial and level with our nature to maintain as firm a hold upon us as the inducements of sensual gratification. It will be our wisdom to employ nature against itself. Thus sorrow, sickness, and care are providential antagonists to our inward disorders; they come upon us as years pass on, and generally produce their natural effects on us, in proportion as we are subjected to their influence. These, however, are God's instruments, not ours; we need a similar remedy, which we can make our own, the object of some legitimate faculty, or the aim of some natural affection, which is capable of resting on the mind, and taking up its familiar lodging with it, and engrossing it, and which thus becomes a match for the besetting power of sensuality, and a sort of homœopathic medicine for the disease. Here then I think is the important aid which intellectual cultivation furnishes to us in rescuing the victims of passion and self-will. It does not supply religious motives; it is not the cause or proper antecedent of any thing supernatural; it is not meritorious of heavenly aid or reward; but it does a work, at least *materially* good (as theologians speak), whatever be its real and formal character. It expels the excitements of sense by the introduction of those of the intellect.

This then is the *primâ facie* advantage of the pursuit of Knowledge; it is the drawing the mind off from things which will harm it to subjects which are worthy a rational being; and though it does not raise it above nature, nor has any tendency to make us pleasing to our Maker, yet is it nothing to substitute what is in itself harmless for what is, to say the least, inexpressibly dangerous? is it a little thing to exchange a circle of ideas which are certainly sinful, for others which are certainly not so? You will say, perhaps, in the words of the Apostle, "Knowledge puffeth up:" and doubtless this mental cultivation, even when it is successful for the purpose for which I am applying it, may be from the first nothing more than the substitution of pride for sensuality. I grant it, I think I shall have something to say on this point presently; but this is not a necessary result, it is but an incidental evil, a danger which may be realized or may be averted, whereas we may in most cases predicate guilt, and guilt of a heinous kind, where the mind is suffered to run wild and indulge its thoughts without training or law of any kind; and surely to turn away a soul from mortal sin is a good and a gain so far, whatever comes of it. And therefore, if a friend in need is twice a friend, I conceive that intellectual employments, though they do no more than occupy the mind with objects naturally noble or innocent, have a special claim upon our consideration and gratitude. . . .

Such are the instruments by which an age of advanced civilization combats those moral disorders, which Reason as well as Revelation denounces; and I have not been backward to express my sense of their serviceableness to Religion. Moreover, they are but the foremost of a series of influences, which intellectual culture exerts upon our moral nature, and all upon the type of Christianity, manifesting themselves in veracity, probity, equity, fairness, gentleness, benevolence, and amiableness; so much so, that a character more noble to look at, more beautiful, more winning, in the various relations of life and in personal duties, is hardly conceivable, than may, or might be, its result, when that culture is bestowed upon a soil naturally adapted to virtue. If you would obtain a picture for contemplation which may seem to fulfil the ideal, which the Apostle has delineated under the name of charity, in its sweetness and harmony, its generosity, its courtesy to others, and its depreciation of self, you could not have recourse to a better furnished *studio* than to that of Philosophy, with the specimens of it, which with greater or less exactness are scattered through society in a civilized age. It is enough to refer you, Gentlemen, to the various Biographies and Remains of contemporaries and others, which from time to time issue from the press, to see how striking is the action of our intellectual upon our moral nature, where the moral material is rich, and the intellectual cast is perfect. Individuals will occur to all of us, who deservedly attract our love and admiration, and whom the world almost worships as the work of its own hands. Religious principle, indeed,—that is, faith,—is, to all appearance, simply away; the work is as certainly not supernatural as it is certainly noble and beautiful. This must be insisted on, that the intellect may have its due; but it also must be insisted on for the sake of conclusions to which I wish to conduct our investigation. The radical difference indeed of this mental refinement from genuine religion, in spite of its seeming relationship, is the very cardinal point on which my present discussion turns; yet, on the other hand, such refinement may readily be assigned to a Christian origin by hasty or distant observers, or by those who view it in a particular light. And as this is the case, I think it advisable, before proceeding with the delineation of its characteristic features, to point out to you distinctly the elementary principles on which its morality is based.

You will bear in mind then, Gentlemen, that I spoke just now of the scorn and hatred which a cultivated mind feels for some kinds of vice, and the utter disgust and profound humiliation which may come over it, if it should happen in any degree to be betrayed into them. Now this feeling may have its root in faith and love, but it may not; there is nothing really religious in it, considered by itself. Conscience indeed is implanted in the breast by nature, but it inflicts upon us fear as well as shame; when the mind is simply angry with itself and nothing more, surely the true import of the voice of nature and the depth of its intimations have been forgotten, and a false philosophy has misinterpreted emotions which ought to lead to God. Fear implies the transgression of a law, and a law implies a lawgiver and judge; but the tendency of intellectual culture is to swallow up the fear in the self-reproach, and self-reproach is directed and limited to our mere sense of what is fitting and becoming. Fear carries us out of ourselves, whereas shame may act upon us only within the round of our own thoughts. Such, I say, is the danger which awaits a civilized age; such is its besetting sin (not inevitable, God forbid! or we must abandon the use of God's own gifts), but still the ordinary sin of the Intellect; conscience tends to become what is called a moral sense; the command of duty is a sort of taste; sin is not an offence against God, but against human nature.

The less amiable specimens of this spurious religion are those which we meet not unfrequently in my own country. . . .

We find there men possessed of many virtues, but proud, bashful, fastidious, and reserved. Why is this? it is because they think and act as if there were really nothing objective in their religion; it is because conscience to them is not the word of a lawgiver, as it ought to to be, but the dictate of their own minds and nothing more; it is because they do not look out of themselves, because they do not look through and beyond their own minds to their Maker, but are engrossed in notions of what is due to themselves, to their own dignity and their own consistency. Their conscience has become a mere self-respect. Instead of doing one thing and then another, as each is called for, in faith and obedience, careless of what may be called the *keeping* of deed with deed, and leaving Him who gives the command to blend the portions of their conduct into a whole, their one object, however unconscious to themselves, is to paint a smooth and perfect surface, and to be able to say to themselves that they have done their duty. When they do wrong, they feel, not contrition, of which God is the object, but remorse, and a sense of degradation. They call themselves fools, not sinners; they are angry and impatient, not humble. They shut themselves up in themselves; it is misery to them to think or to speak of their own feelings; it is misery to suppose that others see them, and their shyness and sensitiveness often become morbid. As to confession, which is so natural to the Catholic, to them it is impossible; unless indeed, in cases where they have been guilty, an apology is due to their own character, is expected of them, and will be satisfactory to look back upon. They are victims of an intense self-contemplation. . . .

. . . And thus at length we find, surprising as the change may be, that that very refinement of Intellectualism, which began by repelling sensuality, ends by excusing it. Under the shadow indeed of the Church, and in its due development, Philosophy does service to the cause of morality; but, when it is strong enough to have a will of its own, and is lifted up with an idea of its own importance, and attempts to form a theory, and to lay down a principle, and to carry out a system of ethics, and undertakes the moral education of the man, then it does but abet evils to which at first it seemed instinctively opposed. True Religion is slow in growth, and, when once planted, is difficult of dislodgement; but its intellectual counterfeit has no root in itself: it springs up suddenly, it suddenly withers. It appeals to what is in nature, and it falls under the dominion of the old Adam. Then, like dethroned princes, it keeps up a state and majesty, when it has lost the real power. Deformity is its abhorrence; accordingly, since it cannot dissuade men from vice, therefore in order to escape the sight of its deformity, it embellishes it. . . .

And from this shallowness of philosophical Religion it comes to pass that its disciples seem able to fulfil certain precepts of Christianity more readily and exactly than Christians themselves. . . . At this day the "gentleman" is the creation, not of Christianity, but of civilization. But the reason is obvious. The world is content with setting right the surface of things; the Church aims at regenerating the very depths of the heart. She ever begins with the beginning; and, as regards the multitude of her children, is never able to get beyond the beginning, but is continually employed in laying the foundation. She is engaged with what is essential, as previous and as introductory to the ornamental and the attractive. She is curing men and keeping them clear of mortal sin; she is "treating of justice and chastity, and the judgment to come:" she is insisting on faith and hope, and devotion, and honesty, and the elements of charity; and has so much to do with precept, that she almost leaves it to inspirations from Heaven to suggest what is of counsel and perfection. She aims at what is necessary rather than at what is desirable. She is for the many as well as for the few. She is putting souls in the way of salvation, that they may then be in a condition, if they shall be called upon, to aspire to the heroic, and to attain the full proportions, as well as the rudiments, of the beautiful.

Such is the method, or the policy (so to call it), of the Church; but Philosophy looks at the matter from a very different point of view: what have Philosophers to do with the terror of judgment or the saving of the soul? . . .

The embellishment of the exterior is almost the beginning and the end of philosophical morality. This is why it aims at being modest rather than humble; this is how it can be proud at the very time that it is unassuming. To humility indeed it does not even aspire; humility is one of the most difficult of virtues both to attain and to ascertain. It lies close upon the heart itself, and its tests are exceedingly delicate and subtle. Its counterfeits abound; however, we are little concerned with them here, for, I repeat, it is hardly professed even by name in the code of ethics which we are reviewing. As has been often observed, ancient civilization had not the idea, and had no word to express it: or rather, it had the idea, and considered it a defect of mind, not a virtue, so that the word which denoted it conveyed a reproach. As to the modern world, you may gather its ignorance of it by its perversion of the somewhat parallel term "condescension." Humility or condescension, viewed as a virtue of conduct, may be said to consist, as in other things, so in our placing ourselves in our thoughts on a level with our inferiors; it is not only a voluntary relinquishment of the privileges of our own station, but an actual participation or assumption of the condition of those to whom we stoop. This is true humility, to feel and to behave as if we were low; not, to cherish a notion of our importance, while we affect a low position. Such was St. Paul's humility, when he called himself "the least of the saints;" such the humility of those many holy men who have considered themselves the greatest of sinners. It is an abdication, as far as their own thoughts are concerned, of those prerogatives or privileges to which others deem them entitled. Now it is not a little instructive to contrast with this idea, Gentlemen,—with this theological meaning of the word "condescension,"—its proper English sense; put them in juxta-position, and you will at once see the difference between the world's humility and the humility of the Gospel. As the world uses the word, "condescension" is a stooping indeed of the person, but a bending forward, unattended with any the slightest effort to leave by a single inch the seat in which it is so firmly established. It is the act of a superior, who protests to himself, while he commits it, that he is superior still, and that he is doing nothing else but an act of grace towards those on whose level, in theory, he is placing himself. And this is the nearest idea which the philosopher can form of the virtue of self-abasement; to do more than this is to his mind a meanness or an hypocrisy, and at once excites his suspicion and disgust. What the world is, such it has ever been; we know the contempt which the educated pagans had for the martyrs and confessors of the Church; and it is shared by the anti-Catholic bodies of this day.

Such are the ethics of Philosophy, when faithfully represented; but an age like this, not pagan, but professedly Christian, cannot venture to reprobate humility in set terms, or to make a boast of pride. Accordingly, it looks out for some expedient by which it may blind itself to the real state of the case. Humility, with its grave and self-denying attributes, it cannot love: but what is more beautiful, what more winning, than modesty? what virtue, at first sight, simulates humility so well? though what in fact is more radically distinct from it? In truth, great as is its charm, modesty is not the deepest or the most religious of virtues. Rather it is the advanced guard or sentinel of the soul militant, and watches continually over its nascent intercourse with the world about it. It goes the round of the senses; it mounts up into the countenance; it protects the eye and ear; it reigns in the voice and gesture. Its province is the outward deportment, as other virtues have relation to matters theological, others to society, and others to the mind itself. And being more superficial than other virtues, it is more easily disjoined from their company; it admits of being associated with principles or qualities naturally foreign to it, and is often made the cloak of feelings or ends for which it was never given to us. So little is it the necessary index of humility, that it is even compatible with pride. The better for the purpose of Philosophy; humble it cannot be, so forthwith modesty becomes its humility.

Pride, under such training, instead of running to waste in the education of the mind, is turned to account; it gets a new name; it is called self-respect; and ceases to be the disagreeable, uncompanionable quality which it is in itself. Though it be the motive principle of the soul, it seldom comes to view; and when it shows itself, then delicacy and gentleness are its attire, and good sense and sense of honour direct its motions. It is no longer a restless agent, without definite aim; it has a large field of exertion assigned to it, and its subserves those social interests which it would naturally trouble. It is directed into the channel of industry, frugality, honesty, and obedience; and it becomes the very staple of the religion and morality held in honour in a day like our own. It becomes the safeguard of chastity, the guarantee of veracity, in high and low; it is the very household god of society, as at present constituted, inspiring neatness and decency in the servant girl, propriety of carriage and refined manners in her mistress, uprightness, manliness, and generosity in the head of the family. It diffuses a light over town and country; it covers the soil with handsome edifices and smiling gardens; it tills the field, it stocks and embellishes the shop. It is the stimulating principle of providence on the one hand, and of free expenditure on the other; of an honourable ambition, and of elegant enjoyment. It breathes upon the face of the community, and the hollow sepulchre is forthwith beautiful to look upon.

Refined by the civilization which has brought it into activity, this self-respect infuses into the mind an intense horror of exposure, and a keen sensitiveness of notoriety and ridicule. It becomes the enemy of extravagances of any kind; it shrinks from what are called scenes; it has no mercy on the mock-heroic, on pretence or egotism, on verbosity in language, or what is called prosiness in conversation. It detests gross adulation; not that it tends at all to the eradication of the appetite to which the flatterer ministers, but it sees the absurdity of indulging it, it understands the annoyance thereby given to others, and if a tribute must be paid to the wealthy or the powerful, it demands greater subtlety and art in the preparation. Thus vanity is changed into a more dangerous self-conceit, as being checked in its natural eruption. It teaches men to suppress their feelings, and to control their tempers, and to mitigate both the severity and the tone of their judgments. As Lord Shaftesbury would desire, it prefers playful wit and satire in putting down what is objectionable, as a more refined and good-natured, as well as a more effectual method, than the expedient which is natural to uneducated minds. It is from this impatience of the tragic and the bombastic that it is now quietly but energetically opposing itself to the unchristian practice of duelling, which it brands as simply out of taste, and as the remnant of a barbarous age; and certainly it seems likely to effect what Religion has aimed at abolishing in vain.

Hence it is that it is almost a definition of a gentleman to say he is one who never inflicts pain. This description is both refined and, as far as it goes, accurate. He is mainly occupied in merely removing the obstacles which hinder the free and unembarrassed action of those about him; and he concurs with their movements rather than takes the initiative himself. His benefits may be considered as parallel to what are called comforts or conveniences in arrangements of a personal nature: like an easy chair or a good fire, which do their part in dispelling cold and fatigue, though nature provides both means of rest and animal heat without them. The true gentleman in like manner carefully avoids whatever may cause a jar or a jolt in the minds of those with whom he is cast;—all clashing of opinion, or collision of feeling, all restraint, or suspicion, or gloom, or resentment; his great concern being to make every one at their ease and at home. He has his eyes on all his company; he is tender towards the bashful, gentle towards the distant, and merciful towards the absurd; he can recollect to whom he is speaking; he guards against unseasonable allusions, or topics which may irritate; he is seldom prominent in conversation, and never wearisome. He makes light of favours while he does them, and seems to be receiving when he is conferring. He never speaks of himself except when compelled, never defends himself by a mere retort, he has no ears for slander or gossip, is scrupulous in imputing motives to those who interfere with him, and interprets every thing for the best. He is never mean or little in his disputes, never takes unfair advantage, never mistakes personalities or sharp sayings for arguments, or insinuates evil which he dare not say out. For a long-sighted prudence, he observes the maxim of the ancient sage, that we should ever conduct ourselves towards our enemy as if he were one day to be our friend. He has too much good sense to be affronted at insults, he is too well employed to remember injuries, and too indolent to bear malice. He is patient, forbearing, and resigned, on philosophical principles; he submits to pain, because it is inevitable, to bereavement, because it is irreparable, and to death, because it is his destiny. If he engages in controversy of any kind, his disciplined intellect preserves him from the blundering discourtesy of better, perhaps, but less educated minds; who, like blunt weapons, tear and hack instead of cutting clean, who mistake the point in argument, waste their strength on trifles, misconceive their adversary, and leave the question more involved than they find it. He may be right or wrong in his opinion, but he is too clear-headed to be unjust; he is as simple as he is forcible, and as brief as he is decisive. Nowhere shall we find greater candour, consideration, indulgence: he throws himself into the minds of his opponents, he accounts for their mistakes. He knows the weakness of human reason as well as its strength, its province and its limits. If he be an unbeliever, he will be too profound and large-minded to ridicule religion or to act against it; he is too wise to be a dogmatist or fanatic in his infidelity. He respects piety and devotion; he even supports institutions as venerable, beautiful, or useful, to which he does not assent; he honours the ministers of religion, and it contents him to decline its mysteries without assailing or denouncing them. He is a friend of religious toleration, and that, not only because his philosophy has taught him to look on all forms of faith with an impartial eye, but also from the gentleness and effeminacy of feeling, which is the attendant on civilization.

Not that he may not hold a religion too, in his own way, even when he is not a Christian. In that case his religion is one of imagination and sentiment; it is the embodiment of those ideas of the sublime, majestic, and beautiful, without which there can be no large philosophy. Sometimes he acknowledges the being of God, sometimes he invests an unknown principle or quality with the attributes of perfection. And this deduction of his reason, or creation of his fancy, he makes the occasion of such excellent thoughts, and the starting-point of so varied and systematic a teaching, that he even seems like a disciple of Christianity itself. From the very accuracy and steadiness of his logical powers, he is able to see what sentiments are consistent in those who hold any religious doctrine at all, and he appears to others to feel and to hold a whole circle of theological truths, which exist in his mind no otherwise than as a number of deductions.

Such are some of the lineaments of the ethical character, which the cultivated intellect will form, apart from religious principle. They are seen within the pale of the Church and without it, in holy men, and in profligate; they form the *beau-ideal* of the world; they partly assist and partly distort the development of the Catholic. They may subserve the education of a St. Francis de Sales or a Cardinal Pole; they may be the limits of the contemplation of a Shaftesbury or a Gibbon. Basil and Julian were fellow-students at the schools of Athens; and one became the Saint and Doctor of the Church, the other her scoffing and relentless foe.

[Discourse 8]

2

THE PHILOSOPHER

Newman's *Sermons, Chiefly on the Theory of Religious Belief, preached before the University of Oxford* (1843) were delivered from the pulpit of the University Church of St Mary the Virgin between 1826 and 1843. Their main theme is the relation of faith to reason. Newman argues that far from faith being opposed to reason, faith is an example of reasoning based upon 'presumptions' or 'antecedent probabilities' as opposed to empirical evidence or logic. But this is true, too, of other kinds of reasoning which are also likely to be criticized as irrational. Faith, then, in this sense is no different from other kinds of intellectual activity where proof follows from unproved assumptions. But religious faith, like unbelief, rests upon probabilities which are determined by moral principles: what seems probable to a good person will not seem probable to a bad person. Faith, like other kinds of reasoning which it resembles, may be either of an explicit or an implicit nature depending on whether or not the actual grounds and reasons are recognized and understood. The originality of these sermons, usually referred to as the *Oxford University Sermons*, lies in the way Newman replaces the Englightenment's narrow conception of reason not with non-intellectual experiences and feelings but with a wider understanding of the rational or reasonable.

Newman's concern to provide a philosophical justification for religious belief remained with him through the years while society became more and more secularized. Above all he wanted to show that it is as reasonable for an uneducated person to be certain about their faith as a learned theologian. His various attempts to write on the problem of certainty came to nothing, until 1866, when, on a holiday in Switzerland, it suddenly struck him that he must consider first the nature of 'assent' in general. The moment marked the conception of *An Essay in Aid of a Grammar of Assent* (1870). It is the one work of Newman which, far from being, like so many of his books, a controversial writing for a particular occasion, was on the contrary long meditated and premeditated, and, even if not his best book, it is in a real sense his *magnum opus*. On its appearance it was criticized by both Catholic scholastic philosophers and also by secular critics for its failure to provide a theory of truth which would enable true certitudes to be distinguished from false certitudes. Newman did not profess to be able to furnish such a rationale.

The first part of the book is concerned mainly with the celebrated distinction between 'notional' and 'real' apprehension and assent. In the second half Newman maintains that certitude is generally 'indefectible', but that there are no sure tests and exceptions are always possible. In discussing how we attain to certitude in concrete matters where strictly logical reasoning is not applicable, he concludes that it is 'the cumulation of probabilities' that leads the 'illative sense' or 'right judgment' in informal reasoning to certainty. If the *Grammar of Assent* is misunderstood to be a 'metaphysical' work, it is bound to disappoint. But that does not mean that it is exactly a 'psychological' study. Rather, it is a philosophical analysis of that state of mind which we ordinarily call certitude or certainty and of the intellectual acts associated with it; and, as such, it has come to be recognized as a classic of the philosophy of religion. Religious certainty becomes one of many kinds of certainty which are not 'proved' in the formal sense of the word. The justification of religious belief merges into a much more general justification of the validity of those ordinary processes of thought which normally lead to conviction.

OXFORD UNIVERSITY SERMONS

Faith and Reason, Contrasted as Habits of Mind
(Preached on the Epiphany, 1839)

Now Faith is the substance of things hoped for, the evidence of things
not seen.

Heb. 11: 1

... It is scarcely necessary to prove from Scripture, the especial dignity
and influence of Faith, under the Gospel Dispensation, as regards both
our spiritual and moral condition. Whatever be the particular faculty or
frame of mind denoted by the word, certainly Faith is regarded in
Scripture as the chosen instrument connecting heaven and earth, as a
novel principle of action most powerful in the influence which it exerts
both on the heart and on the Divine view of us, and yet in itself of a
nature to excite the contempt or ridicule of the world. ...

... in the minds of the sacred writers, Faith is an instrument of
knowledge and action, unknown to the world before, a principle *sui*
generis, distinct from those which nature supplies, and in particular
(which is the point into which I mean to inquire) independent of what is
commonly understood by Reason. Certainly if, after all that is said
about Faith in the New Testament, as if it were what may be called a
discovery of the Gospel, and a special divine method of salvation; if, after
all, it turns out merely to be a believing upon evidence, or a sort of
conclusion upon a process of reasoning, a resolve formed upon a
calculation, the inspired text is not level to the understanding, or
adapted to the instruction, of the unlearned reader. If Faith be such a
principle, how is it novel and strange? ...

Such is the question which presents itself to readers of Scripture, as to
the relation of Faith to Reason: and it is usual at this day to settle it in
disparagement of Faith,—to say that Faith is but a moral quality,
dependent upon Reason,—that Reason judges both of the evidence on
which Scripture is to be received, and of the meaning of Scripture; and
then Faith follows or not, according to the state of the heart; that we
make up our minds by Reason without Faith, and then we proceed to
adore and to obey by Faith apart from Reason; that, though Faith rests
on testimony, not on reasonings, yet that testimony, in its turn, depends
on Reason for the proof of its pretensions, so that Reason is an
indispensable preliminary.

Now, in attempting to investigate what are the distinct offices of Faith
and Reason in religious matters, and the relation of the one to the other,

I observe, first, that undeniable though it be, that Reason has a power of analysis and criticism in all opinion and conduct, and that nothing is true or right but what may be justified, and, in a certain sense, proved by it, and undeniable, in consequence, that, unless the doctrines received by Faith are approvable by Reason, they have no claim to be regarded as true, it does not therefore follow that Faith is actually grounded on Reason in the believing mind itself; unless, indeed, to take a parallel case, a judge can be called the origin, as well as the justifier, of the innocence or truth of those who are brought before him. A judge does not make men honest, but acquits and vindicates them: in like manner, Reason need not be the origin of Faith, as Faith exists in the very persons believing, though it does test and verify it. This, then, is one confusion, which must be cleared up in the question,—the assumption that Reason must be the inward principle of action in religious inquiries or conduct in the case of this or that individual, because, like a spectator, it acknowledges and concurs in what goes on;—the mistake of a critical for a creative power.

This distinction we cannot fail to recognize as true in itself, and applicable to the matter in hand. It is what we all admit at once as regards the principle of Conscience. No one will say that Conscience is against Reason, or that its dictates cannot be thrown into an argumentative form; yet who will, therefore, maintain that it is not an original principle, but must depend, before it acts, upon some previous processes of Reason? Reason analyzes the grounds and motives of action: a reason is an analysis, but is not the motive itself. As, then, Conscience is a simple element in our nature, yet its operations admit of being surveyed and scrutinized by Reason; so may faith be cognizable, and its acts be justified, by Reason, without therefore being, in matter of fact, dependent upon it; and as we reprobate, under the name of Utilitarianism, the substitution of Reason for Conscience, so perchance it is a parallel error to teach that a process of Reason is the *sine quâ non* for true religious Faith. When the Gospel is said to require a rational Faith, this need not mean more than that Faith is accordant to right Reason in the abstract, not that it results from it in the particular case.

A parallel and familiar instance is presented by the generally-acknowledged contrast between poetical or similar powers, and the art of criticism. That art is the sovereign awarder of praise and blame, and constitutes a court of appeal in matters of taste; as then the critic ascertains what he cannot himself create, so Reason may put its sanction upon the acts of Faith, without in consequence being the source from which Faith springs.

On the other hand, Faith certainly does seem, in matter of fact, to exist and operate quite independently of Reason. Will any one say that a child or uneducated person may not savingly act on Faith, without being able to produce reasons why he so acts? What sufficient view has he of the Evidences of Christianity? What logical proof of its divinity? If he has none, Faith, viewed as an internal habit or act, does not depend upon inquiry and examination, but has its own special basis, whatever that is, as truly as Conscience has. We see, then, that Reason may be the judge, without being the origin, of Faith; and that Faith may be justified by Reason, without making use of it. This is what it occurs to mention at first sight.

Next, I observe, that, whatever be the real distinction and relation existing between Faith and Reason, which it is not to our purpose at once to determine, the contrast that would be made between them, on a popular view, is this,—that Reason requires strong evidence before it assents, and Faith is content with weaker evidence.

For instance: when a well-known infidel of the last century argues, that the divinity of Christianity is founded on the testimony of the Apostles, in opposition to the experience of nature, and that the laws of nature are uniform, those of testimony variable, and scoffingly adds that Christianity is founded on Faith, not on Reason, what is this but saying that Reason is severer in its demands of evidence than Faith?

Again, the founder of the recent Utilitarian School insists, that all evidence for miracles, before it can be received, should be brought into a court of law, and subjected to its searching forms:—this too is to imply that Reason demands exact proofs, but that Faith accepts inaccurate ones.

The same thing is implied in the notion which men of the world entertain, that Faith is but credulity, superstition, or fanaticism; these principles being notoriously such as are contented with insufficient evidence concerning their objects. On the other hand, scepticism, which shows itself in a dissatisfaction with evidence of whatever kind, is often called by the name of Reason. What Faith, then, and Reason are, when compared together, may be determined from their counterfeits,—from the mutual relation of credulity and scepticism, which no one can doubt about. . . .

When, then, Reason and Faith are contrasted together, Faith means easiness, Reason, difficulty of conviction. Reason is called either strong sense or scepticism, according to the bias of the speaker; and Faith, either teachableness or credulity. . . .

Faith, then, as I have said, does not demand evidence so strong as is necessary for what is commonly considered a rational conviction, or belief on the ground of Reason; and why? For this reason, because it is mainly swayed by antecedent considerations. In this way it is, that the two principles are opposed to one another: Faith is influenced by previous notices, prepossessions, and (in a good sense of the word) prejudices; but Reason, by direct and definite proof. The mind that believes is acted upon by its own hopes, fears, and existing opinions; whereas it is supposed to reason severely, when it rejects antecedent proof of a fact,—rejects every thing but the actual evidence producible in its favour. This will appear from a very few words.

Faith is a principle of action, and action does not allow time for minute and finished investigations. We may (if we will) think that such investigations are of high value; though, in truth, they have a tendency to blunt the practical energy of the mind, while they improve its scientific exactness; but, whatever be their character and consequences, they do not answer the needs of daily life. Diligent collection of evidence, sifting of arguments, and balancing of rival testimonies, may be suited to persons who have leisure and opportunity to act when and how they will; they are not suited to the multitude. Faith, then, as being a principle for the multitude and for conduct, is influenced more by what (in language familiar to us of this place) are called εἰκότα than by σημεῖα,—less by evidence, more by previously-entertained principles, views, and wishes.

This is the case with all Faith, and not merely religious. We hear a report in the streets, or read it in the public journals. We know nothing of the evidence; we do not know the witnesses, or any thing about them: yet sometimes we believe implicitly, sometimes not; sometimes we believe without asking for evidence, sometimes we disbelieve till we receive it. Did a rumour circulate of a destructive earthquake in Syria or the south of Europe, we should readily credit it; both because it might easily be true, and because it was nothing to us though it were. Did the report relate to countries nearer home, we should try to trace and authenticate it. We do not call for evidence till antecedent probabilities fail.

Again, it is scarcely necessary to point out how much our inclinations have to do with our belief. It is almost a proverb, that persons believe what they wish to be true. . . .

Such are the inducements to belief which prevail with all of us, by a law of our nature, and whether they are in the particular case reasonable or not. When the probabilities we assume do not really exist, or our wishes are inordinate, or our opinions are wrong, our Faith degenerates into weakness, extravagance, superstition, enthusiasm, bigotry, prejudice, as the case may be; but when our prepossessions are unexceptionable, then we are right in believing or not believing, not indeed without, but upon slender evidence.

Whereas Reason then (as the word is commonly used) rests on the evidence, Faith is influenced by presumptions; and hence, while Reason requires rigid proofs, Faith is satisfied with vague or defective ones. . . .

. . . Next it is plain in what sense Faith is a moral principle. It is created in the mind, not so much by facts, as by probabilities; and since probabilities have no definite ascertained value, and are reducible to no scientific standard, what are such to each individual, depends on his moral temperament. A good and a bad man will think very different things probable. In the judgment of a rightly disposed mind, objects are desirable and attainable which irreligious men will consider to be but fancies. Such a correct moral judgment and view of things is the very medium in which the argument for Christianity has its constraining influence; a faint proof under circumstances being more availing than a strong one, apart from those circumstances.

This holds good as regards the matter as well as the evidence of the Gospel. It is difficult to say where the evidence, whether for Scripture or the Creed, would be found, if it were deprived of those adventitious illustrations which it extracts and absorbs from the mind of the inquirer, and which a merciful Providence places there for that very purpose. Texts have their illuminating power, from the atmosphere of habit, opinion, usage, tradition, through which we see them. On the other hand, irreligious men are adequate judges of the value of mere evidence, when the decision turns upon it; for evidence is addressed to the Reason, compels the Reason to assent so far as it is strong, and allows the Reason to doubt or disbelieve so far as it is weak. The blood on Joseph's coat of many colours was as perceptible to enemy as to friend; miracles appeal to the senses of all men, good and bad; and, while their supernatural character is learned from that experience of nature which is common to the just and to the unjust, the fact of their occurrence depends on considerations about testimony, enthusiasm, imposture, and the like, in which there is nothing inward, nothing personal. It is a sort of proof which a man does not make for himself, but which is made for him. It exists independently of him, and is apprehended from its own clear and objective character. It is its very boast that it does but require a candid hearing; nay, it especially addresses itself to the unbeliever, and engages to convert him as if against his will. There is no room for choice; there is no merit, no praise or blame, in believing or disbelieving; no test of character in the one or the other. But a man *is* responsible for his faith, because he is responsible for his likings and dislikings, his hopes and his opinions, on all of which his faith depends. And whereas unbelievers do not see this distinction, they persist in saying that a man is as little responsible for his faith as for his bodily functions; that both are from nature; that the will cannot make a weak proof a strong one; that if a person thinks a certain reason goes only a certain way, he is dishonest in attempting to make it go farther; that if he is after all wrong in his judgment, it is only his misfortune, not his fault; that he is acted on by certain principles from without, and must obey the laws of evidence, which are necessary and constant. But in truth, though a given evidence does not vary in force, the antecedent probability attending it does vary without limit, according to the temper of the mind surveying it. . . .

And here, again, we see what is meant by saying that Faith is a supernatural principle. The laws of evidence are the same in regard to the Gospel as to profane matters. If they were the sole arbiters of Faith, of course Faith could have nothing supernatural in it. But love of the great Object of Faith, watchful attention to Him, readiness to believe Him near, easiness to believe Him interposing in human affairs, fear of the risk of slighting or missing what may really come from Him; these are feelings not natural to fallen man, and they come only of supernatural grace; and these are the feelings which make us think evidence sufficient, which falls short of a proof in itself. The natural man has no heart for the promises of the Gospel, and dissects its evidence without reverence, without hope, without suspense, without misgivings; and, while he analyzes that evidence perhaps more philosophically than another, and treats it more luminously, and sums up its result with the precision and propriety of a legal tribunal, he rests in it as an end, and neither attains the farther truths at which it points, nor inhales the spirit which it breathes.

And this remark bears upon a fact which has sometimes perplexed Christians,—that those philosophers, ancient and modern, who have been eminent in physical science, have not unfrequently shown a tendency to infidelity. The system of physical causes is so much more tangible and satisfying than that of final, that unless there be a pre-existent and independent interest in the inquirer's mind, leading him to dwell on the phenomena which betoken an Intelligent Creator, he will certainly follow out those which terminate in the hypothesis of a settled order of nature and self-sustained laws. It is indeed a great question whether Atheism is not as philosophically consistent with the phen-omena of the physical world, taken by themselves, as the doctrine of a creative and governing Power. But, however this be, the practical safeguard against Atheism in the case of scientific inquirers is the inward need and desire, the inward experience of that Power, existing in the mind before and independently of their examination of His material world.

And in this lies the main fallacy of the celebrated argument against miracles, already alluded to, of a Scotch philosopher, whose depth and subtlety all must acknowledge. Let us grant (at least for argument's sake) that judging from the experience of life, it is more likely that witnesses should deceive, than that the laws of nature should be suspended. Still there may be considerations distinct from this view of the question which turn the main probability the other way,—viz. the likelihood, *à priori*, that a Revelation should be given. Here, then, we see how Faith is and is not according to Reason: taken together with the antecedent probability that Providence will reveal Himself to mankind, such evidence of the fact, as is otherwise deficient, may be enough for conviction, even in the judgment of Reason. But enough need not be enough, apart from that probability. That is, Reason, weighing evidence only, or arguing from external experience, is counter to Faith; but, admitting the legitimate influence and logical import of the moral feelings, it concurs with it. . . .

. . . As miracles, according to the common saying, are not wrought to convince Atheists, and, when they claim to be evidence of a Revelation, presuppose the being of an Intelligent Agent to whom they may be referred, so Evidences in general are grounded on the admission that the doctrine they are brought to prove is, not merely not inconsistent, but actually accordant with the laws of His moral governance. Miracles, though they contravene the physical laws of the universe, tend to the due fulfilment of its moral laws. And in matter of fact, when they were wrought, they addressed persons who were already believers, not in the mere probability, but even in the truth of supernatural revelations. This appears from the preaching of our Lord and His Apostles, who are accustomed to appeal to the religious feelings of their hearers; and who, though they might fail with the many, did thus persuade those who were persuaded—not, indeed, the sophists of Athens or the politicians of Rome, yet men of very different states of mind one from another, the pious, the superstitious, and the dissolute, different, indeed, but all agreeing in this, in the acknowledgement of truths beyond this world, whether or not their knowledge was clear, or their lives consistent,—the devout Jew, the proselyte of the gate, the untaught fisherman, the outcast Publican, and the pagan idolater.

And last of all, we here see what divines mean, who have been led to depreciate what are called the Evidences of Religion. The last century, a time when love was cold, is noted as being especially the Age of Evidences; and now, when more devout and zealous feelings have been excited, there is, I need scarcely say, a disposition manifested in various quarters, to think lightly, as of the eighteenth century, so of its boasted demonstrations. I have not here to make any formal comparison of the last century with the present, or to say whether they are nearer the truth, who in these matters advance with the present age, or who loiter behind with the preceding. I will only state what seems to me meant when persons disparage the Evidences,—viz. they consider that, as a general rule, religious minds embrace the Gospel mainly on the great antecedent probability of a Revelation, and the suitableness of the Gospel to their needs; on the other hand, that on men of irreligious minds Evidences are thrown away. Further, they perhaps would say, that to insist much on matters which are for the most part so useless for any practical purpose, draws men away from the true view of Christianity, and leads them to think that Faith is mainly the result of argument, that religious Truth is a legitimate matter of disputation, and that they who reject it rather err in judgment than commit sin. They think they see in the study in question a tendency to betray the sacredness and dignity of Religion, when those who profess themselves as champions allow themselves to stand on the same ground as philosophers of the world, admit the same principles, and only aim at drawing different conclusions.

For is not this the error, the common and fatal error, of the world, to think itself a judge of Religious Truth without preparation of heart? "I am the good Shepherd, and know My sheep, and am known of Mine." "He goeth before them, and the sheep follow Him, for they know His voice." "The pure in heart shall see God:" "to the meek mysteries are revealed;" "he that is spiritual judgeth all things." " The darkness comprehendeth it not." Gross eyes see not; heavy ears hear not. But in the schools of the world the ways towards Truth are considered high roads open to all men, however disposed, at all times. Truth is to be approached without homage. Every one is considered on a level with his neighbour; or rather the powers of the intellect, acuteness, sagacity, subtlety, and depth, are thought the guides into Truth. Men consider that they have as full a right to discuss religious subjects, as if they were themselves religious. They will enter upon the most sacred points of Faith at the moment, at their pleasure,—if it so happen, in a careless frame of mind, in their hours of recreation, over the wine cup. Is it wonderful that they so frequently end in becoming indifferentists, and conclude that Religious Truth is but a name, that all men are right and all wrong, from witnessing externally the multitude of sects and parties, and from the clear consciousness they possess within, that their own inquiries end in darkness?

Yet, serious as these dangers may be, it does not therefore follow that the Evidences may not be of great service to persons in particular frames of mind. Careless persons may be startled by them as they might be startled by a miracle, which is no necessary condition of believing, notwithstanding. Again, they often serve as a test of honesty of mind; their rejection being the condemnation of unbelievers. Again, religious persons sometimes get perplexed and lose their way; are harassed by objections; see difficulties which they cannot surmount; are a prey to subtlety of mind or over-anxiety. Under these circumstances the varied proofs of Christianity will be a stay, a refuge, an encouragement, a rallying point for Faith, a gracious economy; and even in the case of the most established Christian they are a source of gratitude and reverent admiration, and a means of confirming faith and hope. Nothing need be detracted from the use of the Evidences on this score; much less can any sober mind run into the wild notion that actually no proof at all is implied in the maintenance, or may be exacted for the profession of Christianity. I would only maintain that that proof need not be the subject of analysis, or take a methodical form, or be complete and symmetrical, in the believing mind; and that probability is its life. I do but say that it is antecedent probability that gives meaning to those arguments from facts which are commonly called the Evidences of Revelation; that, whereas mere probability proves nothing, mere facts persuade no one; that probability is to fact, as the soul to the body; that mere presumptions may have no force, but that mere facts have no warmth. A mutilated and defective evidence suffices for persuasion where the heart is alive; but dead evidences, however perfect, can but create a dead faith.

To conclude: It will be observed, I have not yet said what Reason really is, or what is its relation to Faith, but have merely contrasted the two together, taking Reason in the sense popularly ascribed to the word. Nor do I aim at more than ascertaining the sense in which the words Faith and Reason are used by Christian and Catholic writers. If I shall succeed in this, I shall be content, without attempting to defend it. Half the controversies in the world are verbal ones; and could they be brought to a plain issue, they would be brought to a prompt termination. Parties engaged in them would then perceive, either that in substance they agreed together, or that their difference was one of first principles. This is the great object to be aimed at in the present age, though confessedly a very arduous one. We need not dispute, we need not prove,—we need but define. At all events, let us, if we can, do this first of all; and then see who are left for us to dispute with, what is left for us to prove. Controversy, at least in this age, does not lie between the hosts of heaven, Michael and his Angels on the one side, and the powers of evil on the other; but it is a sort of night battle, where each fights for himself, and friend and foe stand together. When men understand each other's meaning, they see, for the most part, that controversy is either superfluous or hopeless.

[Sermon 10]

The Nature of Faith in Relation to Reason
(Preached 13 January 1839)

> God hath chosen the foolish things of the world to confound the wise,
> and God hath chosen the weak things of the world to confound the
> things which are mighty.
>
> 1 Cor. 1:27

It is usual at this day to speak as if Faith were simply of a moral nature,
and depended and followed upon a distinct act of Reason beforehand,—
Reason warranting, on the ground of evidence, both ample and
carefully examined, that the Gospel comes from God, and *then* Faith
embracing it. On the other hand, the more Scriptural representation
seems to be this, which is obviously more agreeable to facts also, that,
instead of there being really any such united process of reasoning first,
and then believing, the act of Faith is sole and elementary, and complete
in itself, and depends on no process of mind previous to it: and this
doctrine is borne out by the common opinion of men, who, though they
contrast Faith and Reason, yet rather consider Faith to be weak Reason,
than a moral quality or act following upon Reason. The Word of Life is
offered to a man; and, on its being offered, he has Faith in it. Why? On
these two grounds,—the word of its human messenger, and the
likelihood of the message. And why does he feel the message to be
probable? Because he has a love for it, his love being strong, though the
testimony is weak. He has a keen sense of the intrinsic excellence of the
message, of its desirableness, of its likeness to what it seems to him Divine
Goodness would vouchsafe did He vouchsafe any, of the need of a
Revelation, and its probability. Thus Faith is the reasoning of a religious
mind, or of what Scripture calls a right or renewed heart, which acts
upon presumptions rather than evidence, which speculates and ventures
on the future when it cannot make sure of it.

Thus, to take the instance of St. Paul preaching at Athens: he told his
hearers that he came as a messenger from that God whom they
worshipped already, though ignorantly, and of whom their poets spoke.
He appealed to the conviction that was lodged within them of the
spiritual nature and the unity of God; and he exhorted them to turn to
Him who had appointed One to judge the whole world hereafter. This
was an appeal to the antecedent probability of a Revelation, which
would be estimated variously according to the desire of it existing in each
breast. Now, what was the evidence he gave, in order to concentrate
those various antecedent presumptions, to which he referred, in behalf of
the message which he brought? Very slight, yet something; not a
miracle, but his own word that God had raised Christ from the dead;

very like the evidence given to the mass of men now, or rather not so much. No one will say it was strong evidence; yet, aided by the novelty, and what may be called originality, of the claim, its strangeness and improbability considered as a mere invention, and the personal bearing of the Apostle, and supported by the full force of the antecedent probabilities which existed, and which he stirred within them, it was enough. It was enough, for some did believe,—enough, not indeed in itself, but enough for those who had love, and therefore were inclined to believe. To those who had no fears, wishes, longings, or expectations, of another world, he was but "a babbler;" those who had such, or, in the Evangelist's words in another place, were "ordained to eternal life," "clave unto him, and believed."

This instance, then seems very full to justify the view of Faith which I have been taking, that it is an act of Reason, but of what the world would call weak, bad, or insufficient Reason; and that, because it rests on presumption more, and on evidence less. On the other hand, I conceive that this passage of Scripture does not fit in at all with the modern theory now in esteem that Faith is a mere moral act, dependent on a previous process of clear and cautious Reason. If so, one would think that St. Paul had no claim upon the faith of his hearers, till he had first wrought a miracle, such as Reason might approve, in token that his message was to be handed over to the acceptance of Faith.

Now, that this difference of theories as regards the nature of religious Faith is not a trifling one, is evident, perhaps, from the conclusions which I drew from it last week, which, if legitimate, are certainly important: and as feeling it to be a serious difference, I now proceed to state distinctly what I conceive to be the relation of Faith to Reason. I observe, then, as follows:—

We are surrounded by beings which exist quite independently of us,— exist whether we exist, or cease to exist, whether we have cognizance of them or no. These we commonly separate into two great divisions, material and immaterial. Of the material we have direct knowledge through the senses; we are sensible of the existence of persons and things, of their properties and modes, of their relations towards each other, and the courses of action which they carry on. Of all these we are directly cognizant through the senses; we see and hear what passes, and that immediately. As to immaterial beings, that we have faculties analogous to sense by which we have direct knowledge of their presence, does not appear, except indeed as regards our own soul and its acts. But so far is certain at least, that we are not conscious of possessing them; and we account it, and rightly, to be enthusiasm to profess such consciousness. At times, indeed, that consciousness has been imparted, as in some of the appearances of God to man contained in Scripture: but, in the ordinary course of things, whatever direct intercourse goes on between the soul and immaterial beings, whether we perceive them or not, and are influenced by them or not, certainly we have no consciousness of that perception or influence, such as our senses convey to us in the perception of things material. The senses, then, are the only instruments which we know to be granted to us for direct and immediate acquaintance with things external to us. Moreover, it is obvious that even our senses convey us but a little way out of ourselves, and introduce us to the external world only under circumstances, under conditions of time and place, and of certain media through which they act. We must be near things to touch them; we must be interrupted by no simultaneous sounds to hear them; we must have light to see them; we can neither see, hear, nor touch things past or future.

Now, Reason is that faculty of the mind by which this deficiency is supplied; by which knowledge of things external to us, of beings, facts, and events, is attained beyond the range of sense. It ascertains for us not natural things only, or immaterial only, or present only, or past, or future; but, even if limited in its power, it is unlimited in its range, viewed as a faculty, though, of course, in individuals it varies in range also. It reaches to the ends of the universe, and to the throne of God beyond them; it brings us knowledge, whether clear or uncertain, still knowledge, in whatever degree of perfection, from every side; but, at the same time, with this characteristic, that it obtains it indirectly, not directly.

Reason does not really perceive any thing; but it is a faculty of proceeding from things that are perceived to things which are not; the existence of which it certifies to us on the hypothesis of something else being known to exist, in other words, being assumed to be true.

Such is Reason, simply considered; and hence the fitness of a number of words which are commonly used to denote it and its acts. For instance: its act is usually considered a process, which, of course, a progress of thought from one idea to the other must be; an exercise of mind, which perception through the senses can hardly be called; or, again, an investigation, or an analysis; or it is said to compare, discriminate, judge, and decide: all which words imply, not simply assent to the reality of certain external facts, but a search into grounds, and an assent upon grounds. It is, then, the faculty of gaining knowledge upon grounds given; and its exercise lies in asserting one thing, because of some other thing; and, when its exercise is conducted rightly, it leads to knowledge; when wrongly, to apparent knowledge, to opinion, and error.

Now, if this be Reason, an act or process of Faith, simply considered, is certainly an exercise of Reason; whether a right exercise or not is a farther question; and, whether so to call it, is a sufficient account of it, is a farther question. It is an acceptance of things as real, which the senses do not convey, upon certain previous grounds; it is an instrument of indirect knowledge concerning things external to us,—the process being such as the following: "I assent to this doctrine as true, because I have been taught it;" or, "because superiors tell me so;" or, "because good men think so;" or, "because very different men think so;" or, "because all men;" or, "most men;" or, "because it is established;" or, "because persons whom I trust say that it was once guaranteed by miracles;" or, "because one who is said to have wrought miracles," or "who says he wrought them," "has taught it;" or, "because I have seen one who saw the miracles;" or, "because I saw what I took to be a miracle;" or for all or some of these reasons together. Some such exercise of Reason is the act of Faith, considered in its nature.

On the other hand, Faith plainly lies exposed to the popular charge of being a faulty exercise of Reason, as being conducted on insufficient grounds; and, I suppose, so much must be allowed on all hands, either that it is illogical, or that the mind has some grounds which are not fully brought out, when the process is thus exhibited. In other words, that when the mind savingly believes, the reasoning which that belief involves, if it be logical, does not merely proceed from the actual evidence, but from other grounds besides.

I say, there is this alternative in viewing the particular process of Reason which is involved in Faith;—to say either that the process is illogical, or the subject-matter more or less special and recondite; the act of inference faulty, or the premisses undeveloped; that Faith is weak, or that it is unearthly. Scripture says that it is unearthly, and the world says that it is weak.

This, then, being the imputation brought against Faith, that it is the reasoning of a weak mind, whereas it is in truth the reasoning of a divinely enlightened one, let me now, in a few words, attempt to show the analogy of this state of things, with what takes place in regard to other exercises of Reason also; that is, I shall attempt to show that Faith is not the only exercise of Reason, which, when critically examined, would be called unreasonable, and yet is not so.

In truth, nothing is more common among men of a reasoning turn than to consider that no one reasons well but themselves. All men of course think that they themselves are right and others wrong, who differ from them; and so far all men must find fault with the reasoning of others, since no one proposes to act without reasons of some kind. Accordingly, so far as men are accustomed to analyze the opinions of others and to contemplate their processes of thought, they are tempted to despise them as illogical. If any one sets about examining why his neighbours are on one side in political questions, not on another; why for or against certain measures, of a social, economical, or civil nature; why they belong to this religious party, not to that; why they hold this or that doctrine; why they have certain tastes in literature; or why they hold certain views in matters of opinion; it is needless to say that, if he measures their grounds merely by the reasons which they produce, he will have no difficulty in holding them up to ridicule, or even to censure.

And so again as to the deductions made from definite facts common to all. From the sight of the same sky one may augur fine weather, another bad; from the signs of the times one the coming in of good, another of evil; from the same actions of individuals one infers moral greatness, another depravity or perversity, one simplicity, another craft; upon the same evidence one justifies, another condemns. The miracles of Christianity were in early times imputed by some to magic, others they converted; the union of its professors were ascribed to seditious and traitorous aims by some, while others it moved to say, "See how these Christians love one another." The phenomena of the physical world have given rise to a variety of theories, that is, of alleged facts, at which they are supposed to point; theories of astronomy, chemistry, and physiology; theories religious and atheistical. The same events are considered to prove a particular providence, and not; to attest the divinity of one religion or of another. The downfall of the Roman Empire was to Pagans a refutation, to Christians an evidence, of Christianity. Such is the diversity with which men reason, showing us that Faith is not the only exercise of Reason, which approves itself to some and not to others, or which is, in the common sense of the word, irrational.

Nor can it fairly be said that such varieties do arise from deficiency in the power of reasoning in the multitude; and that Faith, such as I have described it, is but proved thereby to be a specimen of such deficiency. This is what men of clear intellects are not slow to imagine. Clear, strong, steady intellects, if they are not deep, will look on these differences in deduction chiefly as failures in the reasoning faculty, and will despise them or excuse them accordingly. Such are the men who are commonly latitudinarians in religion on the one hand, or innovators on the other; men of exact or acute but shallow minds, who consider all men wrong but themselves, yet think it no matter though they be; who regard the pursuit of truth only as a syllogistic process, and failure in attaining it as arising merely from a want of mental conformity with the laws on which just reasoning is conducted. But surely there is no greater mistake than this. For the experience of life contains abundant evidence that in practical matters, when their minds are really roused, men commonly are not bad reasoners. Men do not mistake when their interest is concerned. They have an instinctive sense in which direction their path lies towards it, and how they must act consistently with self-preservation or self-aggrandisement. And so in the case of questions in which party spirit, or political opinion, or ethical principle, or personal feelings, is concerned, men have a surprising sagacity, often unknown to themselves, in finding their own place. However remote the connexion between the point in question and their own creed, or habits, or feelings, the principles which they profess guide them unerringly to their legitimate issues; and thus it often happens that in apparently indifferent practices or usages or sentiments, or in questions of science, or politics, or literature, we can almost prophesy beforehand, from their religious or moral views, where certain persons will stand, and often can defend them far better than they defend themselves. The same thing is proved from the internal consistency of such religious creeds as are allowed time and space to develope freely; such as Primitive Christianity, or the Medieval system, or Calvinism—a consistency which nevertheless is wrought out in and through the rude and inaccurate minds of the multitude. Again, it is proved from the uniformity observable in the course of the same doctrine in different ages and countries, whether it be political, religious, or philosophical; the laws of Reason forcing it on into the same developments, the same successive phases, the same rise, and the same decay, so that its recorded history in one century will almost suit its prospective course in the next.

All this shows, that in spite of the inaccuracy in expression, or (if we will) in thought, which prevails in the world, men on the whole do not reason incorrectly. If their reason itself were in fault, they would reason each in his own way: whereas they form into schools, and that not merely from imitation and sympathy, but certainly from internal compulsion, from the constraining influence of their several principles. They may argue badly, but they reason well; that is, their professed grounds are no sufficient measures of their real ones. And in like manner, though the evidence with which Faith is content is apparently inadequate to its purpose, yet this is no proof of real weakness or imperfection in its reasoning. It seems to be contrary to Reason, yet is not; it is but independent of and distinct from what are called philosophical inquiries, intellectual systems, courses of argument, and the like.

So much on the general phenomena which attend the exercise of this great faculty, one of the characteristics of human over brute natures. Whether we consider processes of Faith or other exercise of Reason, men advance forward on ground which they do not, or cannot produce, or if they could, yet could not prove to be true, on latent or antecedent grounds which they take for granted.

Next, let it be observed, that however full and however precise our producible grounds may be, however systematic our method, however clear and tangible our evidence, yet when our argument is traced down to its simple elements, there must ever be something assumed ultimately which is incapable of proof, and without which our conclusion will be as illogical as faith is apt to seem to men of the world.

To take the case of actual evidence, and that of the strongest kind. Now, whatever it be, its cogency must be a thing taken for granted; so far it is its own evidence, and can only be received on instinct or prejudice. For instance, we trust our senses, and that in spite of their often deceiving us. They even contradict each other at times, yet we trust them. But even were they ever consistent, never unfaithful, still their fidelity would not be thereby proved. We consider that there is so strong an antecedent probability that they are faithful, that we dispense with proof. We take the point for granted; or, if we have grounds for it, these either lie in our secret belief in the stability of nature, or in the preserving presence and uniformity of Divine Providence,—which, again, are points assumed. As, then, the senses may and do deceive us, and yet we trust them from a secret instinct, so it need not be weakness or rashness, if upon a certain presentiment of mind we trust to the fidelity of testimony offered for a Revelation.

Again: we rely implicitly on our memory, and that, too, in spite of its being obviously unstable and treacherous. And we trust to memory for the truth of most of our opinions; the grounds on which we hold them not being at a given moment all present to our minds. We trust to memory to inform us what we do hold and what we do not. It may be said, that without such assumption the world could not go on: true; and in the same way the Church could not go on without Faith. Acquiescence in testimony, or in evidence not stronger than testimony, is the only method, as far as we see, by which the next world can be revealed to us.

The same remarks apply to our assumption of the fidelity of our reasoning powers; which in certain instances we implicitly believe, though we know they have deceived us in others.

Were it not for these instincts, it cannot be doubted but our experience of the deceivableness of Senses, Memory, and Reason, would perplex us much as to our practical reliance on them in matters of this world. And so, as regards the matters of another, they who have not that instinctive apprehension of the Omnipresence of God and His unwearied and minute Providence which holiness and love create within us, must not be surprised to find that the evidence of Christianity does not perform an office which was never intended for it,—viz. that of recommending itself as well as the Revelation. Nothing, then, which Scripture says about

Faith, however startling it may be at first sight, is inconsistent with the state in which we find ourselves by nature with reference to the acquisition of knowledge generally,—a state in which we must assume something to prove anything, and can gain nothing without a venture.

To proceed. Next let it be considered, that the following law seems to hold in our attainment of knowledge, that according to its desirableness, whether in point of excellence, or range, or intricacy, so is the subtlety of the evidence on which it is received. We are so constituted, that if we insist upon being as sure as is conceivable, in every step of our course, we must be content to creep along the ground, and can never soar. If we are intended for great ends, we are called to great hazards; and, whereas we are given absolute certainty in nothing, we must in all things choose between doubt and inactivity, and the conviction that we are under the eye of One who, for whatever reason, exercises us with the less evidence when He might give us the greater. He has put it into our hands, who loves us; and He bids us examine it, indeed, with our best judgment, reject this and accept that, but still all the while as loving Him in our turn; not coldly and critically, but with the thought of His presence, and the reflection that perchance by the defects of the evidence He is trying our love of its matter; and that perchance it is a law of His Providence to speak less loudly the more He promises. For instance, the touch is the most certain and cautious, but it is the most circumscribed of our senses, and reaches but an arm's length. The eye, which takes in a far wider range, acts only in the light. Reason, which extends beyond the province of sense or the present time, is circuitous and indirect in its conveyance of knowledge, which, even when distinct, is traced out pale and faint, as distant objects on the horizon. And Faith, again, by which we get to know divine things, rests on the evidence of testimony, weak in proportion to the excellence of the blessing attested. And as Reason, with its great conclusions, is confessedly a higher instrument than Sense with its secure premises, so Faith rises above Reason, in its subject-matter, more than it falls below it in the obscurity of its process. And it is, I say, but agreeable to analogy, that Divine Truth should be attained by so subtle and indirect a method, a method less tangible than others, less open to analysis, reducible but partially to the forms of Reason, and the ready sport of objection and cavil.

Further, much might be observed concerning the special delicacy and abstruseness of such reasoning processes as attend the acquisition of all higher knowledge. It is not too much to say that there is no one of the greater achievements of the Reason, which would show to advantage, which would be apparently justified and protected from criticism, if thrown into the technical forms which the science of argument requires. The most remarkable victories of genius, remarkable both in their originality and the confidence with which they have been pursued, have been gained, as though by invisible weapons, by ways of thought so recondite and intricate that the mass of men are obliged to take them on trust, till the event or other evidence confirms them. Such are the methods which penetrating intellects have invented in mathematical science, which look like sophisms till they issue in truths. Here, even in the severest of disciplines, and in absolutely demonstrative processes, the instrument of discovery is so subtle, that technical expressions and formulæ are of necessity substituted for it, to thread the labyrinth withal, by way of tempering its difficulties to the grosser reason of the many. Or, let it be considered how rare and immaterial (if I may use the words) is metaphysical proof: how difficult to embrace, even when presented to us by philosophers in whose clearness of mind and good sense we fully confide; and what a vain system of words without ideas such men seem to be piling up, while perhaps we are obliged to confess that it must be we who are dull, not they who are fanciful; and that, whatever be the character of their investigations, we want the vigour or flexibility of mind to judge of them. Or let us attempt to ascertain the passage of the mind, when slight indications in things present are made the informants of what is to be. Consider the preternatural sagacity with which a great general knows what his friends and enemies are about, and what will be the final result, and where, of their combined movements,—and then say whether, if he were required to argue the matter in word or on paper, all his most brilliant conjectures might not be refuted, and all his producible reasons exposed as illogical.

And, in an analogous way, Faith is a process of the Reason, in which so much of the grounds of inference cannot be exhibited, so much lies in the character of the mind itself, in its general view of things, its estimate of the probable and the improbable, its impressions concerning God's will, and its anticipations derived from its own inbred wishes, that it will ever seem to the world irrational and despicable;—till, that is, the event confirms it. The act of mind, for instance, by which an unlearned person savingly believes the Gospel, on the word of his teacher, may be analogous to the exercise of sagacity in a great statesman or general, supernatural grace doing for the uncultivated reason what genius does for them.

Now it is a singular confirmation of this view of the subject, that the reasonings of inspired men in Scripture, nay, of God Himself, are of this recondite nature; so much so, that irreverent minds scarcely hesitate to treat them with the same contempt which they manifest towards the faith of ordinary Christians. St. Paul's arguments have been long ago abandoned even by men who professed to be defenders of Christianity. Nor can it be said surely that the line of thought (if I may dare so to speak), on which some of our Ever-blessed Saviour's discourses proceed, is more intelligible to our feeble minds. And here, moreover, let it be noted that, supposing the kind of reasoning which we call Faith to be of the subtle character which I am maintaining, and the instances of professed reasoning found in Scripture to be of a like subtlety, light is thrown upon another remarkable circumstance, which no one can deny, and which some have made an objection,—I mean, the indirectness of the Scripture proof on which the Catholic doctrines rest. It may be, that such a peculiarity in the inspired text is the proper correlative of Faith; such a text the proper matter for Faith to work upon; so that a Scripture such as we have, and not such as the Pentateuch was to the Jews, may be implied in our being under Faith and not under the Law.

Lastly, it should be observed that the analogy which I have been pursuing extends to moral actions, and their properties and objects, as well as to intellectual exercises. According as objects are great, the mode of attaining them is extraordinary; and again, according as it is extraordinary, so is the merit of the action. Here, instead of going to Scripture, or to a religious standard, let me appeal to the world's judgment in the matter. Military fame, for instance, power, character for greatness of mind, distinction in experimental science, are all sought and attained by risks and adventures. Courage does not consist in calculation, but in fighting against chances. The statesman whose name endures, is he who ventures upon measures which seem perilous, and yet succeed, and can be only justified on looking back upon them. Firmness and greatness of soul are shown, when a ruler stands his ground on his instinctive perception of a truth which the many scoff at, and which seems failing. The religious enthusiast bows the hearts of men to a voluntary obedience, who has the keenness to see, and the boldness to appeal to, principles and feelings deep buried within them, which they know not themselves, which he himself but by glimpses and at times realizes, and which he pursues from the intensity, not the steadiness of his view of them. And so in all things, great objects exact a venture, and a sacrifice is the condition of honour. And what is true in the world, why should it not be true also in the kingdom of God? We must "launch out into the deep, and let down our nets for a draught;" we must in the morning sow our seed, and in the evening withold not our hand, for we know not whether shall prosper, either this or that. "He that observeth the wind shall not sow, and he that regardeth the clouds shall not reap." He that fails nine times and succeeds the tenth, is a more honourable man than he who hides his talent in a napkin; and so, even though the feelings which prompt us to see God in all things, and to recognize supernatural works in matters of the world, mislead us at times, though they make us trust in evidence which we ought not to admit, and partially incur with justice the imputation of credulity, yet a Faith which generously apprehends Eternal Truth, though at times it degenerates into superstition, is far better than that cold, sceptical, critical tone of mind, which has no inward sense of an overruling, ever-present Providence, no desire to approach its God, but sits at home waiting for the fearful clearness of His visible coming, whom it might seek and find in due measure amid the twilight of the present world.

To conclude: such is Faith as contrasted with Reason;—what it is contrasted with Superstition, how separate from it, and by what principles and laws restrained from falling into it, is a most important question, without settling which any view of the subject of Faith is of course incomplete; but which it does not fall within my present scope to consider.

[Sermon 11]

Implicit and Explicit Reason
(Preached on St Peter's Day, 1840)

> Sanctify the Lord God in your hearts; and be ready always to give an answer to every man that asketh you a reason of the hope that is in you, with meekness and fear.

1 Pet. 3:15

St. Peter's faith was one of his characteristic graces. It was ardent, keen, watchful, and prompt. It dispensed with argument, calculation, deliberation, and delay, whenever it heard the voice of its Lord and Saviour: and it heard that voice even when its accents were low, or when it was unaided by the testimony of the other senses. When Christ appeared walking on the sea, and said, "It is I," Peter answered Him, and said, "Lord, if it be Thou, bid me come unto Thee on the water." When Christ asked His disciples who He was, "Simon Peter answered and said," as we read in the Gospel for this day, "Thou art the Christ, the Son of the Living God," and obtained our Lord's blessing for such clear and ready Faith. At another time, when Christ asked the Twelve whether they would leave Him as others did, St. Peter said, "Lord, to whom shall we go? Thou hast the words of eternal life; and we believe and are sure that Thou art the Christ, the Son of the Living God." And after the Resurrection, when he heard from St. John that it was Christ who stood on the shore, he sprang out of the boat in which he was fishing, and cast himself into the sea, in his impatience to come near Him. Other instances of his faith might be mentioned. If ever Faith forgot self, and was occupied with its Great Object, it was the faith of Peter. If in any one Faith appears in contrast with what we commonly understand by

Reason, and with Evidence, it so appears in the instance of Peter. When he reasoned, it was at times when Faith was lacking. "When he saw the wind boisterous, he was afraid;" and Christ in consequence called him, "Thou of little faith." When He had asked, "Who touched Me?" Peter and others reasoned, "Master," said they, "the multitude throng Thee, and press Thee, and sayest Thou, Who touched Me?" And in like manner, when Christ said that he should one day follow Him in the way of suffering, "Peter said unto Him, Lord, *why* cannot I follow Thee now?"—and we know how his faith gave way soon afterwards.

Faith and Reason, then, stand in strong contrast in the history of Peter: yet it is Peter, and he not the fisherman of Galilee, but the inspired Apostle, who in the text gives us a precept which implies, in order to its due fulfilment, a careful exercise of our Reason, an exercise both upon Faith, considered as an act or habit of mind, and upon the Object of it. We are not only to "sanctify the Lord God in our hearts," not only to prepare a shrine within us in which our Saviour Christ may dwell, and where we may worship Him; but we are so to understand what we do, so to master our thoughts and feelings, so to recognize what we believe, and how we believe, so to trace out our ideas and impressions, and to contemplate the issue of them, that we may be "ready *always* to give an answer to *every* man that asketh us an account of the hope that is in us." In these words, I conceive, we have a clear warrant, or rather an injunction, to cast our religion into the form of Creed and Evidences.

It would seem, then, that though Faith is the characteristic of the Gospel, and Faith is the simple lifting of the mind to the Unseen God, without conscious reasoning or formal argument, still the mind may be allowably, nay, religiously engaged, in reflecting upon its own Faith; investigating the grounds and the Object of it, bringing it out into words, whether to defend, or recommend, or teach it to others. And St. Peter himself, in spite of his ardour and earnestness, gives us in his own case some indications of such an exercise of mind. When he said, "Thou art the Christ, the Son of the Living God," he cast his faith, in a measure, into a dogmatic form: and when he said, "To whom shall we go? Thou hast the words of eternal life," he gave "an account of the hope that was in him," or grounded his faith upon Evidence.

Nothing would be more theoretical and unreal than to suppose that true Faith cannot exist except when moulded upon a Creed, and based upon Evidence; yet nothing would indicate a more shallow philosophy than to say that it ought carefully to be disjoined from dogmatic and argumentative statements. To assert the latter is to discard the science of theology from the service of Religion; to assert the former, is to maintain that every child, every peasant, must be a theologian. Faith cannot exist without grounds or without an object; but it does not follow that all who have faith should recognize, and be able to state what they believe, and why. Nor, on the other hand, because it is not identical with its grounds, and its object, does it therefore cease to be true Faith, on its recognizing them. In proportion as the mind reflects upon itself, it will be able "to give an account" of what it believes and hopes; as far as it has not thus reflected, it will not be able. Such knowledge cannot be wrong, yet cannot be necessary, as long as reflection is at once a natural faculty of our souls, yet not an initial faculty. Scripture gives instances of Faith in each of these states, when attended by a conscious exercise of Reason, and when not. When Nicodemus said, "No man can do these miracles that Thou doest, except God be with him," he investigated. When the Scribe said, "There is One God, and there is none other but He; and to love Him with all the heart . . . is more than all whole burnt offerings and sacrifices," his belief was dogmatical. On the other hand, when the cripple at Lystra believed, on St. Paul's preaching, or the man at the Beautiful gate believed in the Name of Christ, their faith was independent not of objects or grounds (for that is impossible,) but of perceptible, recognized, producible objects and grounds: they believed, they could not say what or why. True Faith, then, admits, but does not require, the exercise of what is commonly understood by Reason. . . .

In the Epistle for this Day we have an account of St. Peter, when awakened by the Angel, obeying him implicitly, yet not understanding, while he obeyed. He girt himself, and bound on his sandals, and cast his garment about him, and "went out and followed him;" yet "wist not that it was true which was done by the Angel, but thought he saw a vision." Afterwards, when he "was come to himself, he said, Now I know of a surety, that the Lord hath sent His Angel, and hath delivered me." First he acted spontaneously, then he contemplated his own acts. This may be taken as an illustration of the difference between the more simple faculties and operations of the mind, and that process of analyzing and describing them, which takes place upon reflection. We not only feel, and think, and reason, but we know that we feel, and think, and reason; not only know, but can inspect and ascertain our thoughts, feelings, and reasonings: not only ascertain, but describe. Children, for a time, do not realize even their material frames, or (as I may say) count their limbs; but, as the mind opens, and is cultivated, they turn their attention to soul as well as body; they contemplate all they are, and all they do; they are no longer beings of impulse, instinct, conscience, imagination, habit, or reason, merely; but they are able to reflect upon their own mind as if it were some external object; they reason upon their reasonings. This is the point on which I shall now enlarge.

Reason, according to the simplest view of it, is the faculty of gaining knowledge without direct perception, or of ascertaining one thing by means of another. In this way it is able, from small beginnings, to create to itself a world of ideas, which do or do not correspond to the things themselves for which they stand, or are true or not, according as it is exercised soundly or otherwise. One fact may suffice for a whole theory; one principle may create and sustain a system; one minute token is a clue to a large discovery. The mind ranges to and fro, and spreads out, and advances forward with a quickness which has become a proverb, and a subtlety and versatility which baffle investigation. It passes on from point to point, gaining one by some indication; another on a probability; then availing itself of an association; then falling back on some received law; next seizing on testimony; then committing itself to some popular impression, or some inward instinct, or some obscure memory; and thus it makes progress not unlike a clamberer on a steep cliff, who, by quick eye, prompt hand, and firm foot, ascends how he knows not himself, by personal endowments and by practice, rather than by rule, leaving no track behind him, and unable to teach another. It is not too much to say that the stepping by which great geniuses scale the mountains of truth is as unsafe and precarious to men in general, as the ascent of a skilful mountaineer up a literal crag. It is a way which they alone can take; and its justification lies in their success. And such mainly is the way in which all men, gifted or not gifted, commonly reason,—not by rule, but by an inward faculty.

Reasoning, then, or the exercise of Reason, is a living spontaneous energy within us, not an art. But when the mind reflects upon itself, it begins to be dissatisfied with the absence of order and method in the exercise, and attempts to analyze the various processes which take place during it, to refer one to another, and to discover the main principles on which they are conducted, as it might contemplate and investigate its faculty of memory or imagination. The boldest, simplest, and most comprehensive theory which has been invented for the analysis of the reasoning process, is the well-known science for which we are indebted to Aristotle, and which is framed upon the principle that every act of reasoning is exercised upon neither more nor less than three terms. Short of this, we have many general words in familiar use to designate particular methods of thought, according to which the mind reasons (that is, proceeds from truth to truth), or to designate particular states of mind which influence its reasonings. Such methods are antecedent probability, analogy, parallel cases, testimony, and circumstantial evidence; and such states of mind are prejudice, deference to authority, party spirit, attachment to such and such principles, and the like. In like manner we distribute the Evidences of Religion into External and Internal; into *à priori* and *à posteriori*; into Evidences of Natural Religion and of Revealed; and so on. Again, we speak of proving doctrines either from the nature of the case, or from Scripture, or from history; and of teaching them in a dogmatic, or a polemical, or a hortatory way. In these and other ways we instance the reflective power of the human mind, contemplating and scrutinizing its own acts.

Here, then, are two processes, distinct from each other,—the original process of reasoning, and next, the process of investigating our reasonings. All men reason, for to reason is nothing more than to gain truth from former truth, without the intervention of sense, to which brutes are limited; but all men do not reflect upon their own reasonings, much less reflect truly and accurately, so as to do justice to their own meaning; but only in proportion to their abilities and attainments. In other words, all men have a reason, but not all men can give a reason. We may denote, then, these two exercises of mind as reasoning and arguing, or as conscious and unconscious reasoning, or as Implicit Reason and Explicit Reason. And to the latter belong the words, science, method, development, analysis, criticism, proof, system, principles, rules, laws, and others of a like nature.

That these two exercises are not to be confounded together would seem too plain for remark, except that they have been confounded. Clearness in argument certainly is not indispensable to reasoning well. Accuracy in stating doctrines or principles is not essential to feeling and acting upon them. The exercise of analysis is not necessary to the integrity of the process analyzed. The process of reasoning is complete in itself, and independent. The analysis is but an account of it; it does not make the conclusion correct; it does not make the inference rational. It does not cause a given individual to reason better. It does but give him a sustained consciousness, for good or for evil, that he is reasoning. How a man reasons is as much a mystery as how he remembers. He remembers better and worse on different subject-matters, and he reasons better and worse. Some men's reason becomes genius in particular subjects, and is less than ordinary in others. The gift or talent of reasoning may be distinct in different subjects, though the process of reasoning is the same. Now a good arguer or clear speaker is but one who excels in analyzing or expressing a process of reason, taken as his subject-matter. He traces out the connexion of facts, detects principles, applies them, supplies deficiencies, till he has reduced the whole into order. But his talent of reasoning, or the gift of reason as possessed by him, may be confined to such an exercise, and he may be as little expert in other exercises, as a mathematician need be an experimentalist; as little creative of the reasoning itself which he analyzes, as a critic need possess the gift of writing poems. . . .

Faith, then, though in all cases a reasonable process, is not necessarily founded on investigation, argument, or proof; these processes being but the explicit form which the reasoning takes in the case of particular minds. Nay, so far from it, that the opposite opinion has, with much more plausibility, been advanced, viz. that Faith is not even compatible with these processes. Such an opinion, indeed, cannot be maintained, particularly considering the light which Scripture casts upon the subject, as in the text; but it may easily take possession of serious minds. When they witness the strife and division to which argument and controversy minister, the proud self-confidence which is fostered by strength of the reasoning powers, the laxity of opinion which often accompanies the study of the Evidences, the coldness, the formality, the secular and carnal spirit which is compatible with an exact adherence to dogmatic formularies; and on the other hand, when they recollect that Scripture represents religion as a divine life, seated in the affections and manifested in spiritual graces, no wonder that they are tempted to rescue Faith from all connexion with faculties and habits which may exist in perfection without Faith, and which too often usurp from Faith its own province, and profess to be a substitute for it. I repeat, such a persuasion is extreme, and will not maintain itself, and cannot be acted on, for any long time; it being as paradoxical to prohibit religious inquiry and inference, as to make it imperative. Yet we should not dismiss the notice of it, on many accounts, without doing justice to it; and therefore I propose now, before considering some of the uses of our critical and analytical powers, in the province of Religion, to state certain of the inconveniences and defects; an undertaking which will fully occupy what remains of our time this morning.

Inquiry and argument may be employed, first, in ascertaining the divine origin of Religion, Natural and Revealed; next, in interpreting Scripture; and thirdly, in determining points of Faith and Morals; that is, in the Evidences, Biblical Exposition, and Dogmatic Theology. In all three departments there is, first of all, an exercise of implicit reason, which is in its degree common to all men; for all men gain a certain impression, right or wrong, from what comes before them, for or against Christianity, for or against certain interpretations of Scripture, for or against certain doctrines. This impression, made upon their minds, whether by the claim itself of Revealed Religion, or by its documents, or by its teaching, it is the object of science to analyze, verify, methodize, and exhibit. We believe certain things, on certain grounds, through certain informants; and the analysis of these three, the why, the how, and the what, seems pretty nearly to constitute the science of divinity.

By the Evidences of Religion I mean the systematic analysis of all the grounds on which we believe Christianity to be true. I say "all", because the word Evidence is often restricted to denote only such arguments as arise out of the thing itself which is to be proved; or, to speak more definitely, facts and circumstances which presuppose the point under inquiry as a condition of their existence, and which are weaker or stronger arguments, according as that point approaches more or less closely to be a necessary condition of them. Thus blood on the clothes is an evidence of a murderer, just so far as a deed of violence is necessary to the fact of the stains, or alone accounts for them. Such are the Evidences as drawn out by Paley and other writers; and though only a secondary part, they are popularly considered the whole of the Evidences, because they can be exhibited and studied with far greater ease than antecedent considerations, presumptions and analogies, which, vague and abstruse as they are, still are more truly the grounds on which religious men receive the Gospel. . . .

No analysis is subtle and delicate enough to represent adequately the state of mind under which we believe, or the subjects of belief, as they are presented to our thoughts. The end proposed is that of delineating, or, as it were, painting what the mind sees and feels: now let us consider what it is to portray duly in form and colour things material, and we shall surely understand the difficulty, or rather the impossibility, of representing the outline and character, the hues and shades, in which any intellectual view really exists in the mind, or of giving it that substance and that exactness in detail in which consists its likeness to the original, or of sufficiently marking those minute differences which attach to the same general state of mind or tone of thought as found in this or that individual respectively. It is probable that a given opinion, as held by several individuals, even when of the most congenial views, is as distinct from itself as are their faces. Now how minute is the defect in imitation which hinders the likeness of a portrait from being successful! how easy it is to recognize who is intended by it, without allowing that really he is represented! Is it not hopeless, then, to expect that the most diligent and anxious investigation can end in more than in giving some very rude description of the living mind, and its feelings, thoughts, and reasonings? And if it be difficult to analyze fully any state, or frame, or opinion of our own minds, is it a less difficulty to delineate, as Theology professes to do, the works, dealings, providences, attributes, or nature of Almighty God?

In this point of view we may, without irreverence, speak even of the words of inspired Scripture as imperfect and defective; and though they are not subjects for our judgment (God forbid), yet they will for that very reason serve to enforce and explain better what I would say, and how far the objection goes. Inspiration is defective, not in itself, but in consequence of the medium it uses and the beings it addresses. It uses human language, and it addresses man; and neither can man compass, nor can his hundred tongues utter, the mysteries of the spiritual world, and God's appointments in this. This vast and intricate scene of things cannot be generalized or represented through or to the mind of man; and inspiration, in undertaking to do so, necessarily lowers what is divine to raise what is human. What, for instance, is the mention made in Scripture of the laws of God's government, of His providences, counsels, designs, anger, and repentance, but a gracious mode (the more gracious because necessarily imperfect) of making man contemplate what is far beyond him? Who shall give method to what is infinitely complex, and measure to the unfathomable? We are as worms in an abyss of divine works; myriads upon myriads of years would it take, were our hearts ever so religious, and our intellects ever so apprehensive, to receive from without the just impression of those works as they really are, and as experience would convey them to us:—sooner, then, than we should know nothing, Almighty God has condescended to speak to us so far as human thought and language will admit, by approximations, in order to give us practical rules for our own conduct amid His infinite and eternal operations.

And herein consists one great blessing of the Gospel Covenant, that in Christ's death on the Cross, and in other parts of that all-gracious Economy, are concentrated, as it were, and so presented to us those attributes and works which fill eternity. And with a like graciousness we are also told, in human language, things concerning God Himself, concerning His Son and His Spirit, and concerning His Son's incarnation, and the union of two natures in His One Person—truths which even a peasant holds implicitly, but which Almighty God, whether by His Apostles, or by His Church after them, has vouchsafed to bring together and methodize, and to commit to the keeping of science.

Now all such statements are likely at first to strike coldly or harshly upon religious ears, when taken by themselves, for this reason if for no other,—that they express heavenly things under earthly images, which are infinitely below the reality. This applies especially to the doctrine of the Eternal Sonship of our Lord and Saviour, as all know who have turned their minds to the controversies on the subject.

Again, it may so happen, that statements are only possible in the case of certain aspects of a doctrine, and that these seem inconsistent with each other, or mysteries, when contrasted together, apart from what lies between them; just as if one were shown the picture of a little child and an old man, and were told that they represented the same person,—a statement which would be incomprehensible to beings who were unacquainted with the natural changes which take place, in the course of years, in the human frame.

Or doctrinal statements may be introduced, not so much for their own sake, as because many consequences flow from them, and therefore a great variety of errors may, by means of them, be prevented. Such is the doctrine that our Saviour's personality is in His Godhead, not in His manhood; that He has taken the manhood into God. It is evident that such statements, being made for the sake of something beyond, when viewed apart from their end, or in themselves, are abrupt, and may offend hearers.

Again, so it is, however it be explained, that frequently we do not recognize our sensations and ideas, when put into words ever so carefully. The representation seems out of shape and strange, and startles us, even though we know not how to find fault with it. This applies, at least in the case of some persons, to portions of the received theological analysis of the impression made upon the mind by the Scripture notices concerning Christ and the Holy Spirit. In like manner, such phrases as "good works are a condition of eternal life," or "the salvation of the regenerate ultimately depends upon themselves,"— though unexceptionable, are of a nature to offend certain minds.

This difficulty of analyzing our more recondite feelings happily and convincingly, has a most important influence upon the science of the Evidences. Defenders of Christianity naturally select as reasons for belief, not the highest, the truest, the most sacred, the most intimately persuasive, but such as best admit of being exhibited in argument; and these are commonly not the real reasons in the case of religious men.

Nay, they are led for the same reason, to select such arguments as all will allow; that is, such as depend on principles which are a common measure for all minds. A science certainly is, in its very nature, public property; when, then, the grounds of Faith take the shape of a book of Evidences, nothing properly can be assumed but what men in general will grant as true; that is, nothing but what is on a level with all minds, good and bad, rude and refined.

Again, as to the difficulty of detecting and expressing the real reasons on which we believe, let this be considered,—how very differently an argument strikes the mind at one time and another, according to its particular state, or the accident of the moment. At one time it is weak and unmeaning,—at another, it is nothing short of demonstration. We take up a book at one time, and see nothing in it; at another, it is full of weighty remarks and precious thoughts. Sometimes a statement is axiomatic,—sometimes we are at a loss to see what can be said for it. Such, for instance, are the following, many like which are found in controversy;—that true saints cannot but persevere to the end; or that the influences of the Spirit cannot but be effectual; or that there must be an infallible Head of the Church on earth; or that the Roman Church, extending into all lands, is the Catholic Church; or that a Church, which is Catholic abroad, cannot be schismatical in England; or that, if our Lord is the Son of God, He must be God; or that a Revelation is probable; or that, if God is All-powerful, He must be also All-good. Who shall analyze the assemblage of opinions in this or that mind, which occasions it almost instinctively to reject or to accept each of these and similar positions? Far be it from me to seem to insinuate that they are *but* opinions, neither true nor false, and approving themselves or not, according to the humour or prejudice of the individual: so far from it, that I would maintain that the recondite reasons which lead each person to take or decline them, are just the most important portion of the considerations on which his conviction depends; and I say so, by way of showing that the science of controversy, or again the science of Evidences, has done very little, since it cannot analyze and exhibit these momentous reasons; nay, so far has done worse than little, in that it professes to have done much, and leads the student to mistake what are but secondary points in debate, as if they were the most essential.

It often happens, for the same reason, that controversialists or philosophers are spoken of by this or that person as unequal, sometimes profound, sometimes weak. Such cases of inequality, of course, do occur; but we should be sure, when tempted so to speak, that the fault is not with ourselves, who have not entered into an author's meaning, or analyzed the implicit reasonings along which his mind proceeds in those parts of his writings which we not merely dissent from (for that we have a right to do), but criticize as inconsecutive.

These remarks apply especially to the proofs commonly brought, whether for the truth of Christianity, or for certain doctrines from texts of Scripture. Such alleged proofs are commonly strong or slight, not in themselves, but according to the circumstances under which the doctrine professes to come to us, which they are brought to prove; and they will have a great or small effect upon our minds, according as we admit those circumstances or not. Now, the admission of those circumstances involves a variety of antecedent views, presumptions, implications, associations, and the like, many of which it is very difficult to detect and analyze. One person, for instance, is convinced by Paley's argument from the Miracles, another is not; and why? Because the former admits that there is a God, that He governs the world, that He wishes the salvation of man, that the light of nature is not sufficient for man, that there is no other way of introducing a Revelation but miracles, and that men, who were neither enthusiasts nor impostors, could not have acted as the Apostles did, unless they had seen the miracles which they attested; the other denies some one, or more, of these statements, or does not feel the force of some other principle more recondite and latent still than any of these, which is nevertheless necessary to the validity of the argument.

Further, let it be considered, that, even as regards what are commonly called Evidences, that is, arguments *à posteriori*, conviction for the most part follows, not upon any one great and decisive proof or token of the point in debate, but upon a number of very minute circumstances together, which the mind is quite unable to count up and methodize in an argumentative form. Let a person only call to mind the clear impression he has about matters of every day's occurrence, that this man is bent on a certain object, or that that man was displeased, or another suspicious; or that one is happy, and another unhappy; and how much depends in such impressions on manner, voice, accent, words uttered, silence instead of words, and all the many subtle symptoms which are felt by the mind, but cannot be contemplated; and let him consider how very poor an account he is able to give of his impression, if he avows it, and is called upon to justify it. This, indeed, is meant by what is called moral proof, in opposition to legal. We speak of an accused person being guilty without any doubt, even though the evidences of his guilt are none of them broad and definite enough in themselves to admit of being forced upon the notice of those who will not exert themselves to see them.

Now, should the proof of Christianity, or the Scripture proof of its doctrines, be of this subtle nature, of course it cannot be exhibited to advantage in argument: and even if it be not such, but contain strong and almost legal evidences, still there will always be a temptation in the case of writers on Evidence, or on the Scripture proof of doctrine, to over-state and exaggerate, or to systematize in excess; as if they were making a case in a court of law, rather than simply and severely analyzing, as far as is possible, certain existing reasons why the Gospel is true, or why it should be considered of a certain doctrinal character. It is hardly too much to say, that almost all reasons formally adduced in moral inquiries, are rather specimens and symbols of the real grounds, than those grounds themselves. They do but approximate to a representation of the general character of the proof which the writer wishes to convey to another's mind. They cannot, like mathematical proof, be passively followed with an attention confined to what is stated, and with the admission of nothing but what is urged. Rather, they are hints towards, and samples of, the true reasoning, and demand an active, ready, candid, and docile mind, which can throw itself into what is said, neglect verbal difficulties, and pursue and carry out principles. This is the true office of a writer, to excite and direct trains of thought; and this, on the other hand, is the too common practice of readers, to expect every thing to be done for them,—to refuse to think,—to criticize the letter, instead of reaching forwards towards the sense,—and to account every argument as unsound which is illogically worded.

Here is the fertile source of controversy, which may undoubtedly be prolonged without limit by those who desire it, while words are incomplete exponents of ideas, and complex reasons demand study, and involve prolixity. They, then, who wish to shorten the dispute, and to silence a captious opponent, look out for some strong and manifest argument which may be stated tersely, handled conveniently, and urged rhetorically; some one reason, which bears with it a show of vigour and plausibility, or a profession of clearness, simplicity, or originality, and may be easily reduced to mood and figure. Hence the stress often laid upon particular texts, as if decisive of the matter in hand: hence one disputant dismisses all parts of the Bible which relate to the Law,—another finds the high doctrines of Christianity revealed in the Book of Genesis,—another rejects certain portions of the inspired volume, as the Epistle of St. James,—another gives up the Apocrypha,—another rests the defence of Revelation on Miracles only, or the Internal Evidence only,—another sweeps away all Christian teaching but Scripture,—one and all from impatience at being allotted, in the particular case, an evidence which does little more than create an impression on the mind; from dislike of an evidence, varied, minute, complicated, and a desire of something producible, striking, and decisive.

Lastly, since a test is in its very nature of a negative character, and since argumentative forms are mainly a test of reasoning, so far they will be but critical, not creative. They will be useful in raising objections, and in ministering to scepticism; they will pull down, and will not be able to build up.

I have been engaged in proving the following points: that the reasonings and opinions which are involved in the act of Faith are latent and implicit; that the mind reflecting on itself is able to bring them out into some definite and methodical form; that Faith, however, is complete without this reflective faculty, which, in matter of fact, often does interfere with it, and must be used cautiously. . . .

[Sermon 13]

A GRAMMAR OF ASSENT

Conscience

Now certainly the thought of God, as Theists entertain it, is not gained by an instinctive association of His presence with any sensible phenomena; but the office which the senses directly fulfil as regards creation that devolves indirectly on certain of our mental phenomena as regards the Creator. Those phenomena are found in the sense of moral obligation. As from a multitude of instinctive perceptions, acting in particular instances, of something beyond the senses, we generalize the notion of an external world, and then picture that world in and according to those particular phenomena from which we started, so from the perceptive power which identifies the intimations of conscience with the reverberations or echoes (so to say) of an external admonition, we proceed on the notion of a Supreme Ruler and Judge, and then again we image Him and His attributes in those recurring intimations, out of which, as mental phenomena, our recognition of His existence was originally gained. And, if the impressions which His creatures make on us through our senses oblige us to regard those creatures as *sui generis* respectively, it is not wonderful that the notices, which He indirectly gives us through our conscience, of His own nature are such as to make us understand that He is like Himself and like nothing else.

I have already said I am not proposing here to prove the Being of a God; yet I have found it impossible to avoid saying where I look for the proof of it. For I am looking for that proof in the same quarter as that from which I would commence a proof of His attributes and character,—by the same means as those by which I show how we apprehend Him, not merely as a notion, but as a reality. The last indeed of these three investigations alone concerns me here, but I cannot altogether exclude the two former from my consideration. However, I repeat, what I am directly aiming at, is to explain how we gain an image of God and give a real assent to the proposition that He exists. And next, in order to do this, of course I must start from some first principle;—and that first principle, which I assume and shall not attempt to prove, is that which I should also use as a foundation in those other two inquiries, viz. that we have by nature a conscience.

I assume, then, that Conscience has a legitimate place among our mental acts; as really so, as the action of memory, of reasoning, of imagination, or as the sense of the beautiful; that, as there are objects which, when presented to the mind, cause it to feel grief, regret, joy, or desire, so there are things which excite in us approbation or blame, and which we in consequence call right or wrong; and which, experienced in ourselves, kindle in us that specific sense of pleasure or pain, which goes by the name of a good or bad conscience. This being taken for granted, I shall attempt to show that in this special feeling, which follows on the commission of what we call right or wrong, lie the materials for the real apprehension of a Divine Sovereign and Judge.

The feeling of conscience (being, I repeat, a certain keen sensibility, pleasant or painful,—self-approval and hope, or compunction and fear,—attendant on certain of our actions, which in consequence we call right or wrong) is twofold:—it is a moral sense, and a sense of duty; a judgment of the reason and a magisterial dictate. Of course its act is indivisible; still it has these two aspects, distinct from each other, and admitting of a separate consideration. Though I lost my sense of the obligation which I lie under to abstain from acts of dishonesty, I should not in consequence lose my sense that such actions were an outrage offered to my moral nature. Again; though I lost my sense of their moral deformity, I should not therefore lose my sense that they were forbidden to me. Thus conscience has both a critical and a judicial office, and though its promptings, in the breasts of the millions of human beings to whom it is given, are not in all cases correct, that does not necessarily interfere with the force of its testimony and of its sanction: its testimony that there is a right and a wrong, and its sanction to that testimony conveyed in the feelings which attend on right or wrong conduct. Here I have to speak of conscience in the latter point of view, not as supplying us, by means of its various acts, with the elements of morals, such as may be developed by the intellect into an ethical code, but simply as the dictate of an authoritative monitor bearing upon the details of conduct as they come before us, and complete in its several acts, one by one.

Let us then thus consider conscience, not as a rule of right conduct, but as a sanction of right conduct. This is its primary and most authoritative aspect; it is the ordinary sense of the word. Half the world would be puzzled to know what was meant by the moral sense; but every one knows what is meant by a good or bad conscience. Conscience is ever forcing on us by threats and by promises that we must follow the right and avoid the wrong; so far it is one and the same in the mind of every one, whatever be its particular errors in particular minds as to the acts which it orders to be done or to be avoided; and in this respect it corresponds to our perception of the beautiful and deformed. As we have naturally a sense of the beautiful and graceful in nature and art, though tastes proverbially differ, so we have a sense of duty and obligation, whether we all associate it with the same certain actions in particular or not. Here, however, Taste and Conscience part company: for the sense of beautifulness, as indeed the Moral Sense, has no special relations to persons, but contemplates objects in themselves; conscience, on the other hand, is concerned with persons primarily, and with actions mainly as viewed in their doers, or rather with self alone and one's own actions, and with others only indirectly and as if in association with self. And further, taste is its own evidence, appealing to nothing beyond its own sense of the beautiful or the ugly, and enjoying the specimens of the beautiful simply for their own sake; but conscience does not repose on itself, but vaguely reaches forward to something beyond self, and dimly discerns a sanction higher than self for its decisions, as is evidenced in that keen sense of obligation and responsibility which informs them.

And hence it is that we are accustomed to speak of conscience as a voice, a term which we should never think of applying to the sense of the beautiful; and moreover a voice, or the echo of a voice, imperative and constraining, like no other dictate in the whole of our experience.

And again, in consequence of this prerogative of dictating and commanding, which is of its essence, Conscience has an intimate bearing on our affections and emotions, leading us to reverence and awe, hope and fear, especially fear, a feeling which is foreign for the most part, not only to Taste, but even to the Moral Sense, except in consequence of accidental associations. No fear is felt by any one who recognizes that his conduct has not been beautiful, though he may be mortified at himself, if perhaps he has thereby forfeited some advantage; but, if he has been betrayed into any kind of immorality, he has a lively sense of responsibility and guilt, though the act be no offence against society,—of distress and apprehension, even though it may be of present service to him,—of compunction and regret, though in itself it be most pleasurable,—of confusion of face, though it may have no witnesses. These various perturbations of mind which are characteristic of a bad conscience, and may be very considerable,—self-reproach, poignant shame, haunting remorse, chill dismay at the prospect of the future,—and their contraries, when the conscience is good, as real though less forcible, self-approval, inward peace, lightness of heart, and the like,—these emotions constitute a specific difference between conscience and our other intellectual senses,—common sense, good sense, sense of expedience, taste, sense of honour, and the like,—as indeed they would also constitute between conscience and the moral sense, supposing these two were not aspects of one and the same feeling, exercised upon one and the same subject-matter.

So much for the characteristic phenomena, which conscience presents, nor is it difficult to determine what they imply. I refer once more to our sense of the beautiful. This sense is attended by an intellectual enjoyment, and is free from whatever is of the nature of emotion, except in one case, viz. when it is excited by personal objects; then it is that the tranquil feeling of admiration is exchanged for the excitement of affection and passion. Conscience too, considered as a moral sense, an intellectual sentiment, is a sense of admiration and disgust, of approbation and blame: but it is something more than a moral sense; it is always, what the sense of the beautiful is only in certain cases; it is always emotional. No wonder then that it always implies what that sense only sometimes implies; that it always involves the recognition of a living object, towards which it is directed. Inanimate things cannot stir our affections; these are correlative with persons. If, as is the case, we feel responsibility, are ashamed, are frightened, at transgressing the voice of conscience, this implies that there is One to whom we are responsible, before whom we are ashamed, whose claims upon us we fear. If, on doing wrong, we feel the same tearful, broken-hearted sorrow which overwhelms us on hurting a mother; if, on doing right, we enjoy the same sunny serenity of mind, the same soothing, satisfactory delight which follows on our receiving praise from a father, we certainly have within us the image of some person, to whom our love and veneration look, in whose smile we find our happiness, for whom we yearn, towards whom we direct our pleadings, in whose anger we are troubled and waste away. These feelings in us are such as require for their exciting cause an intelligent being: we are not affectionate towards a stone, nor do we feel shame before a horse or a dog; we have no remorse or compunction on breaking mere human law: yet, so it is, conscience excites all these painful emotions, confusion, foreboding, self-condemnation; and on the other hand it sheds upon us a deep peace, a sense of security, a resignation, and a hope, which there is no sensible, no earthly object to elicit. "The wicked flees, when no one pursueth;" then why does he flee? whence his terror? Who is it that he sees in solitude, in darkness, in the hidden chambers of his heart? If the cause of these emotions does not belong to this visible world, the Object to which his perception is directed must be Supernatural and Divine; and thus the phenomena of Conscience, as a dictate, avail to impress the imagination with the picture of a Supreme Governor, a Judge, holy, just, powerful, all-seeing, retributive, and is the creative principle of religion, as the Moral Sense is the principle of ethics. . . .

[ch. 5]

Assent

As apprehension is a concomitant, so inference is ordinarily the antecedent of assent;—on this surely I need not enlarge;—but neither apprehension nor inference interferes with the unconditional character of the assent, viewed in itself. The circumstances of an act, however necessary to it, do not enter into the act; assent is in its nature absolute and unconditional, though it cannot be given except under certain conditions.

This is obvious; but what presents some difficulty is this, how it is that a conditional acceptance of a proposition,—such as is an act of inference,—is able to lead as it does, to an unconditional acceptance of it,—such as is assent; how it is that a proposition which is not, and cannot be, demonstrated, which at the highest can only be proved to be truth-like, not true, such as "I shall die," nevertheless claims and receives our unqualified adhesion. . . .

The doctrine which I have been enunciating requires such careful explanation, that it is not wonderful that writers of great ability and name are to be found who have put it aside in favour of a doctrine of their own; but no doctrine on the subject is without its difficulties, and certainly not theirs, though it carries with it a show of common sense. The authors to whom I refer wish to maintain that there are degrees of assent, and that, as the reasons for a proposition are strong or weak, so is the assent. It follows from this that absolute assent has no legitimate exercise, except as ratifying acts of intuition or demonstration. What is thus brought home to us is indeed to be accepted unconditionally; but, as to reasonings in concrete matters, they are never more than probabilities, and the probability in each conclusion which we draw is the measure of our assent to that conclusion. Thus assent becomes a sort of necessary shadow, following upon inference, which is the substance; and is never without some alloy of doubt, because inference in the concrete never reaches more than probability.

Such is what may be called the *à priori* method of regarding assent in its relation to inference. It condemns an unconditional assent in concrete matters on what may be called the nature of the case. Assent cannot rise higher than its source, inference in such matters is at best conditional, therefore assent is conditional also. Abstract argument is always dangerous, and this instance is no exception to the rule; I prefer to go by facts. The theory to which I have referred cannot be carried out in practice. It may be rightly said to prove too much; for it debars us from unconditional assent in cases in which the common voice of mankind, the advocates of this theory included, would protest against the prohibition. There are many truths in concrete matter, which no one can demonstrate, yet every one unconditionally accepts; and though of course there are innumerable propositions to which it would be absurd to give an absolute assent, still the absurdity lies in the circumstances of each particular case, as it is taken by itself, not in their common violation of the pretentious axiom that probable reasoning can never lead to certitude.

Locke's remarks on the subject are an illustration of what I have been saying. This celebrated writer, after the manner of his school, speaks freely of degrees of assent, and considers that the strength of assent given to each proposition varies with the strength of the inference on which the assent follows; yet he is obliged to make exceptions to his general principle,—exceptions, unintelligible on his abstract doctrine, but demanded by the logic of facts. The practice of mankind is too strong for the antecedent theorem, to which he is desirous to subject it. . . .

. . . He takes a view of the human mind, in relation to inference and assent, which to me seems theoretical and unreal. Reasonings and convictions which I deem natural and legitimate, he apparently would call irrational, enthusiastic, perverse, and immoral; and that, as I think, because he consults his own ideal of how the mind ought to act, instead of interrogating human nature, as an existing thing, as it is found in the world. Instead of going by the testimony of psychological facts, and thereby determining our constitutive faculties and our proper condition, and being content with the mind as God has made it, he would form men as he thinks they ought to be formed, into something better and higher, and calls them irrational and indefensible, if (so to speak) they take to the water, instead of remaining under the narrow wings of his own arbitrary theory. . . .

. . . Assent on reasonings not demonstrative is too widely recognized an act to be irrational, unless man's nature is irrational, too familiar to the prudent and clear-minded to be an infirmity or an extravagance. None of us can think or act without the acceptance of truths, not inituitive, not demonstrated, yet sovereign. If our nature has any constitution, any laws, one of them is this absolute reception of propositions as true, which lie outside the narrow range of conclusions to which logic, formal or virtual, is tethered; nor has any philosophical theory the power to force on us a rule which will not work for a day.

When, then, philosophers lay down principles, on which it follows that our assent, except when given to objects of intuition or demonstration, is conditional, that the assent given to propositions by well-ordered minds necessarily varies with the proof producible for them, and that it does not and cannot remain one and the same while the proof is strengthened or weakened,—are they not to be considered as confusing together two things very distinct from each other, a mental act or state and a scientific rule, an interior assent and a set of logical formulas? When they speak of degrees of assent, surely they have no intention at all of defining the position of the mind itself relative to the adoption of a given conclusion, but they are recording their perception of the relation of that conclusion towards its premises. They are contemplating how representative symbols work, not how the intellect is affected towards the thing which those symbols represent. In real truth they as little mean to assert the principle of measuring our assents by our logic, as they would fancy they could record the refreshment which we receive from the open air by the readings of the graduated scale of a thermometer. There is a connexion doubtless between a logical conclusion and an assent, as there is between the variation of the mercury and our sensations; but the mercury is not the cause of life and health, nor is verbal argumentation the principle of inward belief. If we feel hot or chilly, no one will convince us to the contrary by insisting that the glass is at 60°. It is the mind that reasons and assents, not a diagram on paper. . . .

Acts of Inference are both the antecedents of assent before assenting, and its usual concomitants after assenting. For instance, I hold absolutely that the country which we call India exists, upon trustworthy testimony; and next, I may continue to believe it on the same testimony. In like manner, I have ever believed that Great Britain is an island, for certain sufficient reasons; and on the same reasons I may persist in the belief. But it may happen that I forget my reasons for what I believe to be so absolutely true; or I may never have asked myself about them, or formally marshalled them in order, and have been accustomed to assent without a recognition of my assent or of its grounds, and then perhaps something occurs which leads to my reviewing and completing those grounds, analyzing and arranging them, yet without on that account implying of necessity any suspense, ever so slight, of assent, to the proposition that India is in a certain part of the earth, and that Great Britain is an island. With no suspense of assent at all; any more than the boy in my former illustration had any doubt about the answer set down in his arithmetic-book, when he began working out the question; any more than he would be doubting his eyes and his common sense, that the two sides of a triangle are together greater than the third, because he drew out the geometrical proof of it. He does but repeat, after his formal demonstration, that assent which he made before it, and assents to his previous assenting. This is what I call a reflex or complex assent.

I say, there is no necessary incompatibility between thus assenting and yet proving,—for the conclusiveness of a proposition is not synonymous with its truth. A proposition may be true, yet not admit of being concluded;—it may be a conclusion and yet not a truth. To contemplate it under one aspect, is not to contemplate it under another; and the two aspects may be consistent, from the very fact that they are two *aspects*. Therefore to set about concluding a proposition is not *ipso facto* to doubt its truth; we may aim at inferring a proposition, while all the time we assent to it. We have to do this as a common occurrence, when we take on ourselves to convince another on any point in which he differs from us. We do not deny our own faith, because we become controversialists; and in like manner we may employ ourselves in proving what we already believe to be true, simply in order to ascertain the producible evidence in its favour, and in order to fulfil what is due to ourselves and to the claims and responsibilities of our education and social position.

I have been speaking of investigation, not of inquiry; it is quite true that inquiry is inconsistent with assent, but inquiry is something more than the mere exercise of inference. He who inquires has not found; he is in doubt where the truth lies, and wishes his present profession either proved or disproved. We cannot without absurdity call ourselves at once believers and inquirers also. Thus it is sometimes spoken of as a hardship that a Catholic is not allowed to inquire into the truth of his Creed;—of course he cannot, if he would retain the name of believer. He cannot be both inside and outside of the Church at once. It is merely common sense to tell him that, if he is seeking, he has not found. If seeking includes doubting, and doubting excludes believing, then the Catholic who sets about inquiring, thereby declares that he is not a Catholic. He has already lost faith. And this is his best defence to himself for inquiring, viz. that he is no longer a Catholic, and wishes to become one. They who would forbid him to inquire, would in that case be shutting the stable-door after the steed is stolen. What can he do better than inquire, if he is in doubt? how else can he become a Catholic again? Not to inquire is in his case to be satisfied with disbelief.

However, in thus speaking, I am viewing the matter in the abstract, and without allowing for the manifold inconsistencies of individuals, as they are found in the world, who attempt to unite incompatibilities; who do not doubt, but who act as if they did; who, though they believe, are weak in faith, and put themselves in the way of losing it by unnecessarily listening to objections. Moreover, there are minds, undoubtedly, with whom at all times to question a truth is to make it questionable, and to investigate is equivalent to inquiring; and again, there may be beliefs so sacred or so delicate, that, if I may use the metaphor, they will not wash without shrinking and losing colour. I grant all this; but here I am discussing broad principles, not individual cases; and these principles are, that inquiry implies doubt, and that investigation does not imply it, and that those who assent to a doctrine or fact may without inconsistency investigate its credibility, though they cannot literally inquire about its truth.

Next, I consider that, in the case of educated minds, investigations into the argumentative proof of the things to which they have given their assent, is an obligation, or rather a necessity. Such a trial of their intellects is a law of their nature, like the growth of childhood into manhood, and analogous to the moral ordeal which is the instrument of their spiritual life. The lessons of right and wrong, which are taught them at school, are to be carried out into action amid the good and evil of the world; and so again the intellectual assents, in which they have in like manner been instructed from the first, have to be tested, realized, and developed by the exercise of their mature judgment.

Certainly, such processes of investigation, whether in religious subjects or secular, often issue in the reversal of the assents which they were originally intended to confirm: as the boy who works out an arithmetical problem from his book may end in detecting, or thinking he detects, a false print in the answer. But the question before us is whether acts of assent and of inference are compatible; and my vague consciousness of the possibility of a reversal of my belief in the course of my researches, as little interferes with the honesty and firmness of that belief while those researches proceed, as the recognition of the possibility of my train's oversetting is an evidence of an intention on my part of undergoing so great a calamity. My mind is not moved by a scientific computation of chances, nor can any law of averages affect my particular case. To incur a risk is not to expect reverse; and if my opinions are true, I have a right to think that they will bear examining. Nor, on the other hand, does belief, viewed in its idea, imply a positive resolution in the party believing never to abandon that belief. What belief, as such, does imply is, not an intention never to change, but the utter absence of all thought, or expectation, or fear of changing. A spontaneous resolution never to change is inconsistent with the idea of belief; for the very force and absoluteness of the act of assent precludes any such resolution. We do not commonly determine not to do what we cannot fancy ourselves ever doing. We should readily indeed make such a formal promise if we were called upon to do so; for, since we have the truth, and truth cannot change, how can we possibly change in our belief, except indeed through our own weakness or fickleness? We have no intention whatever of being weak or fickle; so our promise is but the natural guarantee of our sincerity. It is possible then, without disloyalty to our convictions, to examine their grounds, even though in the event they are to fail under the examination, for we have no suspicion of this failure.

And such examination, as I have said, does but fulfil a law of our nature. Our first assents, right or wrong, are often little more than prejudices. The reasonings, which precede and accompany them though sufficient for their purpose, do not rise up to the importance and energy of the assents themselves. As time goes on, by degrees and without set purpose, by reflection and experience, we begin to confirm or to correct the notions and the images to which those assents are given. At times it is a necessity formally to undertake a survey and revision of this or that class of them, of those which relate to religion, or to social duty, or to politics, or to the conduct of life. Sometimes this review begins in doubt as to the matters which we propose to consider, that is, in a suspension of the assents hitherto familiar to us; sometimes those assents are too strong to allow of being lost on the first stirring of the inquisitive intellect, and if, as time goes on, they give way, our change of mind, be it for good or for evil, is owing to the accumulating force of the arguments, sound or unsound, which bear down upon the propositions which we have hitherto received. Objections, indeed, as such, have no direct force to weaken assent; but, when they multiply, they tell against the implicit reasonings or the formal inferences which are its warrant, and suspend its acts and gradually undermine its habit. Then the assent goes; but whether slowly or suddenly, noticeably or imperceptibly, is a matter of circumstance or accident. However, whether the original assent is continued on or not, the new assent differs from the old in this, that it has the strength of explicitness and deliberation, that it is not a mere prejudice, and its strength the strength of prejudice. It is an assent, not only to a given proposition, but to the claim of that proposition on our assent as true; it is an assent to an assent, or what is commonly called a conviction.

Of course these reflex acts may be repeated in a series. As I pronounce that "Great Britain is an island," and then pronounce "That 'Great Britain is an island' has a claim on my assent," or is to "be assented-to," or to be "accepted as true," or to be "believed," or simply "is true" (these predicates being equivalent), so I may proceed, "The proposition 'that *Great-Britain-is-an-island* is to be believed' is to be believed," &c., &c., and so on to *ad infinitum*. But this would be trifling. The mind is like a double mirror, in which reflexions of self within self multiply themselves till they are undistinguishable, and the first reflexion contains all the rest. At the same time, it is worth while to notice two other reflex propositions:—"That 'Great Britain is an island' is probable" is true:— and "That 'Great Britain is an island' is uncertain" is true;—for the former of these is the expression of Opinion, and the latter of formal or theological doubt, as I have already determined.

I have one step farther to make—let the proposition to which the assent is given be as absolutely true as the reflex act pronounces it to be, that is, objectively true as well as subjectively:—then the assent may be called a *perception*, the conviction a *certitude*, the proposition or truth a *certainty*, or thing known, or a matter of *knowledge*, and to assent to it is to *know*.

Of course, in thus speaking, I open the all-important question, what is truth, and what apparent truth? what is genuine knowledge, and what is its counterfeit? what are the tests for discriminating certitude from mere persuasion or delusion? Whatever a man holds to be true, he will say he holds for certain; and for the present I must allow him in his assumption, hoping in one way or another, as I proceed, to lessen the difficulties which lie in the way of calling him to account for so doing. And I have the less scruple in taking this course, as believing that, among fairly prudent and circumspect men, there are far fewer instances of false certitude than at first sight might be supposed. Men are often doubtful about propositions which are really true; they are not commonly certain of such as are simply false. What they judge to be a certainty is in matter of fact for the most part a truth. Not that there is not a great deal of rash talking even among the educated portion of the community, and many a man makes professions of certitude, for which he has no warrant; but that such off-hand, confident language is no token how these persons will express themselves when brought to book. No one will with justice consider himself certain of any matter, unless he has sufficient reasons for so considering; and it is rare that what is not true should be so free from every circumstance and token of falsity as to create no suspicion in his mind to its disadvantage, no reason for suspense of judgment. However, I shall have to remark on this difficulty by and by; here I will mention two conditions of certitude, in close connexion with that necessary preliminary of investigation and proof of which I have been speaking, which will throw some light upon it. The one, which is *à priori*, or from the nature of the case, will tell us what is not certitude; the other, which is *à posteriori*, or from experience, will tell us in a measure what certitude is.

Certitude, as I have said, is the perception of a truth with the perception that it is a truth, or the consciousness of knowing, as expressed in the phrase, "I know that I know," or "I know that I know that I know,"—or simply "I know;" for one reflex assertion of the mind about self sums up the series of self-consciousnesses without the need of any actual evolution of them.

But if so, if by certitude about a thing is to be understood the knowledge of its truth, let it be considered that what is once true is always true, and cannot fail, whereas what is once known need not always be known, and is capable of failing. It follows, that if I am certain of a thing, I believe it will remain what I now hold it to be, even though my mind should have the bad fortune to let it drop. Since mere argument is not the measure of assent, no one can be called certain of a proposition, whose mind does not spontaneously and promptly reject, on their first suggestion, as idle, as impertinent, as sophistical, any objections which are directed against its truth. No man is certain of a truth who can endure the thought of the fact of its contradictory existing or occurring; and that not from any set purpose or effort to reject that thought, but, as I have said, by the spontaneous action of the intellect. What is contradictory to the truth, with its apparatus of argument, fades out of the mind as fast as it enters it; and though it be brought back to the mind ever so often by the pertinacity of an opponent, or by a voluntary or involuntary act of imagination, still that contradictory proposition and its arguments are mere phantoms and dreams, in the light of our certitude, and their very entering into the mind is the first step of their going out of it. . . .

We may indeed say, if we please, that a man ought not to have so supreme a conviction in a given case, or in any case whatever; and that he is therefore wrong in treating opinions which he does not himself hold, with this even involuntary contempt;—certainly, we have a right to say so, if we will; but if, in matter of fact, a man has such a conviction, if he is sure that Ireland is to the West of England, or that the Pope is the Vicar of Christ, nothing is left to him, if he would be consistent, but to carry his conviction out into this magisterial intolerance of any contrary assertion; and if he were in his own mind tolerant, I do not say patient (for patience and gentleness are moral duties, but I mean intellectually tolerant), of objections as objections, he would virtually be giving countenance to the views which those objections represented. I say I certainly should be very intolerant of such a notion as that I shall one day be Emperor of the French; I should think it too absurd even to be ridiculous, and that I must be mad before I could entertain it. And did a man try to persuade me that treachery, cruelty, or ingratitude was as praiseworthy as honesty and temperance, and that a man who lived the life of a knave and died the death of a brute had nothing to fear from future retribution, I should think there was no call on me to listen to his arguments, except with the hope of converting him, though he called me a bigot and a coward for refusing to inquire into his speculations. And if, in a matter in which my temporal interests were concerned, he attempted to reconcile me to fraudulent acts by what he called philosophical views, I should say to him, "Retro Satana," and that, not from any suspicion of his ability to reverse immutable principles, but from a consciousness of my own moral changeableness, and a fear, on that account, that I might not be intellectually true to the truth. This, then, from the nature of the case, is a main characteristic of certitude in any matter, to be confident indeed that that certitude will last, but to be confident of this also, that, if it did fail, nevertheless, the thing itself, whatever it is, of which we are certain, will remain just as it is, true and irreversible. . . .

Now to consider what Certitude is, not simply as it must be, but in our actual experience of it.

It is accompanied, as a state of mind, by a specific feeling, proper to it, and discriminating it from other states, intellectual and moral, I do not say, as its practical test or as its *differentia*, but as its token, and in a certain sense its form. When a man says he is certain, he means he is conscious to himself of having this specific feeling. It is a feeling of satisfaction and self-gratulation, of intellectual security, arising out of a sense of success, attainment, possession, finality, as regards the matter which has been in question. As a conscientious deed is attended by a self-approval which nothing but itself can create, so certitude is united to a sentiment *sui generis* in which it lives and is manifested. These two parallel sentiments indeed have no relationship with each other, the enjoyable self-repose of certitude being as foreign to a good deed, as the self-approving glow of conscience is to the perception of a truth; yet knowledge, as well as virtue, is an end, and both knowledge and virtue, when reflected on, carry with them respectively their own reward in the characteristic sentiment, which, as I have said, is proper to each. And, as the performance of what is right is distinguished by this religious peace, so the attainment of what is true is attested by this intellectual security.

[ch. 6]

Certitude

It is the characteristic of certitude that its object is a truth, a truth as such, a proposition as true. There are right and wrong convictions, and certitude is a right conviction; if it is not right with a consciousness of being right, it is not certitude. Now truth cannot change; what is once truth is always truth; and the human mind is made for truth, and so rests in truth, as it cannot rest in falsehood. When then it once becomes possessed of a truth, what is to dispossess it? but this is to be certain; therefore once certitude, always certitude. If certitude in any matter be the termination of all doubt or fear about its truth, and an unconditional conscious adherence to it, it carries with it an inward assurance, strong though implicit, that it shall never fail. Indefectibility almost enters into its very idea, enters into it at least so far as this, that its failure, if of frequent occurrence, would prove that certitude was after all and in fact an impossible act, and that what looked like it was a mere extravagance of the intellect. Truth would still be truth, but the knowledge of it would be beyond us and unattainable. It is of great importance then to show, that, as a general rule, certitude does not fail; that failures of what was taken for certitude are the exception; that the intellect, which is made for truth, can attain truth, and, having attained it, can keep it, can recognize it, and preserve the recognition.

This is on the whole reasonable; yet are the stipulations, thus obviously necessary for an act or state of certitude, ever fulfilled? We know what conjecture is, and what opinion, and what assent is, can we point out any specific state or habit of thought, of which the distinguishing mark is unchangeableness? On the contrary, any conviction, false as well as true, may last; and any conviction, true as well as false, may be lost. A conviction in favour of a proposition may be exchanged for a conviction of its contradictory; and each of them may be attended, while they last, by that sense of security and repose, which a true object alone can legitimately impart. No line can be drawn between such real certitudes as have truth for their object, and apparent certitudes. No distinct test can be named, sufficient to discriminate between what may be called the false prophet and the true. What looks like certitude always is exposed to the chance of turning out to be a mistake. If our intimate, deliberate conviction may be counterfeit in the case of one proposition, why not in the case of another? if in the case of one man, why not in the case of a hundred? Is certitude then ever possible without the attendant gift of infallibility? can we know what is right in one case, unless we are secured against error in any? Further, if one man is infallible, why is he different from his brethren? unless indeed he is distinctly marked out for the prerogative. Must not all men be infallible by consequence, if any man is to be considered as certain?

The difficulty, thus stated argumentatively, has only too accurate a response in what actually goes on in the world. It is a fact of daily occurrence that men change their certitudes, that is, what they consider to be such, and are as confident and well-established in their new opinions as they were once in their old. They take up forms of religion only to leave them for their contradictories. They risk their fortunes and their lives on impossible adventures. They commit themselves by word and deed, in reputation and position, to schemes which in the event they bitterly repent of and renounce; they set out in youth with intemperate confidence in prospects which fail them, and in friends who betray them, ere they come to middle age; and they end their days in cynical disbelief of truth and virtue any where;—and often, the more absurd are their means and their ends, so much the longer do they cling to them, and then again so much the more passionate is their eventual disgust and contempt of them. How then can certitude be theirs, how is certitude possible at all, considering it is so often misplaced, so often fickle and inconsistent, so deficient in available criteria? And, as to the feeling of finality and security, ought it ever to be indulged? Is it not a mere weakness or extravagance, a deceit, to be eschewed by every clear and prudent mind? With the countless instances, on all sides of us, of human fallibility, with the constant exhibitions of antagonist certitudes, who can so sin against modesty and sobriety of mind, as not to be content with probability, as the true guide of life, renouncing ambitious thoughts, which are sure either to delude him, or to disappoint?

This is what may be objected: now let us see what can be said in answer, particularly as regards religious certitude.

First, as to fallibility and infallibility. It is very common, doubtless, especially in religious controversy, to confuse infallibility with certitude, and to argue that, since we have not the one, we have not the other, for that no one can claim to be certain on any point, who is not infallible about all; but the two words stand for things quite distinct from each other. For example, I remember for certain what I did yesterday, but still my memory is not infallible; I am quite clear that two and two make four, but I often make mistakes in long addition sums. I have no doubt whatever that John or Richard is my true friend, but I have before now trusted those who failed me, and I may do so again before I die. A certitude is directed to this or that particular proposition; it is not a faculty or gift, but a disposition of mind relatively to a definite case which is before me. Infallibility, on the contrary, is just that which certitude is not; it *is* a faculty or gift, and relates, not to some one truth in particular, but to all possible propositions in a given subject-matter. We ought in strict propriety, to speak, not of infallible acts, but of acts of infallibility. A belief or opinion as little admits of being called infallible, as a deed can correctly be called immortal. A deed is done and over; it may be great, momentous, effective, anything but immortal; it is its fame, it is the work which it brings to pass, which is immortal, not the deed itself. And as a deed is good or bad, but never immortal, so a belief, opinion, or certitude is true or false, but never infallible. We cannot speak of things which exist or things which once were, as if they were someting *in posse*. It is persons and rules that are infallible, not what is brought out into act, or committed to paper. A man is infallible, whose words are always true; a rule is infallible, if it is unerring in all its possible applications. An infallible authority is certain in every particular case that may arise; but a man who is certain in some one definite case, is not on that account infallible.

I am quite certain that Victoria is our Sovereign, and not her father, the late Duke of Kent, without laying any claim to the gift of infallibility; as I may do a virtuous action, without being impeccable. I may be certain that the Church is infallible, while I am myself a fallible mortal; otherwise, I cannot be certain that the Supreme Being is infallible, until I am infallible myself. . . .

It must be recollected that certitude is a deliberate assent given expressly after reasoning. If then my certitude is unfounded, it is the reasoning that is in fault, not my assent to it. It is the law of my mind to seal up the conclusions to which ratiocination has brought me, by that formal assent which I have called a certitude. I could indeed have withheld my assent, but I should have acted against my nature, had I done so when there was what I considered a proof; and I did only what was fitting, what was incumbent on me, upon those existing conditions, in giving it. This is the process by which knowledge accumulates and is stored up both in the individual and in the world. It has sometimes been remarked, when men have boasted of the knowledge of modern times, that no wonder we see more than the ancients, because we are mounted upon their shoulders. The conclusions of one generation are the truths of the next. We are able, it is our duty, deliberately to take things for granted which our forefathers had a duty to doubt about; and unless we summarily put down disputation on points which have been already proved and ruled, we shall waste our time, and make no advances. Circumstances indeed may arise, when a question may legitimately be revived, which has already been definitely determined; but a re-consideration of such a question need not abruptly unsettle the existing certitude of those who engage in it, or throw them into a scepticism about things in general, even though eventually they find they have been wrong in a particular matter. It would have been absurd to prohibit the controversy which has lately been held concerning the obligations of Newton to Pascal; and supposing it had issued in their being established, the partisans of Newton would not have thought it necessary to renounce their certitude of the law of gravitation itself, on the ground that they had been mistaken in their certitude that Newton discovered it.

If we are never to be certain, after having been once certain wrongly, then we ought never to attempt a proof because we have once made a bad one. Errors in reasoning are lessons and warnings, not to give up reasoning, but to reason with greater caution. It is absurd to break up the whole structure of our knowledge, which is the glory of the human intellect, because the intellect is not infallible in its conclusions. If in any particular case we have been mistaken in our inferences and the certitudes which followed upon them, we are bound of course to take the fact of this mistake into account, in making up our minds on any new question, before we proceed to decide upon it. But if, while weighing the arguments on one side and the other and drawing our conclusion, that old mistake has already been allowed for, or has been, to use a familiar mode of speaking, discounted, then it has no outstanding claim against our acceptance of that conclusion, after it has actually been drawn. Whatever be the legitimate weight of the fact of that mistake in our inquiry, justice has been done to it, before we have allowed ourselves to be certain again. Suppose I am walking out in the moonlight, and see dimly the outlines of some figure among the trees;—it is a man. I draw nearer,—it is still a man; nearer still, and all hesitation is at an end,—I am certain it is a man. But he neither moves, nor speaks when I address him; and then I ask myself what can be his purpose in hiding among the trees at such an hour. I come quite close to him, and put out my arm. Then I find for certain that what I took for a man is but a singular shadow, formed by the falling of the moonlight on the interstices of some branches or their foliage. Am I not to indulge my second certitude, because I was wrong in my first? does not any objection, which lies against my second from the failure of my first, fade away before the evidence on which my second is founded?

Or again: I depose on my oath in a court of justice, to the best of my knowledge and belief, that I was robbed by the prisoner at the bar. Then, when the real offender is brought before me, I am obliged, to my great confusion, to retract. Because I have been mistaken in my certitude, may I not at least be certain that I have been mistaken? And further, in spite of the shock which that mistake gives me, is it impossible that the sight of the real culprit may give me so luminous a conviction that at length I have got the right man, that, were it decent towards the court, or consistent with self-respect, I may find myself prepared to swear to the identity of the second, as I have already solemnly committed myself to the identity of the first? It is manifest that the two certitudes stand each on its own basis, and the antecedent objection to my admission of a truth which was brought home to me second, drawn from a hallucination which came first, is a mere abstract argument, impotent when directed against good evidence lying in the concrete.

If in the criminal case which I have been supposing, the second certitude, felt by a witness, was a legitimate state of mind, so was the first. An act, viewed in itself, is not wrong because it is done wrongly. False certitudes are faults because they are false, not because they are (supposed) certitudes. They are, or may be, the attempts and the failures of an intellect insufficiently trained, or off its guard. Assent is an act of the mind, congenial to its nature; and it, as other acts, may be made both when it ought to be made, and when it ought not. It is a free act, a personal act for which the doer is responsible, and the actual mistakes in making it, be they ever so numerous or serious, have no force whatever to prohibit the act itself. We are accustomed in such cases, to appeal to the maxim, "Usum non tollit abusus;" and it is plain that, if what may be called functional disarrangements of the intellect are to be considered fatal to the recognition of the functions themselves, then the mind has no laws whatever and no normal constitution. I just now spoke of the growth of knowledge; there is also a growth in the use of those faculties by which knowledge is acquired. The intellect admits of an education; man is a being of progress; he has to learn how to fulfil his end, and to be what facts show that he is intended to be. His mind is in the first instance in disorder, and runs wild; his faculties have their rudimental and inchoate state, and are gradually carried on by practice and experience to their perfection. No instances then whatever of mistaken certitude are sufficient to constitute a proof, that certitude itself is a perversion or extravagance of his nature.

We do not dispense with clocks, because from time to time they go wrong, and tell untruly. A clock, organically considered, may be perfect, yet it may require regulating. Till that needful work is done, the moment-hand perhaps marks the half-minute, when the minute-hand is at the quarter-past, and the hour hand is just at noon, and the quarter-bell strikes the three-quarters, and the hour-bell strikes four, while the sun-dial precisely tells two o'clock. The sense of certitude may be called the bell of the intellect; and that it strikes when it should not is a proof that the clock is out of order, no proof that the bell will be untrustworthy and useless, when it comes to us adjusted and regulated from the hands of the clock-maker.

Our conscience too may be said to strike the hours, and will strike them wrongly, unless it be duly regulated for the performance of its proper function. It is the loud announcement of the principle of right in the details of conduct, as the sense of certitude is the clear witness to what is true. Both certitude and conscience have a place in the normal condition of the mind. As a human being, I am unable, if I were to try, to live without some kind of conscience; and I am as little able to live without these landmarks of thought which certitude secures for me; still, as the hammer of a clock may tell untruly, so may my conscience and my sense of certitude be attached to mental acts, whether of consent or of assent, which have no claim to be thus sanctioned. Both the moral and the intellectual sanction are liable to be biassed by personal inclinations and motives; both require and admit of discipline; and, as it is no disproof of the authority of conscience that false consciences abound, neither does it destroy the importance and the uses of certitude, because even educated minds, who are earnest in their inquiries after the truth, in many cases remain under the power of prejudice or delusion.

To this deficiency in mental training a wider error is to be attributed,—the mistaking for conviction and certitude states and frames of mind which make no pretence to the fundamental condition on which conviction rests as distinct from assent. The multitude of men confuse together the probable, the possible, and the certain, and apply these terms to doctrines and statements almost at random. They have no clear view what it is they know, what they presume, what they suppose, and what they only assert. They make little distinction between credence, opinion, and profession; at various times they give them all perhaps the name of certitude, and accordingly, when they change their minds, they fancy they have given up points of which they had a true conviction. Or at least bystanders thus speak of them, and the very idea of certitude falls into disrepute. . . .

. . . Now a religion is not a proposition, but a system; it is a rite, a creed, a philosophy, a rule of duty, all at once; and to accept a religion is neither a simple assent to it nor a complex, neither a conviction nor a prejudice, neither a notional assent nor a real, not a mere act of profession, nor of credence, nor of opinion, nor of speculation, but it is a collection of all these various kinds of assents, at once and together, some of one description, some of another; but, out of all these different assents, how many are of that kind which I have called certitude? Certitudes indeed do not change, but who shall pretend that assents are indefectible?

For instance: the fundamental dogma of Protestantism is the exclusive authority of Holy Scripture; but in holding this a Protestant holds a host of propositions, explicitly or implicitly, and holds them with assents of various character. Among these propositions, he holds that Scripture is the Divine Revelation itself, that it is inspired, that nothing is known in doctrine but what is there, that the Church has no authority in matters of doctrine, that, as claiming it, it was condemned long ago in the Apocalypse, that St. John wrote the Apocalypse, that justification is by faith only, that our Lord is God, that there are seventy-two generations between Adam and our Lord. Now of which, out of all these propositions, is he certain? and to how many of them is his assent of one and the same description? His belief, that Scripture is commensurate with the Divine Revelation, is perhaps implicit, not conscious; as to inspiration, he does not well know what the word means, and his assent is scarcely more than a profession; that no doctrine is true but what can be proved from Scripture he understands, and his assent to it is what I have called speculative; that the Church has no authority he holds with a real assent or belief; that the Church is condemned in the Apolcalypse is a standing prejudice; that St. John wrote the Apocalypse is his opinion; that justification is by faith only, he accepts, but scarcely can be said to apprehend; that our Lord is God perhaps he is certain; that there are seventy-two generations between Adam and Christ he accepts on credence. Yet, if he were asked the question, he would most probably answer that he was certain of the truth of "Protestantism," though "Protestantism" means these things and a hundred more all at once, and though he believes with actual certitude only one of them all,—that indeed a dogma of most sacred importance, but not the discovery of Luther or Calvin. He would think it enough to say that he was a foe to "Romanism" and "Socinianism," and to avow that he gloried in the Reformation. He looks upon each of these religious professions, Protestantism, Romanism, Socinianism and Theism, merely as units, as if they were not each made up of many elements, as if they had nothing in common, as if a transition from the one to the other involved a simple obliteration of all that had been as yet written on his mind, and would be the reception of a new faith.

When, then, we are told that a man has changed from one religion to another, the first question which we have to ask, is, have the first and the second religions nothing in common? If they have common doctrines, he has changed only a portion of his creed, not the whole: and the next question is, has he ever made much of any doctrines but such as are if otherwise common to his new creed and his old? what doctrines was he certain of among the old, and what among the new?

Thus, of three Protestants, one becomes a Catholic, a second a Unitarian, and a third an unbeliever: how is this? The first becomes a Catholic, because he assented, as a Protestant, to the doctrine of our Lord's divinity, with a real assent and a genuine conviction, and because this certitude, taking possession of his mind, led him on to welcome the Catholic doctrines of the Real Presence and of the Theotocos, till his Protestantism fell off from him, and he submitted himself to the Church. The second became a Unitarian, because, proceeding on the principle that Scripture was the rule of faith and that a man's private judgment was its rule of interpretation, and finding that the doctrine of the Nicene and Athanasian Creeds did not follow by logical necessity from the text of Scripture, he said to himself, "The word of God has been made of none effect by the traditions of men," and therefore nothing was left for him but to profess what he considered primitive Christianity, and to become a Humanitarian. The third gradually subsided into infidelity, because he started with the Protestant dogma, cherished in the depths of his nature, that a priesthood was a corruption of the simplicity of the Gospel. First, then, he would protest against the sacrifice of the Mass; next he gave up baptismal regeneration, and the sacramental principle; then he asked himself whether dogmas were not a restraint on Christian liberty as well as sacraments; then came the question, what after all was the use of teachers of religion? why should any one stand between him and his Maker? After a time it struck him, that this obvious question had to be answered by the Apostles, as well as by the Anglican clergy; so he came to the conclusion that the true and only revelation of God to man is that which is written on the heart. This did for a time, and he remained a Deist. But then it occurred to him, that this inward moral law was there within the breast, whether there was a God or not, and that it was a roundabout way of enforcing that law, to say that it came from God, and simply unnecessary, considering it carried with it its own sacred and sovereign authority, as our feelings instinctively testified; and when he turned to look at the physical world around him, he really did not see what scientific proof there was there of the Being of God at all, and it seemed to him as if all things would go on quite as well as at present, without that hypothesis as with it; so he dropped it, and became a *purus, putus* Atheist.

Now the world will say, that in these three cases old certitudes were lost, and new were gained; but it is not so: each of the three men started with just one certitude, as he would have himself professed, had he examined himself narrowly; and he carried it out and carried it with him into a new system of belief. He was true to that one conviction from first to last; and on looking back on the past, would perhaps insist upon this, and say he had really been consistent all through, when others made much of his great changes in religious opinion. He has indeed made serious additions to his initial ruling principle, but he has lost no conviction of which he was originally possessed.

I will take one more instance. A man is converted to the Catholic Church from his admiration of its religious system, and his disgust with Protestantism. That admiration remains; but, after a time, he leaves his new faith, perhaps returns to his old. The reason, if we may conjecture, may sometimes be this: he has never believed in the Church's infallibility; in her doctrinal truth he has believed, but in her infallibility, no. He was asked, before he was received, whether he held all that the Church taught, he replied he did; but he understood the question to mean, whether he held those particular doctrines "which at that time the Church in matter of fact formally taught," whereas it really meant "whatever the Church then or at any future time should teach." Thus, he never had the indispensable and elementary faith of a Catholic, and was simply no subject for reception into the fold of the Church. This being the case, when the Immaculate Conception is defined, he feels that it is something more than he bargained for when he became a Catholic, and accordingly he gives up his religious profession. The world will say that he has lost his certitude of the divinity of the Catholic Faith, but he never had it.

The first point to be ascertained, then, when we hear of a change of religious certitude in another, is, what the doctrines are on which his so-called certitude before now and at present has respectively fallen. All doctrines besides these were the accidents of his profession, and the indefectibility of certitude would not be disproved, though he changed them every year. There are few religions which have no points in common; and these, whether true or false, when embraced with an absolute conviction, are the pivots on which changes take place in that collection of credences, opinions, prejudices, and other assents, which make up what is called a man's selection and adoption of a form of religion, a denomination, or a Church. There have been Protestants whose idea of enlightened Christianity has been a strenuous antagonism to what they consider the unmanliness and unreasonableness of Catholic morality, an antipathy to the precepts of patience, meekness, forgiveness of injuries, and chastity. All this they have considered a woman's religion, the ornament of monks, of the sick, the feeble, and the old. Lust, revenge, ambition, courage, pride, these, they have fancied, made the man, and want of them the slave. No one could fairly accuse such men of any great change of their convictions, or refer to them in proof of the defectibility of certitude, if they were one day found to have taken up the profession of Islam.

And if this intercommunion of religions holds good, even when the common points between them are but errors held in common, much more natural will be the transition from one religion to another, without injury to existing certitudes, when the common points, the objects of those certitudes, are truths; and still stronger in that case and more constraining will be the sympathy, with which minds that love truth, even when they have surrounded it with error, will yearn towards the Catholic faith, which contains within itself, and claims as its own, all truth that is elsewhere to be found, and more than all, and nothing but truth. This is the secret of the influence, by which the Church draws to herself converts from such various and conflicting religions. They come, not so much to lose what they have, as to gain what they have not; and in order that, by means of what they have, more may be given to them. St. Augustine tells us that there is no false teaching without an intermixture of truth; and it is by the light of those particular truths, contained respectively in the various religions of men, and by our certitudes about them, which are possible wherever those truths are found, that we pick our way, slowly perhaps, but surely, into the One Religion which God has given, taking our certitudes with us, not to lose, but to keep them more securely, and to understand and love their objects more perfectly. . . .

. . . I will not urge (lest I should be accused of quibbling), that certitude is a conviction of what is true, and that these so-called certitudes have come to nought, because, their objects being errors, not truths, they really were not certitudes at all; nor will I insist, as I might, that they ought to be proved first to be something more than mere prejudices, assents without reason and judgment, before they can fairly be taken as instances of the defectibility of certitude; but I simply ask, as regards the zeal of the Jews for the sufficiency of their law, (even though it implied genuine certitude, not a prejudice, not a mere conviction,) still was such zeal, such professed certitude, found in those who were eventually converted, or in those who were not; for, if those who had not that certitude became Christians and those who had it remained Jews, then loss of certitude in the latter is not instanced in the fact of the conversion of the former. St. Paul certainly is an exception, but his conversion, as also his after-life, was miraculous; ordinarily speaking, it was not the zealots who supplied members to the Catholic Church, but those "men of good will," who, instead of considering the law as perfect and eternal, "looked for the redemption of Israel," and for "the knowledge of salvation in the remission of sins." And, in like manner, as to those learned and devout men among the Anglicans at the present day, who come so near the Church without acknowledging her claims, I ask whether there are not two classes among them also,—those who are looking out beyond their own body for the perfect way, and those on the other hand who teach that the Anglican communion is the golden mean between men who believe too much and men who believe too little, the centre of unity to which East and West are destined to gravitate, the instrument and the mould, as the Jews might think of their own moribund institutions, through which the kingdom of Christ is to be established all over the earth. And next I would ask, which of these two classes supplies converts to the Church; for if they come from among those who never professed to be quite certain of the special strength of the Anglican position, such men cannot be quoted as instances of the defectibility of certitude.

There is indeed another class of beliefs, of which I must take notice, the failure of which may be taken at first sight as a proof that certitude may be lost. Yet they clearly deserve no other name than prejudices, as being founded upon reports of facts, or on arguments, which will not bear careful examination. Such was the disgust felt towards our predecessors in primitive times, the Christians of the first centuries, as a secret society, as a conspiracy against the civil power, as a set of mean, sordid, despicable fanatics, as monsters revelling in blood and impurity. Such also is the deep prejudice now existing against the Church among Protestants, who dress her up in the most hideous and loathsome images, which rightly attach, in the prophetic descriptions, to the evil spirit, his agents and instruments. And so of the numberless calumnies directed against individual Catholics, against our religious bodies and men in authority, which serve to feed and sustain the suspicion and dislike with which everything Catholic is regarded in this country. But as a persistence in such prejudices is no evidence of their truth, so an abandonment of them is no evidence that certitude can fail.

There is yet another class of prejudices against the Catholic Religion, which is far more tolerable and intelligible than those on which I have been dwelling, but still in no sense certitudes. Indeed, I doubt whether they would be considered more than presumptive opinions by the persons who entertain them. Such is the idea which has possessed certain philosophers, ancient and modern, that miracles are an infringement and disfigurement of the beautiful order of nature. Such, too, is the persuasion, common among political and literary men, that the Catholic Church is inconsistent with the true interests of the human race, with social progress, with rational freedom, with good government. A renunciation of these imaginations is not a change in certitudes.

So much on this subject. All concrete laws are general, and persons, as such, do not fall under laws. Still, I have gone a good way, as I think, to remove the objections to the doctrine of the indefectibility of certitude in matters of religion, though I cannot assign to it an infallible token.

One further remark may be made. Certitude does not admit of an interior, immediate test, sufficient to discriminate it from false certitude. Such a test is rendered impossible from the circumstance that, when we make the mental act expressed by "I know," we sum up the whole series of reflex judgments which might, each in turn, successively exercise a critical function towards those of the series which precede it. But still, if it is the general rule that certitude is indefectible, will not that indefectibility itself become at least in the event a criterion of the genuineness of the certitude? or is there any rival state or habit of the intellect, which claims to be indefectible also? A few words will suffice to answer these questions.

Premising that all rules are but general, especially those which relate to the mind, I observe that indefectibility may at least serve as a negative test of certitude, or *sine quâ non* condition, so that whoever loses his conviction on a given point is thereby proved not to have been certain of it. Certitude ought to stand all trials, or it is not certitude. Its very office is to cherish and maintain its object, and its very lot and duty is to sustain rude shocks in maintenance of it without being damaged by them.

I will take an example. Let us suppose we are told on an unimpeachable authority, that a man whom we saw die is now alive again and at his work, as it was his wont to be; let us suppose we actually see him and converse with him; what will become of our certitude of his death? I do not think we should give it up; how could we, when we actually saw him die? At first, indeed, we should be thrown into an astonishment and confusion so great, that the world would seem to reel round us, and we should be ready to give up the use of our senses and of our memory, of our reflective powers, and of our reason, and even to deny our power of thinking, and our existence itself. Such confidence have we in the doctrine that when life goes it never returns. Nor would our bewilderment be less, when the first blow was over; but our reason would rally, and with our reason our certitude would come back to us. Whatever came of it, we should never cease to know and to confess to ourselves both of the contrary facts, that we saw him die, and that after dying we saw him alive again. The overpowering strangeness of our experience would have no power to shake our certitude in the facts which created it.

Again, let us suppose, for argument's sake, that ethnologists, philologists, anatomists, and antiquarians agreed together in separate demonstrations that there were half a dozen races of men, and that they were all descended from gorillas, or chimpanzees, or ourang-outangs, or baboons; moreover, that Adam was an historical personage, with a well-ascertained dwelling-place, surroundings and date, in a comparatively modern world. On the other hand, let me believe that the Word of God Himself distinctly declares that there were no men before Adam, that he was immediately made out of the slime of the earth, and that he is the first father of all men that are or ever have been. Here is a contradiction of statements more direct than in the former instance; the two cannot stand together; one or other of them is untrue. But whatever means I might be led to take, for making, if possible, the antagonism tolerable, I conceive I should never give up my certitude in that truth which on sufficient grounds I determined to come from heaven. If I so believed, I should not pretend to argue, or to defend myself to others; I should be patient; I should look for better days; but I should still believe. If, indeed, I had hitherto only half believed, if I believed with an assent short of certitude, or with an acquiescence short of assent, or hastily or on light grounds, then the case would be altered; but if, after full consideration, and availing myself of my best lights, I did think that beyond all question God spoke as I thought He did, philosophers and experimentalists might take their course for me,—I should consider that they and I thought and reasoned in different mediums, and that my certitude was as little in collision with them or damaged by them, as if they attempted to counteract in some great matter chemical action by the force of gravity, or to weigh magnetic influence against capillary attraction. Of course, I am putting an impossible case, for philosophical discoveries cannot really contradict divine revelation.

So much on the indefectibility of certitude; as to the question whether any other assent is indefectible besides it, I think prejudice may be such; but it cannot be confused with certitude, for the one is an assent previous to rational grounds, and the other an assent given expressly after careful examination.

It seems then that on the whole there are three conditions of certitude: that it follows on investigation and proof, that it is accompanied by a specific sense of intellectual satisfaction and repose, and that it is irreversible. If the assent is made without rational grounds, it is a rash judgment, a fancy, or a prejudice; if without the sense of finality, it is scarcely more than an inference; if without permanence, it is a mere conviction.

[ch. 7]

Inference

Inference is the conditional acceptance of a proposition, Assent is the unconditional; the object of Assent is a truth, the object of Inference is the truth-like or a verisimilitude. The problem which I have undertaken is that of ascertaining how it comes to pass that a conditional act leads to an unconditional; and, having now shown that assent really is unconditional, I proceed to show how inferential exercises, as such, always must be conditional.

We reason, when we hold this by virtue of that; whether we hold it as evident or as approximating or tending to be evident, in either case we so hold it because of holding something else to be evident or tending to be evident. In the next place, our reasoning ordinarily presents itself to our mind as a simple act, not a process or series of acts. We apprehend the antecedent and then apprehend the consequent, without explicit recognition of the medium connecting the two, as if by a sort of direct association of the first thought with the second. We proceed by a sort of instinctive perception, from premiss to conclusion. I call it instinctive, not as if the faculty were one and the same to all men in strength and quality (as we generally conceive of instinct), but because ordinarily, or at least often, it acts by a spontaneous impulse, as prompt and inevitable as the exercise of sense and memory. We perceive external objects, and we remember past events, without knowing how we do so; and in like manner we reason without effort and intention, or any necessary consciousness of the path which the mind takes in passing from antecedent to conclusion.

Such is ratiocination, in what may be called a state of nature, as it is found in the uneducated,—nay, in all men, in its ordinary exercise; nor is there any antecedent ground for determining that it will not be as correct in its informations as it is instinctive, as trustworthy as are sensible perception and memory, though its informations are not so immediate and have a wider range. By means of sense we gain knowledge directly; by means of reasoning we gain it indirectly, that is, by virtue of a previous knowledge. And if we may justly regard the universe, according to the meaning of the word, as one whole, we may also believe justly that to know one part of it is necessarily to know much more than that one part. This thought leads us to a further view of ratiocination. . . . That which the mind is able thus variously to bring together into unity, must have some real intrinsic connexion of part with part. But if this *summa rerum* is thus one whole, it must be constructed on definite principles and laws, the knowledge of which will enlarge our capacity of reasoning about it in particulars;—thus we are led on to aim at determining on a large scale and on system, what even gifted or practised intellects are only able by their own personal vigour to reach piecemeal and fitfully, that is, at substituting scientific methods, such as all may use, for the action of individual genius.

There is another reason for attempting to discover an instrument of reasoning (that is, of gaining new truths by means of old), which may be less vague and arbitrary than the talent and experience of the few or the common-sense of the many. As memory is not always accurate, and has on that account led to the adoption of writing, as being a *memoria technica*, unaffected by the failure of mental impressions,—as our senses at times deceive us, and have to be corrected by each other; so is it also with our reasoning faculty. The conclusions of one man are not the conclusions of another; those of the same man do not always agree together; those of ever so many who agree together may differ from the facts themselves, which those conclusions are intended to ascertain. In consequence it becomes a necessity, if it be possible, to analyze the process of reasoning, and to invent a method which may act as a common measure between mind and mind, as a means of joint investigation, and as a recognized intellectual standard,—a standard such as to secure us against hopeless mistakes, and to emancipate us from the capricious *ipse dixit* of authority.

As the index on the dial notes down the sun's course in the heavens, as a key, revolving through the intricate wards of the lock, opens for us a treasure-house, so let us, if we can, provide ourselves with some ready expedient to serve as a true record of the system of objective truth, and an available rule for interpreting its phenomena; or at least let us go as far as we can in providing it. One such experimental key is the science of geometry, which, in a certain department of nature, substitutes a collection of true principles, fruitful and interminable in consequences, for the guesses, *pro re natâ*, of our intellect, and saves it both the labour and the risk of guessing. Another far more subtle and effective instrument is algebraical science, which acts as a spell in unlocking for us, without merit or effort of our own individually, the *arcana* of the concrete physical universe. A more ambitious, because a more comprehensive contrivance still, for interpreting the concrete world is the method of logical inference. What we desiderate is something which may supersede the need of personal gifts by a far-reaching and infallible rule. Now, without external symbols to mark out and to steady its course, the intellect runs wild; but with the aid of symbols, as in algebra, it advances with precision and effect. Let then our symbols be words: let all thought be arrested and embodied in words. Let language have a monopoly of thought; and thought go for only so much as it can show itself to be worth in language. Let every prompting of the intellect be ignored, every *momentum* of argument be disowned, which is unprovided with an equivalent wording, as its ticket for sharing in the common search after truth. Let the authority of nature, commonsense, exper-ience, genius, go for nothing. Ratiocination, thus restricted and put into grooves, is what I have called Inference, and the science, which is its regulating principle, is Logic.

The first step in the inferential method is to throw the question to be decided into the form of a proposition; then to throw the proof itself into propositions, the force of the proof lying in the comparison of these propositions with each other. When the analysis is carried out fully and put into form, it becomes the Aristotelic syllogism. However, an inference need not be expressed thus technically; an enthymeme fulfils the requirements of what I have called Inference. So does any other form of words with the mere grammatical expressions, "for," "therefore," "supposing," "so that," "similarly," and the like. Verbal reasoning, of whatever kind, as opposed to mental, is what I mean by inference, which differs from logic only inasmuch as logic is its scientific form. And it will be more convenient here to use the two words indiscriminately, for I shall say nothing about logic which does not in its substance also apply to inference.

Logical inference, then, being such, and its office such as I have described, the question follows, how far it answers the purpose for which it is used. It proposes to provide both a test and a common measure of reasoning; and I think it will be found partly to succeed and partly to fail; succeeding so far as words can in fact be found for representing the countless varietes and subtleties of human thought, failing on account of the fallacy of the original assumption, that whatever can be thought can be adequately expressed in words.

In the first place, Inference, being conditional, is hampered with other propositions besides that which is especially its own, that is, with the premises as well as the conclusion, and with the rules connecting the latter with the former. It views its own proper proposition in the medium of prior propositions, and measures it by them. It does not hold a proposition for its own sake, but as dependent upon others, and those others it entertains for the sake of the conclusion. Thus it is practically far more concerned with the comparison of propositions, than with the propositions themselves. It is obliged to regard all the propositions, with which it has to do, not so much for their own sake, as for the sake of each other, as regards the identity or likeness, independence or dissimilarity, which has to be mutually predicated of them. It follows from this, that the more simple and definite are the words of a proposition, and the narrower their meaning, and the more that meaning in each proposition is restricted to the relation which it has to the words of the other propositions compared with it,—in other words, the nearer the propositions concerned in the inference approach to being mental abstractions, and the less they have to do with the concrete reality, and the more closely they are made to express exact, intelligible, comprehensible, communicable notions, and the less they stand for objective things, that is, the more they are the subjects, not of real, but of notional apprehension,—so much the more suitable do they become for the purposes of Inference.

Hence it is that no process of argument is so perfect, as that which is conducted by means of symbols. In Arithmetic 1 is 1, and just 1, and never anything else but 1; it never is 2, it has no tendency to change its meaning, and to become 2; it has no portion, quality, admixture of 2 in its meaning. And 6 under all circumstances is 3 times 2, and the sum of 2 and 4; nor can the whole world supply anything to throw doubt upon these elementary positions. It is not so with language. Take, by contrast, the word "inference," which I have been using: it may stand for the act of inferring, as I have used it; or for the connecting principle, or *inferentia*, between premisses and conclusions; or for the conclusion itself. And sometimes it will be difficult, in a particular sentence, to say which it bears of these three senses. And so again in Algebra, *a* is never *x*, or anything but *a*, wherever it is found; and *a* and *b* are always standard quantities, to which *x* and *y* are always to be referred, and by which they are always to be measured. In Geometry again, the subjects of argument, points, lines, and surfaces, are precise creations of the mind, suggested indeed by external objects, but meaning nothing but what they are defined to mean: they have no colour, no motion, no heat, no qualities which address themselves to the ear or to the palate; so that, in whatever combinations or relations the words denoting them occur, and to whomsoever they come, those words never vary in their meaning, but are just of the same measure and weight at one time and at another.

What is true of Arithmetic, Algebra, and Geometry, is true also of Aristotelic argumentation in its typical modes and figures. It compares two given words separately with a third, and then determines how they stand towards each other, in a *bonâ fide* identity of sense. In consequence, its formal process is best conducted by means of symbols, A, B, and C. While it keeps to these, it is safe; it has the cogency of mathematical reasoning, and draws its conclusions by a rule as unerring as it is blind.

Symbolical notation, then, being the perfection of the syllogistic method, it follows that, when words are substituted for symbols, it will be its aim to circumscribe and stint their import as much as possible, lest perchance A should not always exactly mean A, and B mean B; and to make them, as much as possible, the *calculi* of notions, which are in our absolute power, as meaning just what we choose them to mean, and as little as possible the tokens of real things, which are outside of us, and which mean we do not know how much, but so much certainly as, (in proportion as we enter into them,) may run away with us beyond the range of scientific management. The concrete matter of propositions is a constant source of trouble to syllogistic reasoning, as marring the simplicity and perfection of its process. Words, which denote things, have innumerable implications; but in inferential exercises it is the very triumph of that clearness and hardness of head, which is the characteristic talent for the art, to have stripped them of all these connatural senses, to have drained them of that depth and breadth of associations which constitute their poetry, their rhetoric, and their historical life, to have starved each term down till it has become the ghost of itself, and everywhere one and the same ghost, "omnibus umbra locis," so that it may stand for just one unreal aspect of the concrete thing to which it properly belongs, for a relation, a generalization, or other abstraction, for a notion neatly turned out of the laboratory of the mind, and sufficiently tame and subdued, because existing only in a definition.

Thus it is that the logician for his own purposes, and most usefully as far as those purposes are concerned, turns rivers, full, winding, and beautiful, into navigable canals. To him dog or horse is not a thing which he sees, but a mere name suggesting ideas; and by dog or horse universal he means, not the aggregate of all individual dogs or horses brought together, but a common aspect, meagre but precise, of all existing or possible dogs or horses, which all the while does not really correspond to any one single dog or horse out of the whole aggregate. Such minute fidelity in the representation of individuals is neither necessary nor possible to his art; his business is not to ascertain facts in the concrete, but to find and dress up middle terms; and, provided they and the extremes which they go between are not equivocal, either in themselves or in their use, and he can enable his pupils to show well in a *vivâ voce* disputation, or in a popular harangue, or in a written dissertation, he has achieved the main purpose of his profession.

Such are the characteristics of reasoning, viewed as a science or scientific art, or inferential process, and we might anticipate that, narrow as by necessity is its field of view, for that reason its pretensions to be demonstrative were incontrovertible. In a certain sense they really are so; while we talk logic, we are unanswerable; but then, on the other hand, this universal living scene of things is after all as little a logical world as it is a poetical; and, as it cannot without violence be exalted into poetical perfection, neither can it be attenuated into a logical formula. Abstract can only conduct to abstract; but we have need to attain by our reasonings to what is concrete; and the margin between the abstract conclusions of the science, and the concrete facts which we wish to ascertain, will be found to reduce the force of the inferential method from demonstration to the mere determination of the probable. Thus, whereas (as I have already said) Inference starts with conditions, as starting with premisses, here are two reasons why, when employed upon questions of fact, it can only conclude probabilities: first, because its premisses are assumed, not proved; and secondly, because its conclusions are abstract, and not concrete. I will now consider these two points separately.

Inference comes short of proof in concrete matters, because it has not a full command over the objects to which it relates, but merely assumes its premisses. In order to complete the proof, we are thrown upon some previous syllogism or syllogisms, in which the assumptions may be proved; and then, still farther back, we are thrown upon others again, to prove the new assumptions of that second order of syllogisms. Where is this process to stop? especially since it must run upon separated, divergent, and multiplied lines of argument, the farther the investigation is carried back. At length a score of propositions present themselves, all to be proved by propositions more evident than themselves, in order to enable them respectively to become premisses to that series of inferences which terminates in the conclusion which we originally drew. But even now the difficulty is not at an end; it would be something to arrive at length at premisses which are undeniable, however long we might be in arriving at them; but in this case the long retrospection lodges us at length at what are called first principles, the recondite sources of all knowledge, as to which logic provides no common measure of minds,—which are accepted by some, rejected by others,—in which, and not in the syllogistic exhibitions, lies the whole problem of attaining to truth,—and which are called self-evident by their respective advocates because they are evident in no other way. One of the two uses contemplated in reasoning by rule, or in verbal argumentation, was, as I have said, to establish a standard of truth and to supersede the *ipse dixit* of authority: how does it fulfil this end, if it only leads us back to first principles, about which there is interminable controversy? We are not able to prove by syllogism that there are any self-evident propositions at all; but supposing there are (as of course I hold there are), still who can determine these by logic? Syllogism, then, though of course it has its use, still does only the minutest and easiest part of the work, in the investigation of truth, for when there is any difficulty, that difficulty commonly lies in determining first principles, not in the arrangement of proofs.

Even when argument is the most direct and severe of its kind, there must be those assumptions in the process which resolve themselves into the conditions of human nature; but how many more assumptions does that process in ordinary concrete matters involve, subtle assumptions not directly arising out of these primary conditions, but accompanying the course of reasoning, step by step, and traceable to the sentiments of the age, country, religion, social habits and ideas, of the particular inquires or disputants, and passing current without detection, because admitted equally on all hands! And to these must be added the assumptions which are made from the necessity of the case, in consequence of the prolixity and elaborateness of any argument which should faithfully note down all the propositions which go to make it up. We recognize this tediousness even in the case of the theorems of Euclid, though mathematical proof is comparatively simple.

Logic then does not really prove; it enables us to join issue with others; it suggests ideas; it opens views; it maps out for us the lines of thought; it verifies negatively; it determines when differences of opinion are hopeless; and when and how far conclusions are probable; but for genuine proof in concrete matter we require an *organon* more delicate, versatile, and elastic than verbal argumentation. . . .

This is what I have to say on formal Inference, when taken to represent Ratiocination. Science in all its departments has too much simplicity and exactness, from the nature of the case, to be the measure of fact. In its very perfection lies its incompetency to settle particulars and details. As to Logic, its chain of conclusions hangs loose at both ends; both the point from which the proof should start, and the points at which it should arrive, are beyond its reach; it comes short both of first principles and of concrete issues. Even its most elaborate exhibitions fail to represent adequately the sum-total of considerations by which an individual mind is determined in its judgment of things; even its most careful combinations made to bear on a conclusion want that steadiness of aim which is necessary for hitting it. As I said when I began, thought is too keen and manifold, its sources are too remote and hidden, its path too personal, delicate, and circuitous, its subject-matter too various and intricate, to admit of the trammels of any language, of whatever subtlety and of whatever compass.

Nor is it any disparagement of the proper value of formal reasonings thus to speak of them. That they cannot proceed beyond probabilities is most readily allowed by those who use them most. Philosophers, experimentalists, lawyers, in their several ways, have commonly the reputation of being, at least on moral and religious subjects, hard of belief; because, proceeding in the necessary investigation by the analytical method of verbal inference, they find within its limits no sufficient resources for attaining a conclusion. Nay, they do not always find it possible in their own special province severally; for, even when in their hearts they have no doubt about a conclusion, still often, from the habit of their minds, they are reluctant to own it, and dwell upon the deficiencies of the evidence, or the possibility of error, because they speak by rule and by book, though they judge and determine by common-sense.

Every exercise of nature or of art is good in its place; and the uses of this logical inference are manifold. It is the great principle of order in our thinking; it reduces a chaos into harmony; it catalogues the accumulations of knowledge; it maps out for us the relations of its separate departments; it puts us in the way to correct its own mistakes. It enables the independent intellects of many, acting and re-acting on each other, to bring their collective force to bear upon one and the same subject-matter, or the same question. If language is an inestimable gift to man, the logical faculty prepares it for our use. Though it does not go so far as to ascertain truth, still it teaches us the direction in which truth lies, and how propositions lie towards each other. Nor is it a slight benefit to know what is probable, and what is not so, what is needed for the proof of a point, what is wanting in a theory, how a theory hangs together, and what will follow, if it be admitted. Though it does not itself discover the unknown, it is one principal way by which discoveries are made. Moreover, a course of argument, which is simply conditional, will point out when and where experiment and observation should be applied, or testimony sought for, as often happens both in physical and legal questions. A logical hypothesis is the means of holding facts together, explaining difficulties, and reconciling the imagination to what is strange. And, again, processes of logic are useful as enabling us to get over particular stages of an investigation speedily and surely, as on a journey we now and then gain time by travelling by night, make short cuts when the high-road winds, or adopt water-carriage to avoid fatigue.

But reasoning by rule and in words is too natural to us, to admit of being regarded merely in the light of utility. Our inquiries spontaneously fall into scientific sequence, and we think in logic, as we talk in prose, without aiming at doing so. However sure we are of the accuracy of our instinctive conclusions, we as instinctively put them into words, as far as we can; as preferring, if possible, to have them in an objective shape which we can fall back upon,—first for our own satisfaction, then for our justification with others. Such a tangible defence of what we hold, inadequate as it necessarily is, considered as an analysis of our ratiocination in its length and breadth, nevertheless is in such sense associated with our holdings, and so fortifies and illustrates them, that it acts as a vivid apprehension acts, giving them luminousness and force. Thus inference becomes a sort of symbol of assent, and even bears upon action. . . .

It is plain that formal logical sequence is not in fact the method by which we are enabled to become certain of what is concrete; and it is equally plain, from what has been already suggested, what the real and necessary method is. It is the cumulation of probabilities, independent of each other, arising out of the nature and circumstances of the particular case which is under review; probabilities too fine to avail separately, too subtle and circuitous to be convertible into syllogisms, too numerous and various for such conversion, even were they convertible. As a man's portrait differs from a sketch of him, in having, not merely a continuous outline, but all its details filled in, and shades and colours laid on and harmonized together, such is the multiform and intricate process of ratiocination, necessary for our reaching him as a concrete fact, compared with the rude operation of syllogistic treatment.

Let us suppose I wish to convert an educated, thoughtful Protestant, and accordingly present for his acceptance a syllogism of the following kind:—"All Protestants are bound to join the Church; you are a Protestant: ergo." He answers, we will say, by denying both premises; and he does so by means of arguments, which branch out into other arguments, and those into others, and all of them severally requiring to be considered by him on their own merits, before the syllogism reaches him, and in consequence mounting up, taken altogether, into an array of inferential exercises large and various beyond calculation. Moreover, he is bound to submit himself to this complicated process from the nature of the case; he would act rashly, if he did not; for he is a concrete individual unit, and being so is under so many laws, and is the subject of so many predications all at once, that he cannot determine, offhand, his position and his duty by the law and the predication of one syllogism in particular. I mean he may fairly say, "Distinguo," to each of its premises: he says, "Protestants are bound to join the Church,—under circumstances," and "I am a Protestant—in a certain sense;" and therefore the syllogism, at first sight, does not touch him at all.

Before, then, he grants the major, he asks whether all Protestants really are bound to join the Church—are they bound in case they do not feel themselves bound; if they are satisfied that their present religion is a safe one; if they are sure it is true; if, on the other hand, they have grave doubts as to the doctrinal fidelity and purity of the Church; if they are convinced that the Church is corrupt; if their conscience instinctively rejects certain of its doctrines; if history convinces them that the Pope's power is not *jure divino*, but merely in the order of Providence? if, again, they are in a heathen country where priests are not? or where the only priest who is to be found exacts of them as a condition of their reception, a profession, which the Creed of Pope Pius IV says nothing about; for instance, that the Holy See is fallible even when it teaches, or that the Temporal Power is an anti-Christian corruption? On one or other of such grounds he thinks he need not change his religion; but presently he asks himself, Can a Protestant be in such a state as to be really satisfied with his religion, as he has just now been professing? Can he possibly believe Protestantism came from above, as a whole? how much of it can he believe came from above? and, as to that portion which he feels did come from above, has it not all been derived to him from the Church, when traced to its source? Is not Protestantism in itself a negation? Did not the Church exist before it? and can he be sure, on the other hand, that any one of the Church's doctrines is not from above? Further, he finds he has to make up his mind what is a corruption, and what are the tests of it; what he means by a religion; whether it is obligatory to profess any religion in particular; what are the standards of truth and falsehood in religion; and what are the special claims of the Church.

And so, again, as to the minor premiss, perhaps he will answer, that he is not a Protestant; that he is a Catholic of the early undivided Church; that he is a Catholic, but not a Papist. Then he has to determine questions about division, schism, visible unity, what is essential, what is desirable; about provisional states; as to the adjustment of the Church's claims with those of personal judgment and responsibility; as to the soul of the Church contrasted with the body; as to degrees of proof, and the degree necessary for his conversion; as to what is called his providential position, and the responsibility of change; as to the sincerity of his purpose to follow the Divine Will, whithersoever it may lead him; as to his intellectual capacity of investigating such questions at all.

None of these questions, as they come before him, admit of simple demonstration; but each carries with it a number of independent probable arguments, sufficient, when united, for a reasonable conclusion about itself. And first he determines that the questions are such as he personally, with such talents or attainments as he has, may fairly entertain; and then he goes on, after deliberation, to form a definite judgment upon them; and determines them, one way or another, in their bearing on the bald syllogism which was originally offered to his acceptance. And, we will say, he comes to the conclusion, that he ought to accept it as true in his case; that he is a Protestant in such a sense, of such a complexion, of such knowledge, under such circumstances, as to be called upon by duty to join the Church; that this is a conclusion of which he can be certain, and ought to be certain, and that he will be incurring grave responsibility, if he does not accept it as certain, and act upon the certainty of it. And to this conclusion he comes, as is plain, not by any possible verbal enumeration of all the considerations, minute but abundant, delicate but effective, which unite to bring him to it; but by a mental comprehension of the whole case, and a discernment of its upshot, sometimes after much deliberation, but, it may be, by a clear and rapid act of the intellect, always, however, by an unwritten summing-up, something like the summation of the terms, *plus* and *minus* of an algebraical series.

This I conceive to be the real method of reasoning in concrete matters;

and it has these characteristics:—First, it does not supersede the logical form of inference, but is one and the same with it; only it is no longer an abstraction, but carried out into the realities of life, its premises being instinct with the substance and the momentum of that mass of probabilities, which, acting upon each other in correction and confirmation, carry it home definitely to the individual case, which is its original scope.

Next, from what has been said it is plain, that such a process of reasoning is more or less implicit, and without the direct and full advertence of the mind exercising it. As by the use of our eyesight we recognize two brothers, yet without being able to express what it is by which we distinguish them; as at first sight we perhaps confuse them together, but, on better knowledge, we see no likeness between them at all; as it requires an artist's eye to determine what lines and shades make a countenance look young or old, amiable, thoughtful, angry or conceited, the principle of discrimination being in each case real, but implicit;—so is the mind unequal to a complete analysis of the motives which carry it on to a particular conclusion, and is swayed and determined by a body of proof, which it recognizes only as a body, and not in its constituent parts.

And thirdly, it is plain, that, in this investigation of the method of concrete inference, we have not advanced one step towards depriving inference of its conditional character; for it is still as dependent on premises as it is in its elementary idea. On the contrary, we have rather added to the obscurity of the problem; for a syllogism is at least a demonstration, when the premises are granted, but a cumulation of probabilities, over and above their implicit character, will vary both in their number and their separate estimated value, according to the particular intellect which is employed upon it. It follows that what to one intellect is a proof is not so to another, and that the certainty of a proposition does properly consist in the certitude of the mind which contemplates it. And this of course may be said without prejudice to the objective truth or falsehood of propositions, since it does not follow that these propositions on the one hand are not true, and based on right reason, and those on the other not false, and based on false reason, because not all men discriminate them in the same way. . . .

. . . I think it is the fact that many of our most obstinate and most reasonable certitudes depend on proofs which are informal and personal, which baffle our powers of analysis, and cannot be brought under logical rule, because they cannot be submitted to logical statistics. If we must speak of Law, this recognition of a correlation between certitude and implicit proof seems to me a law of our minds.

I said just now that an object of sense presents itself to our view as one whole, and not in its separate details: we take it in, recognize it, and discriminate it from other objects, all at once. Such too is the intellectual view we take of the *momenta* of proof for a concrete truth; we grasp the full tale of premises and the conclusion, *per modum unius*,—by a sort of instinctive perception of the legitimate conclusion in and through the premises, not by a formal juxta-position of propositions; though of course such a juxta-position is useful and natural, both to direct and to verify, just as in objects of sight our notice of bodily peculiarities, or the remarks of others may aid us in establishing a case of disputed identity. And, as this man or that will receive his own impression of one and the same person, and judge differently from others about his countenance, its expression, its moral significance, its physical contour and complexion, so an intellectual question may strike two minds very differently, may awaken in them distinct associations, may be invested by them in contrary characteristics, and lead them to opposite conclusions;— and so, again, a body of proof, or a line of argument, may produce a distinct, nay, a dissimilar effect, as addressed to one or to the other.

Thus in concrete reasonings we are in a great measure thrown back into that condition, from which logic proposed to rescue us. We judge for ourselves, by our own lights, and on our own principles; and our criterion of truth is not so much the manipulation of propositions, as the intellectual and moral character of the person maintaining them, and the ultimate silent effect of his arguments or conclusions upon our minds.

It is this distinction between ratiocination as the exercise of a living faculty in the individual intellect, and mere skill in argumentative science, which is the true interpretation of the prejudice which exists against logic in the popular mind, and of the animadversions which are levelled against it, as that its formulas make a pedant and a *doctrinaire*, that it never makes converts, that it leads to rationalism, that Englishmen are too practical to be logical, that an ounce of common-sense goes farther than many cartloads of logic, that Laputa is the land of logicians, and the like. Such maxims mean, when analyzed, that the processes of reasoning which legitimately lead to assent, to action, to certitude, are in fact too multiform, subtle, omnigenous, too implicit, to allow of being measured by rule, that they are after all personal. . .

That there are cases, in which evidence, not sufficient for a scientific proof, is nevertheless sufficient for assent and certitude, is the doctrine of Locke, as of most men. . . . The only question is, what these propositions are: this he does not tell us, but he seems to think that they are few in number, and will be without any trouble recognized at once by common-sense; whereas, unless I am mistaken, they are to be found throughout the range of concrete matter, and that supra-logical judgment, which is the warrant for our certitude about them, is not mere common-sense, but the true healthy action of our ratiocinative powers, an action more subtle and more comprehensive than the mere appreciation of a syllogistic argument. It is often called the "judicium prudentis viri," a standard of certitude which holds good in all concrete matter, not only in those cases of practice and duty, in which we are more familiar with it, but in questions of truth and falsehood generally, or in what are called "speculative" questions, and that, not indeed to the exclusion, but as the supplement of logic. Thus a proof, except in abstract demonstration, has always in it, more or less, an element of the personal, because "prudence" is not a constituent part of our nature, but a personal endowment.

And the language in common use, when concrete conclusions are in question, implies the presence of this personal element in the proof of them. We are considered to feel, rather than to see, its cogency; and we decide, not that the conclusion must be, but that it cannot be otherwise. We say, that we do not see our way to doubt it, that it is impossible to doubt, that we are bound to believe it, that we should be idiots, if we did not believe. We never should say, in abstract science, that we could not escape the conclusion that 25 was a mean proportional between 5 and 125; or that a man had no right to say that a tangent to a circle at the extremity of the radius makes an acute angle with it. Yet, though our certitude of the fact is quite as clear, we should not think it unnatural to say that the insularity of Great Britain is as good as demonstrated, or that none but a fool expects never to die. Phrases indeed such as these are sometimes used to express a shade of doubt, but it is enough for my purpose if they are also used when doubt is altogether absent. What, then, they signify, is, what I have so much insisted on, that we have arrived at these conclusions—not *ex opere operato*, by a scientific necessity independent of ourselves,—but by the action of our own minds, by our own individual perception of the truth in question, under a sense of duty to those conclusions and with an intellectual conscientiousness.

This certitude and this evidence are often called moral; a word which I avoid, as having a very vague meaning; but using it here for once, I observe that moral evidence and moral certitude are all that we can attain, not only in the case of ethical and spiritual subjects, such as religion, but of terrestrial and cosmical questions also. . . .

This being the state of the case, the question arises, whether, granting that the personality (so to speak) of the parties reasoning is an important element in proving propositions in concrete matter; any account can be given of the ratiocinative method in such proofs, over and above that analysis into syllogism which is possible in each of its steps in detail. I think there can; though I fear, lest to some minds it may appear far-fetched or fanciful; however, I will hazard this imputation. I consider, then, that the principle of concrete reasoning is parallel to the method of proof which is the foundation of modern mathematical science, as contained in the celebrated lemma with which Newton opens his "Principia." We know that a regular polygon, inscribed in a circle, its sides being continually diminished, tends to become that circle, as its limit; but it vanishes before it has coincided with the circle, so that its tendency to be the circle, though ever nearer fulfilment, never in fact gets beyond a tendency. In like manner, the conclusion in a real or concrete question is foreseen and predicted rather than actually attained; foreseen in the number and direction of accumulated premisses, which all converge to it, and as the result of their combination, approach it more nearly than any assignable difference, yet do not touch it logically (though only not touching it,) on account of the nature of its subject-matter, and the delicate and implicit character of at least part of the reasonings on which it depends. It is by the strength, variety, or multiplicity of premisses, which are only probable, not by invincible syllogisms,—by objections overcome, by adverse theories neutralized, by difficulties gradually clearing up, by exceptions proving the rule, by unlooked-for correlations found with received truths, by suspense and delay in the process issuing in triumphant reactions,—by all these ways, and many others, it is that the practised and experienced mind is able to make a sure divination that a conclusion is inevitable, of which his lines of reasoning do not actually put him in possession. This is what is meant by a proposition being "as good as proved," a conclusion as undeniable "as if it were proved," and by the reasons for it "amounting to a proof," for a proof is the limit of converging probabilities. . . .

I commenced my remarks upon Inference by saying that reasoning ordinarily shows as a simple act, not as a process, as if there were no medium interposed between antecedent and consequent, and the transition from one to the other were of the nature of an instinct,—that is, the process is altogether unconscious and implicit. It is necessary, then, to take some notice of this natural or material Inference, as an existing phenomenon of mind; and that the more, because I shall thereby be illustrating and supporting what I have been saying of the characteristics of inferential processes as carried on in concrete matter, and especially of their being the action of the mind itself, that is, by its ratiocinative or illative faculty, not a mere operation as in the rules of arithmetic.

I say, then, that our most natural mode of reasoning is, not from propositions to propositions, but from things to things, from concrete to concrete, from wholes to wholes. Whether the consequents, at which we arrive from the antecedents with which we start, lead us to assent or only towards assent, those antecedents commonly are not recognized by us as subjects for analysis; nay, often are only indirectly recognized as antecedents at all. Not only is the inference with its process ignored, but the antecedent also. To the mind itself the reasoning is a simple divination or predication; as it literally is in the instance of enthusiasts, who mistake their own thoughts for inspirations.

This is the mode in which we ordinarily reason, dealing with things directly, and as they stand, one by one, in the concrete, with an intrinsic and personal power, not a conscious adoption of an artificial instrument or expedient; and it is especially exemplified both in uneducated men, and in men of genius,—in those who know nothing of intellectual aids and rules, and in those who care nothing for them,—in those who are either without or above mental discipline. As true poetry is a spontaneous outpouring of thought, and therefore belongs to rude as well as to gifted minds, whereas no one becomes a poet merely by the canons of criticism, so this unscientific reasoning, being sometimes a natural, uncultivated faculty, sometimes approaching to a gift, sometimes an acquired habit and second nature, has a higher source than logical rule . . . When it is characterized by precision, subtlety, promptitude, and truth, it is of course a gift and a rarity: in ordinary minds it is biassed and degraded by prejudice, passion, and self-interest; but still, after all, this divination comes by nature, and belongs to all of us in a measure, to women more than to men, hitting or missing, as the case may be, but with a success on the whole sufficient to show that there is a method in it, though it be implicit. . . .

What I have been saying of Ratiocination, may be said of Taste, and is confirmed by the obvious analogy between the two. Taste, skill, invention in the fine arts—and so, again, discretion or judgment in conduct—are exerted spontaneously, when once acquired, and could not give a clear account of themselves, or of their mode of proceeding. They do not go by rule, though to a certain point their exercise may be analyzed, and may take the shape of an art or method. But these parallels will come before us presently.

And now I come to a further peculiarity of this natural and spontaneous ratiocination. This faculty, as it is actually found in us, proceeding from concrete to concrete, is attached to a definite subject-matter, according to the individual. . . . No one would for a moment expect that because Newton and Napoleon both had a genius for ratiocination, that, in consequence, Napoleon could have generalized the principle of gravitation, or Newton have seen how to concentrate a hundred thousand men at Austerlitz. The ratiocinative faculty, then, as found in individuals, is not a general instrument of knowledge, but has its province, or is what may be called departmental. It is not so much one faculty, as a collection of similar or analogous faculties under one name, there being really as many faculties as there are distinct subject-matters, though in the same person some of them may, if it so happen, be united,—nay, though some men have a sort of literary power in arguing in all subject-matters . . . a power extensive, but not deep or real. . . .

. . . Instead of trusting logical science, we must trust persons, namely, those who by long acquaintance with their subject have a right to judge. And if we wish ourselves to share in their convictions and the grounds of them, we must follow their history, and learn as they have learned. We must take up their particular subject as they took it up, beginning at the beginning, give ourselves to it, depend on practice and experience more than on reasoning, and thus gain that mental insight into truth, whatever its subject-matter may be, which our masters have gained before us. By following this course, we may make ourselves of their number, and then we rightly lean upon ourselves, directing ourselves by our own moral or intellectual judgment, not by our skill in argumentation.

. . . Judgment then in all concrete matter is the architectonic faculty; and what may be called the Illative Sense, or right judgment in ratiocination, is one branch of it.

[ch. 8]

Illative Sense

My object in the foregoing pages has been, not to form a theory which may account for those phenomena of the intellect of which they treat, viz. those which characterize inference and assent, but to ascertain what is the matter of fact as regards them, that is, when it is that assent is given to propositions which are inferred, and under what circumstances. I have never had the thought of an attempt which in me would be ambitious and which has failed in the hands of others,—if that attempt may fairly be called unsuccessful, which, though made by the acutest minds, has not succeeded in convincing opponents. Especially have I found myself unequal to antecedent reasonings in the instance of a matter of fact. There are those, who, arguing *à priori*, maintain, that, since experience leads by syllogism only to probabilities, certitude is ever a mistake. There are others, who, while they deny this conclusion, grant the *à priori* principle assumed in the argument, and in consequence are obliged, in order to vindicate the certainty of our knowledge, to have recourse to the hypothesis of institutions, intellectual forms, and the like, which belong to us by nature, and may be considered to elevate our experience into something more than it is in itself. Earnestly maintaining, as I would, with this latter school of philosophers, the certainty of knowledge, I think it enough to appeal to the common voice of mankind in proof of it. That is to be accounted a normal operation of our nature, which men in general do actually instance. That is a law of our minds, which is exemplified in action on a large scale, whether *à priori* it ought to be a law or no. Our hoping is a proof that hope, as such, is not an extravagance; and our possession of certitude is a proof that it is not a weakness or an absurdity to be certain. How it comes about that we can be certain is not my business to determine; for me it is sufficient that certitude is felt. This is what the schoolmen, I believe, call treating a subject *in facto esse*, in contrast with *in fieri*. Had I attempted the latter, I should have been falling into metaphysics; but my aim is of a practical character, such as that of Butler in his *Analogy*, with this difference, that he treats of probability, doubt, expedience, and duty, whereas in these pages, without excluding, far from it, the question of duty, I would confine myself to the truth of things, and to the mind's certitude of that truth.

Certitude is a mental state: certainty is a quality of propositions. Those propositions I call certain, which are such that I am certain of them. Certitude is not a passive impression made upon the mind from without, by argumentative compulsion, but in all concrete questions (nay, even in abstract, for though the reasoning is abstract, the mind which judges of it is concrete) it is an active recognition of propositions as true, such as it is the duty of each individual himself to exercise at the bidding of reason, and, when reason forbids, to withhold. And reason never bids us be certain except on an absolute proof; and such a proof can never be furnished to us by the logic of words, for as certitude is of the mind, so is the act of inference which leads to it. Every one who reasons, is his own centre; and no expedient for attaining a common measure of minds can reverse this truth;—but then the question follows, is there any *criterion* of the accuracy of an inference, such as may be our warrant that certitude is rightly elicited in favour of the proposition inferred, since our warrant cannot, as I have said, be scientific? I have already said that the sole and final judgment on the validity of an inference in concrete matter is committed to the personal action of the ratiocinative faculty, the perfection or virtue of which I have called the Illative Sense, a use of the word "sense" parallel to our use of it in "good sense," "common sense," a "sense of beauty," &c.;— and I own I do not see any way to go farther than this in answer to the question. However, I can at least explain my meaning more fully; and therefore I will now speak, first of the sanction of the Illative Sense, next of its nature, and then of its range.

We are in a world of facts, and we use them; for there is nothing else to use. We do not quarrel with them, but we take them as they are, and avail ourselves of what they can do for us. It would be out of place to demand of fire, water, earth, and air their credentials, so to say, for acting upon us, or ministering to us. We call them elements, and turn them to account, and make the most of them. We speculate on them at our leisure. But what we are still less able to doubt about or annul, at our leisure or not, is that which is at once their counterpart and their witness, I mean, ourselves. We are conscious of the objects of external nature, and we reflect and act upon them, and this consciousness, reflection, and action we call our rationality. And as we use the (so called) elements without first criticizing what we have no command over, so is it much more unmeaning in us to criticize or find fault with our own nature, which is nothing else than we ourselves, instead of using it according to the use of which it ordinarily admits. Our being, with its faculties, mind and body, is a fact not admitting of question, all things being of necessity referred to it, not it to other things.

If I may not assume that I exist, and in a particular way, that is, with a particular mental constitution, I have nothing to speculate about, and had better let speculation alone. Such as I am, it is my all; this is my essential stand-point, and must be taken for granted; otherwise, thought is but an idle amusement, not worth the trouble. There is no medium between using my faculties, as I have them, and flinging myself upon the external world according to the random impulse of the moment, as spray upon the surface of the waves, and simply forgetting that I am.

I am what I am, or I am nothing. I cannot think, reflect, or judge about my being, without starting from the very point which I aim at concluding. My ideas are all assumptions, and I am ever moving in a circle. I cannot avoid being sufficient for myself, for I cannot make myself anything else, and to change me is to destroy me. If I do not use myself, I have no other self to use. My only business is to ascertain what I am, in order to put it to use. It is enough for the proof of the value and authority of any function which I possess, to be able to pronounce that it is natural. What I have to ascertain is the laws under which I live. My first elementary lesson of duty is that of resignation to the laws of my nature, whatever they are; my first disobedience is to be impatient at what I am, and to indulge an ambitious aspiration after what I cannot be, to cherish a distrust of my powers, and to desire to change laws which are identical with myself. . . .

. . . there is no ultimate test of truth besides the testimony born to truth by the mind itself, and . . . this phenomenon, perplexing as we may find it, is a normal and inevitable characteristic of the mental constitution of a being like man on a stage such as the world. His progress is a living growth, not a mechanism; and its instruments are mental acts, not the formulas and contrivances of language. . . .

It is the mind that reasons, and that controls its own reasonings, not any technical apparatus of words and propositions. This power of judging and concluding, when in its perfection, I call the Illative Sense, and I shall best illustrate it by referring to parallel faculties, which we commonly recognize without difficulty.

For instance, how does the mind fulfil its function of supreme direction and control, in matters of duty, social intercourse, and taste? In all of these separate actions of the intellect, the individual is supreme, and responsible to himself, nay, under circumstances, may be justified in opposing himself to the judgment of the whole world; though he uses rules to his great advantage, as far as they go, and is in consequence bound to use them. As regards moral duty, the subject is fully considered in the well-known ethical treatises of Aristotle. He calls the faculty which guides the mind in matters of conduct, by the name of *phronesis*, or judgment. This is the directing, controlling, and determining principle in such matters, personal and social. What it is to be virtuous, how we are to gain the just idea and standard of virtue, how we are to approximate in practice to our own standard, what is right and wrong in a particular case, for the answers in fulness and accuracy to these and similar questions, the philosopher refers us to no code of laws, to no moral treatise, because no science of life, applicable to the case of an individual, has been or can be written. Such is Aristotle's doctrine, and it is undoubtedly true. An ethical system may supply laws, general rules, guiding principles, a number of examples, suggestions, landmarks, limitations, cautions, distinctions, solutions of critical or anxious difficulties; but who is to apply them to a particular case? whither can we go, except to the living intellect, our own, or another's? What is written is too vague, too negative for our need. It bids us avoid extremes; but it cannot ascertain for us, according to our personal need, the golden mean. The authoritative oracle, which is to decide our path, is something more searching and manifold than such jejune generalizations as treatises can give, which are most distinct and clear when we least need them. It is seated in the mind of the individual, who is thus his own law, his own teacher, and his own judge in those special cases of duty which are personal to him. It comes of an acquired habit, though it has its first origin in nature itself, and it is formed and matured by practice and experience; and it manifests itself, not in any breadth of view, any philosophical comprehension of the mutual relations of duty towards duty, or any consistency in its teachings, but it is a capacity sufficient for the occasion, deciding what ought to be done here and now, by this given person, under these given circumstances. It decides nothing hypothetical, it does not determine what a man should do ten years hence, or what another should do at this time. It may indeed happen to decide ten years hence as it does now, and to decide a second case now as it now decides a first; still its present act is for the present, not for the distant or the future. . . .

In this respect of course the law of truth differs from the law of duty, that duties change, but truths never; but, though truth is ever one and the same, and the assent of certitude is immutable, still the reasonings which carry us on to truth and certitude are many and distinct, and vary with the inquirer; and it is not with assent, but with the controlling principle in inferences that I am comparing *phronesis*. It is with this drift that I observe that the rule of conduct for one man is not always the rule for another, though the rule is always one and the same in the abstract, and in its principle and scope. To learn his own duty in his own case, each individual must have recourse to his own rule; and if his rule is not sufficiently developed in his intellect for his need, then he goes to some other living, present authority, to supply it for him, not to the dead letter of treatise or a code. A living, present authority, himself or another, is his immediate guide in matters of a personal, social, or political character. In buying and selling, in contracts, in his treatment of others, in giving and receiving, in thinking, speaking, doing, and working, in toil, in danger, in his recreations and pleasures, every one of his acts, to be praiseworthy, must be in accordance with this practical sense. Thus it is, and not by science, that he perfects the virtues of justice, self-command, magnanimity, generosity, gentleness, and all others. *Phronesis* is the regulating principle of every one of them.

These last words lead me to a further remark. I doubt whether it is correct, strictly speaking, to consider this *phronesis* as a general faculty, directing and perfecting all the virtues at once. So understood, it is little better than an abstract term, including under it a circle of analogous faculties severally proper to the separate virtues. Properly speaking, there are as many kinds of *phronesis* as there are virtues; for the judgment, good sense, or tact which is conspicuous in a man's conduct in one subject-matter, is not necessarily traceable in another. As in the parallel cases of memory and reasoning, he may be great in one aspect of his character, and little-minded in another. He may be exemplary in his family, yet commit a fraud on the revenue; he may be just and cruel, brave and sensual, imprudent and patient. And if this be true of the moral virtues, it holds good still more fully when we compare what is called his private character with his public. A good man may make a bad king; profligates have been great statesmen, or magnanimous political leaders.

So, too, I may go on to speak of the various callings and professions which give scope to the exercise of great talents, for these talents also are matured, not by mere rule, but by personal skill and sagacity. They are as diverse as pleading and cross-examining, conducting a debate in Parliament, swaying a public meeting, and commanding an army; and here, too, I observe that, though the directing principle in each case is called by the same name,—sagacity, skill, tact, or prudence,—still there is no one ruling faculty leading to eminence in all these various lines of action in common, but men will excel in one of them, without any talent for the rest.

The parallel may be continued in the case of the Fine Arts, in which, though true and scientific rules may be given, no one would therefore deny that Phidias or Rafael had a far more subtle standard of taste and a more versatile power of embodying it in his works, than any which he could communicate to others in even a series of treatises. And here again genius is indissolubly united to one definite subject-matter; a poet is not therefore a painter, or an architect a musical composer.

And so, again, as regards the useful arts and personal accomplishments, we use the same word "skill," but proficiency in engineering or in ship-building, or again in engraving, or again in singing, in playing instruments, in acting, or in gymnastic exercises, is as simply one with its particular subject-matter, as the human soul with its particular body, and is, in its own department, a sort of instinct or inspiration, not an obedience to external rules of criticism or of science.

It is natural, then, to ask the question, why ratiocination should be an exception to a general law which attaches to the intellectual exercises of the mind; why it is held to be commensurate with logical science; and why logic is made an instrumental art sufficient for determining every sort of truth, while no one would dream of making any one formula, however generalized, a working rule at once for poetry, the art of medicine, and political warfare?

This is what I have to remark concerning the Illative Sense, and in explanation of its nature and claims; and on the whole, I have spoken of it in four respects,—as viewed in itself, in its subject-matter, in the process it uses, and in its function and scope.

First, viewed in its exercise, it is one and the same in all concrete matters, though employed in them in different measures. We do not reason in one way in chemistry or law, in another in morals or religion; but in reasoning on any subject whatever, which is concrete, we proceed, as far indeed as we can, by the logic of language, but we are obliged to supplement it by the more subtle and elastic logic of thought; for forms by themselves prove nothing.

Secondly, it is in fact attached to definite subject-matters, so that a given individual may possess it in one department of thought, for instance, history, and not in another, for instance, philosophy.

Thirdly, in coming to its conclusion, it proceeds always in the same way, by a method of reasoning, which, as I have observed above, is the elementary principle of that mathematical calculus of modern times, which has so wonderfully extended the limits of abstract science.

Fourthly, in no class of concrete reasonings, whether in experimental science, historical research, or theology, is there any ultimate test of truth and error in our inferences besides the trustworthiness of the Illative Sense that gives them its sanction; just as there is no sufficient test of poetical excellence, heroic action, or gentleman-like conduct, other than the particular mental sense, be it genius, taste, sense of propriety, or the moral sense, to which those subject-matters are severally committed. Our duty in each of these is to strengthen and perfect the special faculty which is its living rule, and in every case as it comes to do our best. And such also is our duty and our necessity, as regards the Illative Sense.

Great as are the services of language in enabling us to extend the compass of our inferences, to test their validity, and to communicate them to others, still the mind itself is more versatile and vigorous than any of its works, of which language is one, and it is only under its penetrating and subtle action that the margin disappears, which I have described as intervening between verbal argumentation and conclusions in the concrete. It determines what science cannot determine, the limit of converging probabilities and the reasons sufficient for a proof. It is the ratiocinative mind itself, and no trick of art, however simple in its form and sure in operation, by which we are able to determine, and thereupon to be certain that a moving body left to itself will never stop, and that no man can live without eating.

Nor, again, is it by any diagram that we are able to scrutinize, sort, and combine the same premisses which must be first run together before we answer duly a given question. It is to the living mind that we must look for the means of using correctly principles of whatever kind, facts or doctrines, experiences or testimonies, true or probable, and of discerning what conclusion from these is necessary, suitable, or expedient, when they are taken for granted; and this, either by means of a natural gift, or from mental formation and practice and a long familiarity with those various starting-points. Thus, when Laud said that he did not see his way to come to terms with the Holy See, "till Rome was other than she was," no Catholic would admit the sentiment: but any Catholic may understand that this is just the judgment consistent with Laud's actual condition of thought and cast of opinions, his ecclesiastical position, and the existing state of England.

Nor, lastly, is an action of the mind itself less necessary in relation to those first elements of thought which in all reasoning are assumptions, the principles, tastes, and opinions, very often of a personal character, which are half the battle in the inference with which the reasoning is to terminate. It is the mind itself that detects them in their obscure recesses, illustrates them, establishes them, eliminates them, resolves them into simpler ideas, as the case may be. The mind contemplates them without the use of words, by a process which cannot be analyzed. Thus it was that Bacon separated the physical system of the world from the theological; thus that Butler connected together the moral system with the religious. Logical formulas could never have sustained the reasonings involved in such investigations.

Thus the Illative Sense, that is, the reasoning faculty, as exercised by gifted, or by educated or otherwise well-prepared minds, has its function in the beginning, middle, and end of all verbal discussion and inquiry, and in every stop of the process. It is a rule to itself, and appeals to no judgment beyond its own; and attends upon the whole course of thought from antecedents to consequents, with a minute diligence and unwearied presence, which is impossible to a cumbrous apparatus of verbal reasoning, though, in communicating with others, words are the only instrument we possess, and a serviceable, though imperfect instrument.

One function indeed there is of Logic, to which I have referred in the preceding sentence, which the Illative Sense does not and cannot perform. It supplies no common measure between mind and mind, as being nothing else than a personal gift or acquisition. Few there are, as I said above, who are good reasoners on all subject-matters. Two men, who reason well each in his own province of thought, may, one or both of them, fail and pronounce opposite judgments on a question belonging to some third province. Moreover, all reasoning being from premises, and those premises arising (if it so happen) in their first elements from personal characteristics, in which men are in fact in essential and irremediable variance one with another, the ratiocinative talent can do no more than point out where the difference between them lies, how far it is immaterial, when it is worth while continuing an argument between them, and when not.

[ch. 9]

Proving Christianity

Truth certainly, as such, rests upon grounds intrinsically and objectively and abstractedly demonstrative, but it does not follow from this that the arguments producible in its favour are unanswerable and irresistible. These latter epithets are relative, and bear upon matters of fact; arguments in themselves ought to do, what perhaps in the particular case they cannot do. The fact of revelation is in itself demonstrably true, but it is not therefore true irresistibly; else, how comes it to be resisted? There is a vast distance between what it is in itself, and what it is to us. Light is a quality of matter, as truth is of Christianity; but light is not recognized by the blind, and there are those who do not recognize truth, from the fault, not of truth, but of themselves. I cannot convert men, when I ask for assumptions which they refuse to grant to me; and without assumptions no one can prove anything about anything.

I am suspicious then of scientific demonstrations in a question of concrete fact, in a discussion between fallible men. However, let those demonstrate who have the gift. . . . For me, it is more congenial to my own judgment to attempt to prove Christianity in the same informal way in which I can prove for certain that I have been born into this world, and that I shall die out of it. . . . I prefer to rely on that of an *accumulation* of various probabilities . . . that from probabilities we may construct legitimate proof, sufficient for certitude . . . that, since a Good Providence watches over us, He blesses such means of argument as it has pleased Him to give us, in the nature of man and of the world, if we use them duly for those ends for which He has given them; and that, as in mathematics we are justified by the dictate of nature in withholding our assent from a conclusion of which we have not yet a strict logical demonstration, so by a like dictate we are not justified, in the case of concrete reasoning and especially of religious inquiry, in waiting till such logical demonstration is ours, but on the contrary are bound in conscience to seek truth and to look for certainty by modes of proof, which, when reduced to the shape of formal propositions, fail to satisfy the severe requisitions of science.

Here then at once is one momentous doctrine or principle, which enters into my own reasoning, and which another ignores, viz. the providence and intention of God; and of course there are other principles, explicit or implicit, which are in like circumstances. It is not wonderful then, that, while I can prove Christianity divine to my own satisfaction, I shall not be able to force it upon any one else. Multitudes indeed I ought to succeed in persuading of its truth without any force at all, because they and I start from the same principles, and what is a proof to me is a proof to them; but if any one starts from any other principles but ours, I have not the power to change his principles, or the conclusion which he draws from them, any more than I can make a crooked man straight. Whether his mind will ever grow straight, whether I can do anything towards it becoming straight, whether he is not responsible, responsible to his Maker, for being mentally crooked, is another matter; still the fact remains, that, in any inquiry about things in the concrete, men differ from each other, not so much in the soundness of their reasoning as in the principles which govern its exercise, that those principles are of a personal character, that where there is no common measure of minds, there is no common measure of arguments, and that the validity of proof is determined, not by any scientific test, but by the illative sense.

Accordingly, instead of saying that the truths of Revelation depend on those of Natural Religion, it is more pertinent to say that belief in revealed truths depends on belief in natural. Belief is a state of mind; belief generates belief; states of mind correspond to each other; the habits of thought and the reasonings which lead us on to a higher state of belief than our present, are the very same which we already possess in connexion with the lower state. Those Jews became Christians in Apostolic times who were already what may be called crypto Christians; and those Christians in this day remain Christian only in name, and (if it so happen) at length fall away, who are nothing deeper or better than men of the world, *savants*, literary men, or politicians.

[ch. 10]

3

THE PREACHER

Newman's eight volumes of *Parochial and Plain Sermons* (1834–43) are one of the great classics of Christian spirituality. It is certainly almost as hard to imagine the Oxford Movement without them as without the *Tracts for the Times*. Newman's preaching in the University Church of St Mary the Virgin became legendary and many descriptions were written of it, the most celebrated being Matthew Arnold's nostalgic evocation of 'the charm of that spiritual apparition, gliding in the dim afternoon light through the aisles of St. Mary's, rising into the pulpit, and then, in the most entrancing of voices, breaking the silence with words and thoughts which were a religious music,—subtle, sweet, mournful'. The contrast which was often noted between the mysterious power of Newman's preaching and the eschewal of any kind of pulpit oratory was reflected in the content of the sermons. On the one hand, the doctrine of the 'indwelling' of the Holy Spirit, which was the most fundamental theological rediscovery that Newman had made from his study of Scripture and the Fathers, is prominently present. On the other hand, the sermons deliberately depreciate the role of the Holy Spirit in the Christian life, to the extent that they emphasize the ordinary human channels and means through which the Spirit has to work. Newman felt strongly that Evangelicalism had encouraged the notion that faith only was necessary, whereas obedience to the commandments was also essential.

The awareness, however, of the indwelling of the Holy Spirit is at the heart of the stress the Tractarians placed on 'mystery' as opposed to the 'enthusiasm' of Evangelicals and the 'coldness' and 'dryness' of the liberal and 'high-and-dry' Anglicans. But by contrast there is nothing in the least mysterious about the obligations which the sermons impose on the believer. Not only is the goal which the preacher relentlessly sets before his hearers disturbingly simple—'Be you content with nothing short of perfection'—but the ideal itself, although depicted with all the eloquence of a great rhetorician, is trenchantly translated into a ruthlessly realistic spirituality which refuses to content itself with merely uplifting platitudes and pious aspirations. The paradox is that the object, holiness, is as elevated and lofty as the means of attaining it are humble and mundane. For Newman real spirituality is characterized by its utter unpretentiousness, for it involves above all the minute performance of the duties of every day. Thus the hallmark of the saint is not spiritual ardour but the unexpected quality of '*consistency*'. The sermons are shot through with a sharp realism that can at times be alarming. The penetrating psychology, the practicality of which contrasts with the idealistic spirituality which it informs, owes much to the kind of rigorous self-analysis and self-examination which Newman had imbibed after his adolescent conversion to Evangelical Christianity. The irony is that, although Newman had come to reject the introverted spirituality which he believed Evangelicalism fostered, nevertheless the habit of mind which it encouraged was as important for his own sermons as it was for the probing psychology of the humanistic novels of George Eliot, another former Evangelical. There is another paradox. Newman had also come to the conclusion that the classic Evangelical distinction between so-called 'nominal' and 'real' Christians was itself an unreal distinction, but still a preoccupation with the 'real' and the 'unreal' runs through his sermons (as it does all his writings), showing itself in a constant vigilance against all forms of unreal religion.

Soon after Newman became a Catholic, he ceased reading his sermons from a prepared text in order to conform to the practice of his adopted Church. As a result, he only thereafter published sermons which were preached for special public occasions and where a script was appropriate. Of the two volumes of sermons which he wrote as a Catholic, the first, *Discourses Addressed to Mixed Congregations* (1849), which he published as a recent convert, contains much that is florid and Italianate, even though the sermon 'Mental Sufferings of our Lord in His Passion' is one of the most powerful spiritual pieces he ever wrote. The second volume, *Sermons Preached on Various Occasions* (1857, 1874), also contains some excellent writing. The simple reason why the following selection contains nothing from either volume is that there is a classic quality to the earlier Anglican sermons which sets them in a class apart.

PAROCHIAL AND PLAIN SERMONS

Secret Faults

Who can understand his errors? Cleanse Thou me from secret faults

Ps. 19:12

Strange as it may seem, multitudes called Christians go through life with no effort to obtain a correct knowledge of themselves. They are contented with general and vague impressions concerning their real state; and, if they have more than this, it is merely such accidental information about themselves as the events of life force upon them. But exact systematic knowledge they have none, and do not aim at it.

When I say this is *strange*, I do not mean to imply that to know ourselves is *easy*; it is very difficult to know ourselves even in part, and so far ignorance of ourselves is not a strange thing. But its strangeness consists in this, viz. that men should profess to receive and act upon the great Christian doctrines, while they are thus ignorant of themselves, considering that self-knowledge is a necessary condition for understanding them. Thus it is not too much to say that all those who neglect the duty of habitual self-examination are using words without meaning. The doctrines of the *forgiveness* of sins, and of a *new birth* from sin, cannot be understood without some right knowledge of the *nature* of sin, that is, of our own heart. We may, indeed, assent to a form of words which declares those doctrines; but if such a mere assent, however sincere, is the same as a real *holding of* them, and belief in them, then it is equally possible to believe in a proposition the terms of which belong to some foreign language, which is obviously absurd. Yet nothing is more common than for men to think that because they are familiar with words, they understand the ideas they stand for. Educated persons despise this fault in illiterate men who use hard words as if they comprehended them. Yet they themselves, as well as others, fall into the same error in a more subtle form, when they think they understand terms used in morals and religion, because such are common words, and have been used by them all their lives.

Now (I repeat) unless we have some just idea of our hearts and of sin, we can have no right idea of a Moral Governor, a Saviour or a Sanctifier, that is, in professing to believe in Them, we shall be using words without attaching distinct meaning to them. Thus self-knowledge is at the root of all real religious knowledge; and it is in vain,—worse than vain,—it is a deceit and a mischief, to think to understand the Christian doctrines as a matter of course, merely by being taught by books, or by attending sermons, or by any outward means, however excellent, taken by themselves. For it is in proportion as we search our hearts and understand our own nature, that we understand what is meant by an Infinite Governor and Judge; in proportion as we comprehend the nature of disobedience and our actual sinfulness, that we feel what is the blessing of the removal of sin, redemption, pardon, sanctification, which otherwise are mere words. God speaks to us primarily in our hearts. Self-knowledge is the key to the precepts and doctrines of Scripture. The very utmost any outward notices of religion can do, is to startle us and make us turn inward and search our hearts; and then, when we have experienced what it is to read ourselves, we shall profit by the doctrines of the Church and the Bible.

Of course self-knowledge admits of degrees. No one perhaps, is *entirely* ignorant of himself; and even the most advanced Christian knows himself only "in part." However, most men are contented with a slight acquaintance with their hearts, and therefore a superficial faith. This is the point which it is my purpose to insist upon. Men are satisfied to have numberless secret faults. They do not think about them, either as sins or as obstacles to strength of faith, and live on as if they had nothing to learn.

Now let us consider attentively the strong presumption that exists, that we all have serious secret faults; a fact which, I believe, all are ready to confess in general terms, though few like calmly and practically to dwell upon it; as I now wish to do.

1. Now the most ready method of convincing ourselves of the existence in us of faults unknown to ourselves, is to consider how plainly we see the secret faults of others. At first sight there is of course no reason for supposing that we differ materially from those around us; and if we see sins in them which *they* do not see, it is a presumption that they have their own discoveries about ourselves, which it would surprise us to hear. For instance: how apt is an angry man to fancy that he has the command of himself! The very charge of being angry, if brought against him, will anger him more; and, in the height of his discomposure, he will profess himself able to reason and judge with clearness and impartiality. Now, it may be his turn another day, for what we know, to witness the same failing in us; or, if we are not naturally inclined to violent passion, still at least we may be subject to other sins, equally unknown to ourselves, and equally known to him as his anger was to us. For example: there are persons who act mainly from self-interest at times when they conceive they are doing generous or virtuous actions; they give freely, or put themselves to trouble, and are praised by the world, and by themselves, as if acting on high principle; whereas close observers can detect desire of gain, love of applause, shame, or the mere satisfaction of being busy and active, as the principal cause of their good deeds. This may be our

condition as well as that of others; or, if it be not, still a parallel infirmity, the bondage of some other sin or sins, which others see, and we do not.

But, say there is no human being sees sin in us, of which we are not aware ourselves, (though this is a bold supposition to make,) yet why should man's accidental knowledge of us limit the extent of our imperfections? Should all the world speak well of us, and good men hail us as brothers, after all there is a Judge who trieth the hearts and the reins. He knows our real state; have we earnestly besought Him to teach us the knowledge of our own hearts? If we have not, that very omission is a presumption against us. Though our praise were throughout the Church, we may be sure He sees sins without number in us, sins deep and heinous, of which we have no idea. If man sees so much evil in human nature, what must God see? "If our heart condemn us, God is greater than our heart, and knoweth all things." Not *acts* alone of sin does He set down against us daily, of which we know nothing, but the thoughts of the heart too. The stirrings of pride, vanity, covetousness, impurity, discontent, resentment, these succeed each other through the day in momentary emotions, and are known to Him. We know them not; but how much does it concern us to know them!

2. This consideration is suggested by the first view of the subject. Now reflect upon the *actual disclosures* of our hidden weakness, which accidents occasion. Peter followed Christ boldly, and suspected not his own heart, till it betrayed him in the hour of temptation, and led him to deny his Lord. David lived years of happy obedience while he was in private life. What calm, clear-sighted faith is manifested in his answer to Saul about Goliath:—"The Lord that delivered me out of the paw of the lion, and out of the paw of the bear, He will deliver me out of the hand of this Philistine."[1] Nay, not only in retired life, in severe trial, under ill usage from Saul, he continued faithful to his God; years and years did he go on, fortifying his heart, and learning the fear of the Lord; yet power and wealth weakened his faith, and for a season overcame him. There was a time when a prophet could retort upon him, "Thou art the man"[2] whom thou condemnest. He had kept his principles in words, but lost them in his heart. Hezekiah is another instance of a religious man bearing *trouble* well, but for a season falling back under the temptation of prosperity; and that, after extraordinary mercies had been vouchsafed to him.[3] And if these things be so in the case of the favoured saints of God, what (may we suppose) is our own real spiritual state in His sight? It is a serious thought. The warning to be deduced from it is this:—Never to think we have a due knowledge of ourselves till we have been exposed to various kinds of temptations, and tried on every side. Integrity on one side of our

---

[1] 1 Sam. 17: 37.   [2] 2 Sam. 12: 7.   [3] 2 Kings 20: 12–19.

character is no voucher for integrity on another. We cannot tell how we should act if brought under temptations different from those which we have hitherto experienced. This thought should keep us humble. We are sinners, but we do not know how great. He alone knows who died for our sins.

3. Thus much we cannot but allow; that we do not know ourselves in those respects in which we have not been tried. But farther than this; What if we do not know ourselves even where we *have* been tried, and found faithful? It is a remarkable circumstance which has been often observed, that if we look to some of the most eminent saints of Scripture, we shall find their recorded errors to have occurred in those parts of their duty in which each had had most trial, and generally showed obedience most perfect. *Faithful* Abraham through want of faith denied his wife. Moses, the *meekest* of men, was excluded from the land of promise for a passionate word. The *wisdom* of Solomon was seduced to bow down to idols. Barnabas again, the *son of consolation*, had a sharp contention with St. Paul. If then men, who knew themselves better than we doubtless know ourselves, had so much of hidden infirmity about them, even in those parts of their character which were most free from blame, what are we to think of ourselves? and if our very virtues be so defiled with imperfection, what must be the unknown multiplied circumstances of evil which aggravate the guilt of our sins? This is a third presumption against us.

4. Think of this too. No one begins to examine himself, and to pray to know himself (with David in the text), but he finds within him an abundance of faults which before were either entirely or almost entirely unknown to him. That this is so, we learn from the written lives of good men, and our own experience of others. And hence it is that the best men are ever the most humble; for, having a higher standard of excellence in their minds than others have, and knowing themselves better, they see somewhat of the breadth and depth of their own sinful nature, and are shocked and frightened at themselves. The generality of men cannot understand this; and if at times the habitual self-condemnation of religious men breaks out into words, they think it arises from affectation, or from a strange distempered state of mind, or from accidental melancholy and disquiet. Whereas the confession of a good man against himself, is really a witness against all thoughtless persons who hear it, and a call on them to examine their own hearts. Doubtless the more we examine ourselves, the more imperfect and ignorant we shall find ourselves to be.

5. But let a man persevere in prayer and watchfulness to the day of his death, yet he will never get to the bottom of his heart. Though he know more and more of himself as he becomes more conscientious and earnest,

still the full manifestation of the secrets there lodged, is reserved for another world. And at the last day who can tell the affright and horror of a man who lived to himself on earth, indulging his own evil will, following his own chance notions of truth and falsehood, shunning the cross and the reproach of Christ, when his eyes are at length opened before the throne of God, and all his innumerable sins, his habitual neglect of God, his abuse of his talents, his misapplication and waste of time, and the original unexplored sinfulness of his nature, are brought clearly and fully to his view? Nay, even to the true servants of Christ, the prospect is awful. "The righteous," we are told, "will scarcely be saved."[1] Then will the good man undergo the full sight of his sins, which on earth he was labouring to obtain, and partly succeeded in obtaining, though life was not long enough to learn and subdue them all. Doubtless we must all endure that fierce and terrifying vision of our real selves, that last fiery trial of the soul[2] before its acceptance, a spiritual agony and second death to all who are not then supported by the strength of Him who died to bring them safe through it, and in whom on earth they have believed.

My brethren, I appeal to your reason whether these presumptions are not in their substance fair and just. And if so, next I appeal to your consciences, whether they are *new* to you; for if you have not even thought about your real state, nor even know how little you know of yourselves, how can you in good earnest be purifying yourselves for the next world, or be walking in the narrow way?

And yet how many are the chances that a number of those who now hear me have no sufficient knowledge of themselves, or sense of their ignorance, and are in peril of their souls? Christ's ministers cannot tell who are, and who are not, the true elect: but when the difficulties in the way of knowing yourselves aright are considered, it becomes a most serious and immediate question for each of you to entertain, whether or not he is living a life of self-deceit, and thinking far more comfortably of his spiritual state than he has any right to do. For call to mind the impediments that are in the way of your knowing yourselves, or feeling your ignorance, and then judge.

1. First of all, self-knowledge does not come as a matter of course; it implies an effort and a work. As well may we suppose, that the knowledge of the languages comes by nature, as that acquaintance with our own heart is natural. Now the very effort of steadily reflecting, is itself painful to many men; not to speak of the difficulty of reflecting correctly. To ask ourselves *why* we do this or that, to take account of the principles which govern us, and see whether we act for conscience' sake

[1] 1 Pet. 4: 18.   [2] 1 Cor. 3: 13.

or from some lower inducement, is painful. We are busy in the world, and what leisure time we have we readily devote to a less severe and wearisome employment.

2. And then comes in our self-love. We *hope* the best; this saves us the trouble of examining. Self-love answers for our safety. We think it sufficient caution to allow for certain possible unknown faults at the utmost, and to take them *into* the reckoning when we balance our account with our conscience: whereas, if the truth were known to us, we should find we had nothing but debts, and those greater than we can conceive, and ever increasing.

3. And this favourable judgment of ourselves will especially prevail, if we have the misfortune to have uninterrupted health and high spirits, and domestic comfort. Health of body and mind is a great blessing, if we can bear it; but unless chastened by watchings and fastings,[1] it will commonly seduce a man into the notion that he is much better than he really is. Resistance to our acting rightly, whether it proceed from within or without, tries our principle; but when things go smoothly, and we have but to wish, and we can perform, we cannot tell how far we do or do not act from a sense of duty. When a man's spirits are high, he is pleased with every thing; and with himself especially. He can act with vigour and promptness, and he mistakes this mere constitutional energy for strength of faith. He is cheerful and contented; and he mistakes this for Christian peace. And, if happy in his family, he mistakes mere natural affection for Christian benevolence, and the confirmed temper of Christian love. In short, he is in a dream, from which nothing could have saved him except deep humility, and nothing will ordinarily rescue him except sharp affliction.

Other accidental circumstances are frequently causes of a similar self-deceit. While we remain in retirement from the world, we do not know ourselves; or after any great mercy or trial, which has affected us much, and given a temporary strong impulse to our obedience; or when we are in keen pursuit of some good object, which excites the mind, and for a time deadens it to temptation. Under such circumstances we are ready to think far too well of ourselves. The world is away; or, at least, we are insensible to its seductions; and we mistake our merely temporary tranquillity, or our over-wrought fervour of mind, on the one hand for Christian peace, on the other for Christian zeal.

4. Next we must consider the force of habit. Conscience at first warns us against sin; but if we disregard it, it soon ceases to upbraid us; and thus sins, once known, in time become secret sins. It seems then (and it is a startling reflection), that the more guilty we are, the less we know it;

---

[1] 2 Cor. 11: 27.

for the oftener we sin, the less we are distressed at it. I think many of us may, on reflection, recollect instances, in our experience of ourselves, of our gradually forgetting things to be wrong which once shocked us. Such is the force of habit. By it (for instance) men contrive to allow themselves in various kinds of dishonesty. They bring themselves to affirm what is untrue, or what they are not sure is true, in the course of business. They overreach and cheat; and still more are they likely to fall into low and selfish ways without their observing it, and all the while to continue careful in their attendance on the Christian ordinances, and bear about them a form of religion. Or, again, they will live in self-indulgent habits; eat and drink more than is right; display a needless pomp and splendour in their domestic arrangements, without any misgiving; much less do they think of simplicity of manners and abstinence as Christian duties. Now we cannot suppose they *always* thought their present mode of living to be justifiable, for *others* are still struck with its impropriety; and what others now feel, doubtless they once felt themselves. But such is the force of habit. So again, to take as a third instance, the duty of stated private prayer; at first it is omitted with compunction, but soon with indifference. But it is not the less a sin because we do not feel it to be such. Habit has made it a secret sin.

5. To the force of habit must be added that of custom. Every age has its own wrong ways; and these have such influence, that even good men, from living in the world, are unconsciously misled by them. At one time a fierce persecuting hatred of those who erred in Christian doctrine has prevailed; at another, an odious over-estimation of wealth and the means of wealth; at another an irreligious veneration of the mere intellectual powers; at another, a laxity of morals; at another, disregard of the forms and discipline of the Church. The most religious men, unless they are especially watchful, will feel the sway of the fashion of their age; and suffer from it as Lot in wicked Sodom, though unconsciously. Yet their ignorance of the mischief does not change the nature of their sin;— sin it still is, only custom makes it *secret* sin.

6. Now what is our chief guide amid the evil and seducing customs of the world?—obviously, the Bible. "The world passeth away, but the word of the Lord endureth for ever."[1] How much extended, then, and strengthened, necessarily must be this secret dominion of sin over us, when we consider how little we read Scripture! Our conscience gets corrupted,—true; but the words of truth, though effaced from our minds, remain in Scripture, bright in their eternal youth and purity. Yet, we do not study Scripture to stir up and refresh our minds. Ask

---

[1] Isa. 40: 8; 1 Pet. 1: 24, 25; 1 John 2: 17.

yourselves, my brethren, what do you know of the Bible? Is there any one part of it you have read carefully, and as a whole? One of the Gospels, for instance? Do you know very much more of your Saviour's works and words than you have heard read in church? Have you compared His precepts, or St. Paul's, or any other Apostle's, with your own daily conduct, and prayed and endeavoured to act upon them? If you have, so far is well; go on to do so. If you have not, it is plain you do not possess, for you have not sought to possess, an adequate notion of that perfect Christian character which it is your duty to aim at, nor an adequate notion of your actual sinful state; you are in the number of those who "come not to the light, lest their deeds should be reproved."

These remarks may serve to impress upon us the difficulty of knowing ourselves aright, and the consequent danger to which we are exposed, of speaking peace to our souls, when there is no peace.

Many things are against us; this is plain. Yet is not our future prize worth a struggle? It is not worth present discomfort and pain to accomplish an escape from the fire that never shall be quenched? Can we endure the thought of going down to the grave with a load of sins on our head unknown and unrepented of? Can we content ourselves with such an unreal faith in Christ, as in no sufficient measure includes self-abasement, or thankfulness, or the desire or effort to be holy? for how can we feel our need of His help, or our dependence on Him, or our debt to Him, or the nature of His gift to us, unless we know ourselves? How can we in any sense be said to have that "mind of Christ," to which the Apostle exhorts us, if we cannot follow Him to the height above, or the depth beneath; if we do not in some measure discern the cause and meaning of his sorrows, but regard the world, and man, and the system of Providence, in a light different from that which His words and acts supply? If you receive revealed truth merely through the eyes and ears, you believe words, not things; you deceive yourselves. You may conceive yourselves sound in faith, but you know nothing in any true way. Obedience to God's commandments, which implies knowledge of sin and of holiness, and the desire and endeavour to please Him, this is the only practical interpreter of Scripture doctrine. Without self-knowledge you have no root in yourselves personally; you may endure for a time, but under affliction or persecution your faith will not last. This is why many in this age (and in every age) become infidels, heretics, schismatics, disloyal despisers of the Church. They cast off the form of truth, because it never has been to them more than a form. They endure not, because they never have tasted that the Lord is gracious; and they never have had experience of His power and love, because they have never known their own weakness and need. This *may* be the future condition of some of us, if we harden our hearts to-day,—*apostasy*. Some day, even in this world, we may be found openly among the enemies of God and of His Church.

But, even should we be spared this present shame, what will it ultimately profit a man to profess without understanding? to *say* he has faith, when he has not works?[1] In that case we shall remain in the heavenly vineyard, stunted plants, without the principle of growth in us, barren; and, in the end, we shall be put to shame before Christ and the holy Angels, "as trees of withering fruits, twice dead, plucked up by the roots," even though we die in outward communion with the Church.

To think of these things, and to be alarmed, is the first step towards acceptable obedience; to be at ease, is to be unsafe. We must know what the evil of sin is hereafter, if we do not learn it here. God give us all grace to choose the pain of present repentance before the wrath to come!

[vol. i, no. 4]

Promising without Doing

A certain man had two sons; and he came to the first, and said, Son go work to-day in my vineyard. He answered and said, I will not; but afterward he repented, and went. And he came to the second, and said likewise. And he answered and said, I go, Sir; and went not.

Matt. 21:28–30

Our religious professions are at a far greater distance from our acting upon them, than we ourselves are aware. We know generally that it is our duty to serve God, and we resolve we will do so faithfully. We are sincere in thus generally desiring and purposing to be obedient, and we think we are in earnest; yet we go away, and presently, without any struggle of mind or apparent change of purpose, almost without knowing ourselves what we do,—we go away and do the very contrary to the resolution we have expressed. This inconsistency is exposed by our Blessed Lord in the second part of the parable which I have taken for my text. You will observe, that in the case of the first son, who said he would not go work, and yet did go, it is said, "afterward he repented;" he underwent a positive change of purpose. But in the case of the second, it is merely said, "he answered, I go, Sir; and went not;"—for here there was *no* revolution of sentiment, nothing deliberate; he merely acted according to his habitual frame of mind; he did *not* go work, because it was contrary to his general character to work; only he did not know this. He said, "I go, Sir," sincerely, from the feeling of the moment; but when the words were out of his mouth, then they were forgotten. It was like the

---

[1] James 2: 14.

wind blowing against a stream, which seems for a moment to change its course in consequence, but in fact flows down as before.

To this subject I shall now call your attention, as drawn from the latter part of this parable, passing over the case of the repentant son, which would form a distinct subject in itself. "He answered and said, I go, Sir; and went not." We promise to serve God: we do not perform; and that not from deliberate faithlessness in the particular case, but because it is our nature, our *way* not to obey, and *we* do not know this; we do not know ourselves, or what we are promising. I will give several instances of this kind of weakness.

1. For instance; that of mistaking good feelings for real religious principle. Consider how often this takes place. It is the case with the young necessarily, who have not been exposed to temptation. They have (we will say) been brought up religiously, they wish to be religious, and so are objects of our love and interest; but they think themselves far more religious than they really are. They suppose they hate sin, and understand the Truth, and can resist the world, when they hardly know the meaning of the words they use. Again, how often is a man incited by circumstances to utter a virtuous wish, or propose a generous or valiant deed; and perhaps applauds himself for his own good feeling, and has no suspicion that he is not able to act upon it! In truth, he does not understand where the real difficulty of his duty lies. He thinks that the characteristic of a religious man is his having correct notions. It escapes him that there is a great interval between feeling and acting. He takes it for granted he can do what he wishes. He knows he is a free agent, and can on the whole do what he will; but he is not conscious of the load of corrupt nature and sinful habits which hang upon his will, and clog it in each particular exercise of it. He has borne these so long, that he is insensible to their existence. He knows that in little things, where passion and inclination are excluded, he can perform as soon as he resolves. Should he meet in his walk two paths, to the right and left, he is sure he can take which he will at once, without any difficulty; and he fancies that obedience to God is not much more difficult than to turn to the right instead of the left.

2. One especial case of this self-deception is seen in delaying repentance. A man says to himself, "Of course, if the worst comes to the worst, if illness comes, or at least old age, I can repent." I do not speak of the dreadful presumption of such a mode of quieting conscience (though many persons really use it who do not speak the words out, or are aware that they act upon it), but, merely, of the ignorance it evidences concerning our moral condition, and our power of willing and doing. If men can repent, why do they not do so at once? they answer, that "they intend to do so hereafter;" i.e. they do *not* repent because they *can*. Such

is their argument; whereas, the very fact that they do not now, should make them suspect that there is a greater difference between intending and doing than they know of.

So very difficult is obedience, so hardly won is every step in our Christian course, so sluggish and inert our corrupt nature, that I would have a man disbelieve he can do one jot or tittle beyond what he has already done; refrain from borrowing aught on the hope of the future, however good a security for it he seems to be able to show; and never take his good feelings and wishes in pledge for one single untried deed. Nothing but *past* acts are the vouchers for *future*. Past sacrifices, past labours, past victories over yourselves,—these, my brethren, are the tokens of the like in store, and doubtless of greater in store; for the path of the just is as the shining, growing light.[1] But trust nothing short of these. "Deeds, not words and wishes," this must be the watchword of your warfare and the ground of your assurance. But if you have done nothing firm and manly hitherto, if you are as yet the coward slave of Satan, and the poor creature of your lusts and passions, never suppose you will one day rouse yourselves from your indolence. Alas! there are men who walk the road to hell, always the while looking back at heaven, and trembling as they pace forward towards their place of doom. They hasten on as under a spell, shrinking from the consequences of their own deliberate doings. Such was Balaam. What would he have given if words and feelings might have passed for deeds! See how religious he was so far as profession goes! How did he revere God in speech! How piously express a desire to die the death of the righteous! Yet he died in battle among God's *enemies*; not suddenly overcome by temptation, only on the other hand, not suddenly turned to God by his good thoughts and fair purposes. But in this respect the power of sin differs from any literal spell or fascination, that we are, after all, willing slaves of it, and shall answer for following it. If "our iniquities, like the wind, take us away,"[2] yet we can help this.

Nor is it only among beginners in religious obedience that there is this great interval between promising and performing. We can never answer how we shall act under new circumstances. A very little knowledge of life and of our own hearts will teach us this. Men whom we meet in the world turn out, in the course of their trial, so differently from what their former conduct promised, they view things so differently *before* they are tempted and *after*, that we, who see and wonder at it, have abundant cause to look to ourselves, not to be "high-minded," but to "fear." Even the most matured saints, those who imbibed in largest measure the power and fulness of Christ's Spirit, and worked righteousness most diligently in

---

[1] Prov. 4: 18.   [2] Isa. 64: 6.

their day, could they have been thoroughly scanned even by man, would (I am persuaded) have exhibited inconsistencies such as to surprise and shock their most ardent disciples. After all, one good deed is scarcely the pledge of another, though I just now said it was. The best men are uncertain; they are great, and they are little again; they stand firm, and then fall. Such is human virtue;—reminding us to call no one master on earth, but to look up to our sinless and perfect Lord; reminding us to humble ourselves, each within himself, and to reflect what we must appear to God, if even to ourselves and each other we seem so base and worthless; and showing clearly that all who are saved, even the least inconsistent of us, can be saved only by faith, not by works.

3. Here I am reminded of another plausible form of the same error. It is a mistake concerning what is meant by faith. We know Scripture tells us that God accepts those who have faith in Him. Now the question is, What *is* faith, and how can a man tell that he *has* faith? Some persons answer at once and without hesitation, that "to have faith is to feel oneself to be nothing, and God every thing; it is to be convinced of sin, to be conscious one cannot save oneself, and to wish to be saved by Christ our Lord; and that it is, moreover, to have the love of Him warm in one's heart, and to rejoice in Him, to desire His glory, and to resolve to live to Him and not to the world." But I will answer, with all due seriousness, as speaking on a serious subject, that this is *not* faith. Not that it is not necessary (it is very necessary) to be convinced that we are laden with infirmity and sin, and without health in us, and to look for salvation solely to Christ's blessed sacrifice on the cross; and we may well be thankful if we are thus minded; but that a man may feel all this that I have described, vividly, and still not yet possess one particle of true religious faith. Why? Because there is an immeasurable distance between feeling right and doing right. A man may have all these good thoughts and emotions, yet (if he has not yet hazarded them to the experiment of practice) he cannot promise himself that he has any sound and permanent principle at all. If he has not yet acted upon them, we have no voucher, barely on *account* of them, to believe that they are any thing but words. Though a man spoke like an angel, I would not believe him, on the mere ground of his speaking. Nay, till he acts upon them, he has not even evidence to himself that he has true living faith. Dead faith (as St. James says) profits no man. Of course; the Devils have it. What, on the other hand is *living* faith? Do fervent thoughts make faith *living*? St. James tells us otherwise. He tells us *works*, deeds of obedience, are the life of faith. "As the body without the spirit is dead, so faith without works is dead also."[1] So that those who think they really believe, because they have in word and thought surrendered themselves

---

[1] Jas. 2: 26.

to God, are much too hasty in their judgment. They have done something, indeed, but not at all the most difficult part of their duty, which is to surrender themselves to God in deed and act. They have as yet done nothing to show they will not, after saying "I go," the next moment "go not;" nothing to show they will not act the part of the self-deceiving disciple, who said, "Though I die with Thee, I will not deny Thee," yet straightway went and denied Christ thrice. As far as we know any thing of the matter, justifying faith has no existence independent of its particular definite acts. It may be described to be the temper under which men obey; the humble and earnest desire to please Christ which causes and attends on actual services. He who does one little deed of obedience, whether he denies himself some comfort to relieve the sick and needy, or curbs his temper, or forgives an enemy, or asks forgiveness for an offence committed by him, or resists the clamour or ridicule of the world—such an one (as far as we are given to judge) evinces more true faith than could be shown by the most fluent religious conversation, the most intimate knowledge of Scripture doctrine, or the most remarkable agitation and change of religious sentiments. Yet how many are there who sit still with folded hands, dreaming, doing nothing at all, thinking they have done every thing, or need do nothing, when they merely have had these good *thoughts*, which will save no one!

My object has been, as far as a few words can do it, to lead you to some true notion of the depths and deceitfulness of the heart, which we do not really know. It is easy to speak of human nature as corrupt in the general, to admit it in the general, and then get quit of the subject; as if the doctrine being once admitted, there was nothing more to be done with it. But in truth we can have no real apprehension of the doctrine of our corruption, till we view the structure of our minds, part by part; and dwell upon and draw out the signs of our weakness, inconsistency, and ungodliness, which are such as can arise from nothing else than some strange original defect in our moral nature.

1. Now it will be well if such self-examination as I have suggested leads us to the habit of constant dependence upon the Unseen God, in whom "we live and move and have our being." We are in the dark about ourselves. When we act, we are groping in the dark, and may meet with a fall any moment. Here and there, perhaps, we see a little; or, in our attempts to influence and move our minds, we are making experiments (as it were) with some delicate and dangerous instrument, which works we do not know how, and may produce unexpected and disastrous effects. The management of our hearts is quite above us. Under these circumstances it becomes our comfort to look up to God. "Thou, God, seest me!" Such was the consolation of the forlorn Hagar in the wilderness. He knoweth whereof we are made, and He alone can uphold

us. He sees with most appalling distinctness all our sins, all the windings and recesses of evil within us; yet it is our only comfort to know this and to trust Him for help against ourselves. To those who have a right notion of their weakness, the thought of their Almighty Sanctifier and Guide is continually present. They believe in the necessity of a spiritual influence to change and strengthen them, not as a mere abstract doctrine, but as a practical and most consolatory truth, daily to be fulfilled in their warfare with sin and Satan.

2. And this conviction of our excessive weakness must further lead us to try ourselves continually in little things, in order to prove our own earnestness; ever to be suspicious of ourselves, and not only to refrain from promising much, but actually to put ourselves to the test in order to keep ourselves wakeful. A sober mind never enjoys God's blessings to the full; it draws back and refuses a portion to show its command over itself. It denies itself in trivial circumstances, even if nothing is gained by denying, but an evidence of its own sincerity. It makes trial of its own professions; and if it has been tempted to say any thing noble and great, or to blame another for sloth or cowardice, it takes itself at its word, and resolves to make some sacrifice (if possible) in little things, as a price for the indulgence of fine speaking, or as a penalty on its censoriousness. Much would be gained if we adopted this rule even in our professions of friendship and service one towards another; and never said a thing which we were not willing to do.

There is only one place where the Christian allows himself to profess openly, and that is in Church. Here, under the guidance of Apostles and Prophets, he says many things boldly, as speaking after them, and as before Him who searcheth the reins. There can be no harm in professing much directly to God, because, *while* we speak, we know He sees through our professions, and takes them for what they really are, *prayers*. How much, for instance, do we profess when we say the Creed! and in the Collects we put on the full character of a Christian. We desire and seek the best gifts, and declare our strong purpose to serve God with our whole hearts. By doing this, we remind ourselves of our duty; and withal, we humble ourselves by the taunt (so to call it) of putting upon our dwindled and unhealthy forms those ample and glorious garments which befit the upright and full-grown believer.

Lastly, we see from the parable, what is the course and character of human obedience on the whole. There are two sides of it. I have taken the darker side; the case of profession without practice, of saying "I go, Sir," and of not going. But what is the brighter side? Nothing better than to say, "I go not," and to repent and go. The more *common* condition of men is, not to know their inability to serve God, and readily to answer for themselves; and so they quietly pass through life, as if they had nothing to fear. Their best estate, what is it, but to rise more or less in rebellion against God, to resist His commandments and ordinances, and then poorly to make up for the mischief they have done, by repenting and obeying? Alas! to be alive as a Christian, is nothing better than to struggle against sin, to disobey and repent. There has been but One amongst the sons of men who has said and done *consistently*; who said, "I come to do Thy will, O God," and without delay or hindrance did it. He came to show us what human nature might become, if carried on to its perfection. Thus He teaches us to think highly of our nature as viewed in Him; not (as some do) to speak evil of our nature and exalt ourselves personally, but while we acknowledge *our own* distance from heaven, to view our *nature* as renewed in Him, as glorious and wonderful beyond our thoughts. Thus He teaches us to be hopeful; and encourages us while conscience abases us. Angels seem little in honour and dignity, compared with that nature which the Eternal Word has purified by His own union with it. Henceforth, we dare aspire to enter into the heaven of heavens, and to live for ever in God's presence, because the first-fruits of our race is already there in the Person of His Only-begotten Son.

[vol. i, no. 13]

The Religion of the Day

Let us have grace, whereby we may serve God acceptably with reverence and godly fear. For our God is a consuming fire.

Heb. 12:28, 29

In every age of Christianity, since it was first preached, there has been what may be called a *religion of the world*, which so far imitates the one true religion, as to deceive the unstable and unwary. The world does not oppose religion *as such*. I may say, it never has opposed it. In particular, it has, in all ages, acknowledged in one sense or other the Gospel of Christ, fastened on one or other of its characteristics, and professed to embody this in its practice; while by neglecting the other parts of the holy doctrine, it has, in fact, distorted and corrupted even that portion of it which it has exclusively put forward, and so has contrived to explain away the whole;—for he who cultivates only one precept of the Gospel to the exclusion of the rest, in reality attends to no part at all. Our duties *balance* each other; and though we are too sinful to perform them all perfectly, yet we may in some measure be performing them all, and preserving the balance on the whole; whereas, to give ourselves only to this or that commandment, it to incline our minds in a wrong direction, and at length to pull them down to the earth, which is the aim of our adversary, the Devil.

It is his *aim* to break our strength; to force us down to the earth,—to bind us there. The world is his instrument for this purpose; but he is too wise to set it in open opposition to the Word of God. No! he affects to be a prophet like the prophets of God. He calls his servants also prophets; and they mix with the scattered remnant of the true Church, with the solitary Micaiahs who are left upon the earth, and speak in the name of the Lord. And in one sense they speak the truth; but it is not the whole truth; and we know, even from the common experience of life, that half the truth is often the most gross and mischievous of falsehoods.

Even in the first age of the Church, while persecution still raged, he set up a counter religion among the philosophers of the day, partly like Christianity, but in truth a bitter foe to it; and it deceived and made shipwreck of the faith of those who had not the love of God in their hearts.

Time went on, and he devised a second idol of the true Christ, and it remained in the temple of God for many a year. The age was rude and fierce. Satan took the darker side of the Gospel: its awful mysteriousness, its fearful glory, its sovereign inflexible justice; and here *his* picture of the truth ended, "God is a consuming fire;" so declares the text, and we know it. But we know more, viz. that God is love also; but Satan did not add this to his religion, which became one of *fear*. The religion of the world was then a fearful religion. Superstitions abounded, and cruelties. The noble firmness, the graceful austerity of the true Christian was superseded by forbidding spectres, harsh of eye, and haughty of brow; and these were the patterns or the tyrants of a beguiled people.

What is Satan's device in this day? a far different one; but perhaps a more pernicious. I will attempt to expose it, or rather to suggest some remarks towards its exposure, by those who think it worth while to attempt it; for the subject is too great and too difficult for an occasion such as the present, and, after all, no one can detect falsehood for another;—every man must do it himself; we can but *help* each other.

What is the world's religion now? It has taken the brighter side of the Gospel,—its tidings of comfort, its precepts of love; all darker, deeper views of man's condition and prospects being comparatively forgotten. This is the religion *natural* to a civilized age, and well has Satan dressed and completed it into an idol of the Truth. As the reason is cultivated, the taste formed, the affections and sentiments refined, a general decency and grace will of course spread over the face of society, quite independently of the influence of Revelation. That beauty and delicacy of thought, which is so attractive in books, then extends to the conduct of life, to all we have, all we do, all we are. Our manners are courteous; we avoid giving pain or offence; our words become correct; our relative duties are carefully performed. Our sense of propriety shows itself even in our domestic arrangements, in the embellishments of our houses, in our amusements, and so also in our religious profession. Vice now becomes unseemly and hideous to the imagination, or, as it is sometimes familiarly said, "out of taste." Thus elegance is gradually made the test and standard of virtue, which is no longer thought to possess an intrinsic claim on our hearts, or to exist, *further than* it leads to the quiet and comfort of others. Conscience is no longer recognized as an independent arbiter of actions, its authority is explained away; partly it is superseded in the minds of men by the so-called moral sense, which is regarded merely as the love of the beautiful, partly by the rule of expediency, which is forthwith substituted for it in the details of conduct. Now conscience is a stern, gloomy principle; it tells us of guilt and of prospective punishment. Accordingly, when its terrors disappear, then disappear also, in the creed of the day, those fearful images of Divine wrath with which the Scriptures abound. They are explained away. Every thing is bright and cheerful. Religion is pleasant and easy; benevolence is the chief virtue; intolerance, bigotry, excess of zeal, are the first of sins. Austerity is an absurdity;—even firmness is looked on with an unfriendly, suspicious eye. On the other hand, all open profligacy is discountenanced; drunkenness is accounted a disgrace; cursing and swearing are vulgarities. Moreover, to a cultivated mind, which recreates itself in the varieties of literature and knowledge, and is interested in the ever-accumulating discoveries of science, and the ever-fresh accessions of information, political or otherwise, from foreign countries, religion will commonly seem to be dull, from want of novelty. Hence excitements are eagerly sought out and rewarded. New objects in religion, new systems and plans, new doctrines, new preachers, are necessary to satisfy that craving which the so-called spread of knowledge has created. The mind becomes morbidly sensitive and fastidious; dissatisfied with things as they are, desirous of a change *as such*, as if alteration must of itself be a relief.

Now I would have you put Christianity for an instant out of your thoughts; and consider whether such a state of refinement as I have attempted to describe, is not that to which men might be brought, quite independent of religion, by the mere influence of education and civilization; and then again, whether, nevertheless, this mere refinement of mind is not more or less all that is called religion at this day. In other words, is it not the case, that Satan has so composed and dressed out what is the mere natural produce of the human heart under certain circumstances, as to serve his purposes as the counterfeit of the Truth? I do not at all deny that this spirit of the world uses words, and makes professions, which it would not adopt except for the suggestions of Scripture; nor do I deny that it takes a general colouring from

Christianity, so as really to be modified by it, nay, in a measure enlightened and exalted by it. Again, I fully grant that many persons in whom this bad spirit shows itself, are but partially infected by it, and at bottom, good Christians, though imperfect. Still, after all, here is an existing teaching, only partially evangelical, built upon worldly principle, yet pretending to be the Gospel, dropping one whole side of the Gospel, its austere character, and considering it enough to be benevolent, courteous, candid, correct in conduct, delicate,—though it includes no true fear of God, no fervent zeal for His honour, no deep hatred of sin, no horror at the sight of sinners, no indignation and compassion at the blasphemies of heretics, no jealous adherence to doctrinal truth, no especial sensitiveness about the particular means of gaining ends, provided the ends be good, no loyalty to the Holy Apostolic Church, of which the Creed speaks, no sense of the authority of religion as external to the mind: in a word, no seriousness,—and therefore is neither hot nor cold, but (in Scripture language) *lukewarm*. Thus the present age is the very contrary to what are commonly called the dark ages; and together with the faults of those ages we have lost their virtues. I say their virtues; for even the errors then prevalent, a persecuting spirit, for instance, fear of religious inquiry, bigotry, these were, after all, but perversions and excesses of *real virtues*, such as zeal and reverence; and we, instead of limiting and purifying them, have taken them away root and branch. Why? because we have not acted from a love of the Truth, but from the influence of the Age. The old generation has passed, and its character with it; a new order of things has arisen. Human society has a new framework, and fosters and developes a new character of mind; and this new character is made by the enemy of our souls, to resemble the Christian's obedience as near as it may, its likeness all the time being but accidental. Meanwhile, the Holy Church of God, as from the beginning, continues her course heavenward; despised by the world, yet influencing it, partly correcting it, partly restraining it, and in some happy cases reclaiming its victims, and fixing them firmly and for ever within the lines of the faithful host militant here on earth, which journeys towards the City of the Great King. God give us grace to search our hearts, lest we be blinded by the deceitfulness of sin! lest we serve Satan transformed into an Angel of light, while we think we are pursuing true knowledge; lest, over-looking and ill-treating the elect of Christ here, we have to ask that awful question at the last day, while the truth is bursting upon us, "Lord, *when* saw we Thee a stranger and a prisoner?" when saw we Thy sacred Word and Servants despised and oppressed, "and did not minister unto thee?"[1]

---

[1] Matt. 25: 44.

Nothing shows more strikingly the power of the world's religion, as now described, than to consider the very different classes of men whom it influences. It will be found to extend its sway and its teaching both over the professedly religious and the irreligious.

1. Many religious men, rightly or not, have long been expecting a millennium of purity and peace for the Church. I will not say, whether or not with reason, for good men may well differ on such a subject. But, any how, in the case of those who have expected it, it has become a temptation to take up and recognize the world's religion as I have already delineated it. They have more or less identified their vision of Christ's kingdom with the elegance and refinement of mere human civilization; and have hailed every evidence of improved decency, every wholesome civil regulation, every beneficent and enlightened act of state policy, as signs of their coming Lord. Bent upon achieving their object, an extensive and glorious diffusion and profession of the Gospel, they have been little solicitous about the *means* employed. They have countenanced and acted with men who openly professed unchristian principles. They have accepted and defended what they considered to be reformations and ameliorations of the existing state of things, though injustice must be perpetrated in order to effect them, or long cherished rules of conduct, indifferent perhaps in their origin but consecrated by long usage, must be violated. They have sacrificed Truth to expedience. They have strangely imagined that bad men are to be the immediate instruments of the approaching advent of Christ; and (like the deluded Jews not many years since in a foreign country) they have taken, if not for their Messiah (as the Jews did), at least for their Elijah, their reforming Baptist, the Herald of the Christ, children of this world, and sons of Belial, on whom the anathema of the Apostle lies from the beginning, declaring, "If any man love not the Lord Jesus Christ, let him be Anathema Maran-atha."[1]

2. On the other hand, the form of doctrine, which I have called the religion of the day, is especially adapted to please men of sceptical minds, the opposite extreme to those just mentioned, who have never been careful to obey their conscience, who cultivate the intellect without disciplining the heart, and who allow themselves to speculate freely about what religion *ought to be*, without going to Scripture to discover what it really is. Some persons of this character almost consider religion itself to be an obstacle in the advance of our social and political well-being. But they know human nature requires it; therefore they select the most *rational* form of religion (so they call it) which they can find. Others are far more seriously disposed, but are corrupted by bad example or

[1] 1 Cor. 16: 22.

other cause. But they *all* discard (what they call) gloomy views of religion; they all trust themselves more than God's word, and thus may be classed together; and are ready to embrace the pleasant consoling religion natural to a polished age. They lay much stress on works on *Natural Theology*, and think that all religion is contained in these; whereas, in truth, there is no greater fallacy than to suppose such works to be in themselves in any true sense religious at all. Religion, it has been well observed, is something *relative to us*; a system of commands and promises from God *towards* us. But how are we concerned with the sun, moon, and stars? or with the laws of the universe? how will they teach us our *duty*? how will they speak to *sinners*? They do not speak to sinners at all. They were created *before* Adam fell. They "declare the *glory* of God," but not His *will*. They are all perfect, all harmonious; but that brightness and excellence which they exhibit in their own creation, and the Divine benevolence therein seen, are of little moment to fallen man. We see nothing there of God's *wrath*, of which the conscience of a sinner loudly speaks. So that there cannot be a more dangerous (though a common) device of Satan, than to carry us off from our own secret thoughts, to make us forget our own hearts, which tell us of a God of justice and holiness, and to fix our attention merely on the God who made the heavens; who is *our* God indeed, but not God as manifested to us sinners, but as He shines forth to His Angels, and to His elect hereafter.

When a man has so far deceived himself as to trust his destiny to what the heavens tell him of it, instead of consulting and obeying his conscience, what is the consequence? that at once he misinterprets and perverts the whole tenor of Scripture. It cannot be denied that, pleasant as religious observances are declared in Scripture to be to the holy, yet men in general they are said to be difficult and distasteful; to all men *naturally* impossible, and by few fulfilled even with the assistances of grace, on account of their wilful corruption. Religion is pronounced to be against nature, to be against our original will, to require God's aid to make us love and obey it, and to be commonly refused and opposed in spite of that aid. We are expressly told, that "strait is the gate and narrow the way that leads to life, and few there be that find it;" that we must "*strive*" or struggle "to enter in at the strait gate," for that "many shall *seek* to enter in," but that is not enough, they merely seek and therefore do not find; and further, that they who do not obtain everlasting life, "shall go into everlasting punishment,"[1] This is the dark side of religion; and the men I have been describing cannot bear to think of it. They shrink from it as too terrible. They easily get themselves to believe that those strong declarations of Scripture do not belong to the

---

[1] Matt. 7: 14; Luke 13: 24; Matt. 25: 46.

present day, or that they are figurative. They have no language within their heart responding to them. Conscience has been silenced. The only information they have received concerning God has been from Natural Theology, and that speaks only of benevolence and harmony; so they will not credit the plain word of Scripture. They seize on such parts of Scripture as seem to countenance their own opinions; they insist on its being commanded us to "rejoice evermore;" and they argue that it is our duty to solace ourselves here (in moderation, of course) with the goods of this life,—that we have only to be thankful while we use them,—that we need not alarm ourselves,—that God is a merciful God,—that amendment is quite sufficient to atone for our offences,—that though we have been irregular in our youth, yet that is a thing gone by,—that we forget it, and therefore God forgets it,—that the world is, on the whole, very well disposed towards religion,—that we should avoid enthusiasm,—that we should not be over serious,—that we should have large views on the subject of human nature,—and that we should love all men. This indeed is the creed of shallow men, in *every* age, who reason a little, and feel not at all, and who think themselves enlightened and philosophical. Part of what they say is false, part is true, but misapplied; but why I have noticed it here, is to show how exactly it fits in with what I have already described as the peculiar religion of a civilized age; it fits in with it equally well as does that of the (so called) religious world, which is the opposite extreme.

One further remark I will make about these professedly rational Christians; who, be it observed, often go on to deny the mysteries of the Gospel. Let us take the text;—"Our God is a consuming fire." Now supposing these persons fell upon these words, or heard them urged as an argument against their own doctrine of the unmixed satisfactory character of our prospects in the world to come, and supposing they did not know what part of the Bible they occurred in, what would they say? Doubtless they would confidently say that they applied only to the Jews and not to Christians; that they only described the Divine Author of the Mosaic Law;[1] that God formerly spoke in terrors to the Jews, because they were a gross and brutish people, but that civilization has made us quite other men; that our *reason*, not our *fears*, is appealed to, and that the Gospel is love. And yet, in spite of all this argument, the text occurs in the Epistle to the Hebrews, written by an Apostle of Christ.

I shall conclude with stating more fully what I mean by the dark side of religion; and what judgment ought to be passed on the superstitious and gloomy.

Here I will not shrink from uttering my firm conviction that it would

[1] Deut. 4: 24.

be a gain to this country, were it vastly more supersticious, more bigoted, more gloomy, more fierce in its religion, that at present it shows itself to be. Not, of course, that I think the tempers of mind herein implied desirable, which would be an evident absurdity; but I think them infinitely more desirable and more promising than a heathen obduracy, and a cold, self-sufficient, self-wise tranquillity. Doubtless, peace of mind, a quiet conscience, and a cheerful countenance are the gift of the Gospel, and the sign of a Christian; but the same effects (or, rather, what appear to be the same) may arise from very different causes. Jonah slept in the storm,—so did our Blessed Lord. The one slept in an evil security: the Other in the "peace of God which passeth all understanding." The two states cannot be confounded together, they are perfectly distinct; and as distinct is the calm of the man of the world from that of the Christian. Now take the case of the sailors on board the vessel; they cried to Jonah, "What meanest thou, O sleeper?"—so the Apostles said to Christ; "Lord, we perish." This is the case of the superstitious; they stand between the false peace of Jonah and the true peace of Christ; they are better than the one, though far below the Other. Applying this to the present religion of the educated world, full as it is of security and cheerfulness, and decorum, and benevolence, I observe that these appearances may arise either from a great deal of religion, or from the absence of it; they may be the fruits of shallowness of mind and a blinded conscience, or of that faith which has peace with God through our Lord Jesus Christ. And if this alternative be proposed, I might leave it to the common sense of men to decide (if they could get themselves to think seriously) to which of the two the temper of the age is to be referred. For myself I cannot doubt, seeing what I see of the world, that it arises from the sleep of Jonah; and it is therefore but a dream of religion, far inferior in worth to the well-grounded alarm of the superstitious, who are awakened and see their danger, though they do not attain so far in faith as to embrace the remedy of it.

Think of this, I beseech you, my brethren, and lay it to heart, as far as you go with me, as you will answer for having heard it at the last day. I would not willingly be harsh; but knowing "that the world lieth in wickedness," I think it highly probable that you, so far as you are in it (as you must be, and we all must be in our degree), are, most of you, partially infected with its existing error, that shallowness of religion, which is the result of a blinded conscience; and, therefore, I speak earnestly to you. Believing in the existence of a general plague in the land, I judge that you probably have your share in the sufferings, the voluntary sufferings, which it is spreading among us. The fear of God is the beginning of wisdom; till you see Him to be a consuming fire, and approach Him with reverence and godly fear, as being sinners, you are not even in sight of the strait gate. I do not wish you to be able to point to any particular time when you renounced the world (as it is called), and were converted; this is a deceit. Fear and love must go together; always fear, always love, to your dying day. Doubtless;—still you must know what it is to sow in tears here, if you would reap in joy hereafter. Till you know the weight of your sins, and that not in mere imagination but in practice, not so as merely to confess it in a formal phrase of lamentation, but daily and in your heart in secret, you cannot embrace the offer of mercy held out to you in the Gospel, through the death of Christ. Till you know what it is to fear with the terrified sailors or the Apostles, you cannot sleep with Christ at your Heavenly Father's feet. Miserable as were the superstitions of the dark ages, revolting as are the tortures now in use among the heathen of the East, better, far better is it, to torture the body all one's days, and to make this life a hell upon earth, than to remain in a brief tranquillity here, till the pit at length opens under us, and awakens us to an eternal fruitless consciousness and remorse. Think of Christ's own words: "What shall a man give in exchange for his soul?" Again, He says, "Fear Him, who after He hath killed, hath power to cast into hell; yea, I say unto you, fear Him." Dare not to think you have got to the bottom of your hearts; you do not know what evil lies there. How long and earnestly must you pray, how many years must you pass in careful obedience, before you have any right to lay aside sorrow, and to rejoice in the Lord? In one sense, indeed, you may take comfort from the first; for, though you dare not yet anticipate you are in the number of Christ's true elect, yet from the first you know He desires your salvation, has died for you, has washed away your sins by baptism, and will ever help you; and this thought must cheer you while you go on to examine and review your lives, and to turn to God in self-denial. But, at the same time, you never can be sure of salvation, while you are here; and therefore you must always fear while you hope. Your knowledge of your sins increases with your view of God's mercy in Christ. And this is the true Christian state, and the nearest approach to Christ's calm and placid sleep in the tempest;—not perfect joy and certainty in heaven, but a deep resignation to God's will, a surrender of ourselves, soul and body, to Him; hoping indeed, that we shall be saved, but fixing our eyes more earnestly on Him than on ourselves; that is, acting for His glory, seeking to please Him, devoting ourselves to Him in all manly obedience and strenuous good works; and, when we do look within, thinking of ourselves with a certain abhorence and contempt as being sinners, mortifying our flesh, scourging our appetites, and composedly awaiting that time when, if we be worthy, we shall be stripped of our present selves, and new made in the kingdom of Christ.

[vol. i, no. 24]

Self-Contemplation
(Tuesday in Easter Week)

Looking unto Jesus, the Author and Finisher of our faith.

Heb. 12:2

Surely it is our duty ever to look off ourselves, and to look unto Jesus, that is, to shun the contemplation of our own feelings, emotions, frame and state of mind, as if that were the main business of religion, and to leave these mainly to be secured in their fruits. Some remarks were made yesterday upon this "more excellent" and Scriptural way of conducting ourselves, as it has ever been received in the Church; now let us consider the merits of the rule for holy living, which the fashion of this day would substitute for it.

Instead of looking off to Jesus, and thinking little of ourselves, it is at present thought necessary, among the mixed multitude of religionists, to examine the heart with a view of ascertaining whether it is in a spiritual state or no. A spiritual frame of mind is considered to be one in which the heinousness of sin is perceived, our utter worthlessness, the impossibility of our saving ourselves, the necessity of some Saviour, the sufficiency of our Lord Jesus Christ to be that Saviour, the unbounded riches of His love, the excellence and glory of His work of Atonement, the freeness and fulness of His grace, the high privilege of communion with Him in prayer, and the desirableness of walking with Him in all holy and loving obedience; all of them solemn truths, too solemn to be lightly mentioned, but our hearty reception of which is scarcely ascertainable by a direct inspection of our feelings. Moreover, if one doctrine must be selected above the rest as containing the essence of the truths, which (according to this system) are thus vividly understood by the spiritual Christian, it is that of the necessity of renouncing our own righteousness for the righteousness provided by our Lord and Saviour; which is considered, not as an elementary and simple principle (as it really is), but as rarely and hardly acknowledged by any man, especially repugnant to a certain (so-called) pride of heart, which is supposed to run through the whole race of Adam, and to lead every man instinctively to insist even before God on the proper merit of his good deeds; so that, to trust in Christ, is not merely the work of the Holy Spirit (as all good in our souls is), but is the especial and critical event which marks a man, as issuing from darkness, and sealed unto the privileges and inheritance of the sons of God. In other words, the doctrine of Justification by Faith is accounted to be the one cardinal point of the Gospel; and it is in vain to admit it readily as a clear Scripture truth (which it is), and to attempt to go on unto perfection: the very wish to pass forward is interpreted into a wish to pass over it, and the test of believing it at all, is in fact to insist upon no doctrine but it. And this peculiar mode of inculcating that great doctrine of the Gospel is a proof (if proof were wanting) that the persons who adopt it are not solicitous even about *it* on its own score merely, considered as (what is called) a dogma, but as ascertaining and securing (as they hope) a certain state of heart. For, not content with the simple admission of it on the part of another, they proceed to divide faith into its kinds, living and dead, and to urge against him, that the Truth may be held in a carnal and unrenewed mind, and that men may speak without real feelings and convictions. Thus it is clear they do not contend for the doctrine of Justification as a truth external to the mind, or article of faith, any more than for the doctrine of the Trinity. On the other hand, since they use the same language about dead and living faith, however exemplary the life and conduct be of the individual under their review, they as plainly show that neither are the fruits of righteousness in their system an evidence of spiritual-mindedness, but that a something is to be sought for in the frame of mind itself. All this is not stated at present by way of objection, but in order to settle accurately what they mean to maintain. So now we have the two views of doctrine clearly before us:—the ancient and universal teaching of the Church, which insists on the Objects and fruits of faith, and considers the spiritual character of that faith itself sufficiently secured, if these are as they should be; and the method, now in esteem, of attempting instead to secure directly and primarily that "mind of the Spirit," which may savingly receive the truths, and fulfil the obedience of the Gospel. That such a spiritual temper is indispensable, is agreed on all hands. The simple question is, whether it is formed by the Holy Spirit immediately acting upon our minds, or, on the other hand, by our own particular acts (whether of faith or obedience), prompted, guided, and prospered by Him; whether it is ascertainable otherwise than by its fruits; whether such frames of mind as *are* directly ascertainable and profess to be spiritual, are not rather a delusion, a mere excitement, capricious feeling, fanatic fancy, and the like.—So much then by way of explanation.

1. Now, in the first place, this modern system certainly does disparage the revealed doctrines of the Gospel, however its more moderate advocates may shrink from admitting it. Considering a certain state of heart to be the main thing to be aimed at, they avowedly make the "truth as it is in Jesus," the definite Creed of the Church, secondary in their teaching and profession. They will defend themselves indeed from the appearance of undervaluing it, by maintaining that the existence of right religious affections is a security for sound views of doctrine. And this is abstractedly true;—but not true in the use they

make of it: for they unhappily conceive that they can ascertain in each other the presence of these affections; and when they find men possessed of them (as they conceive), yet not altogether orthodox in their belief, then they relax a little, and argue that an admission of (what they call) the strict and technical niceties of doctrine, whether about the Consubstantiality of the Son or the Hypostatic Union, is scarcely part of the definition of a spiritual believer. In order to support this position, they lay it down as self-evident, that the main purpose of revealed doctrine is to affect the heart,—that that which does not seem to affect it does not affect it,—that what does not affect it, is unnecessary,—and that the circumstance that this or that person's heart seems rightly affected, is a sufficient warrant that such Articles as he may happen to reject, may safely be universally rejected, or at least are only accidentally important. Such principles, when once become familiar to the mind, induce a certain disproportionate attention to the doctrines connected with the work of Christ, in comparison of those which relate to His Person, from their more immediately interesting and exciting character; and carry on the more speculative and philosophical class to view the doctrines of Atonement and Sanctification as the essence of the Gospel, and to advocate them in the place of those "Heavenly Things" altogether, which, as theologically expressed, they have already assailed; and of which they now openly complain as mysteries for bondsmen, not Gospel consolations. The last and most miserable stage of this false wisdom is, to deny that in matters of doctrine there is any one sense of Scripture such, that it is true and all others false; to make the Gospel of Truth (so far) a revelation of words and a dead letter; to consider that inspiration speaks merely of divine operations, not of Persons; and that that is truth to each, which each man thinks to be true, so that one man may say that Christ is God, another deny His pre-existence, yet each have received the Truth according to the peculiar constitution of his own mind, the Scripture doctrine having no real independent substantive meaning. Thus the system under consideration tends legitimately to obliterate the great Objects brought to light in the Gospel, and to darken what I called yesterday the eye of faith,—to throw us back into the vagueness of Heathenism, when men only felt after the Divine Presence, and thus to frustrate the design of Christ's Incarnation, so far as it is a manifestation of the Unseen Creator.

2. On the other hand, the necessity of obedience in order to salvation does not suffer less from the upholders of this modern system than the articles of the Creed. They argue, and truly, that if faith is living, works must follow; but mistaking a following *in order of conception* for a following *in order of time*, they conclude that faith ever comes first, and works afterwards; and therefore, that faith must first be secured, and that, by

some means in which works have no share. Thus, instead of viewing works as the concomitant development and evidence, and instrumental cause, as well as the subsequent result of faith, they lay all the stress upon the direct creation, in their minds, of faith and spiritual-mindedness, which they consider to consist in certain emotions and desires, because they can form abstractedly no better or truer notion of those qualities. Then, instead of being "careful to maintain good works," they proceed to take it for granted, that since they have attained faith (as they consider), works will follow without their trouble as a matter of course. Thus the wise are taken in their own craftiness; they attempt to reason, and are overcome by sophisms. Had they kept to the Inspired Record, instead of reasoning, their way would have been clear; and, considering the serious exhortations to keeping God's commandments, with which all Scripture abounds, from Genesis to the Apocalypse, is it not a very grave question, which the most charitable among Churchmen must put to himself, whether these random expounders of the Blessed Gospel are not risking a participation in the woe denounced against those who preach any other doctrine beisdes that delivered unto us, or who "take away from the words of the Book" of revealed Truth?

3. But still more evidently do they fall into this last imputation, when we consider how they are obliged to treat the Sacred Volume altogether, in order to support the system they have adopted. Is it too much to say that, instead of attempting to harmonize Scripture with Scripture, much less referring to Antiquity to enable them to do so, they either drop altogether, or explain away, whole portions of the Bible, and those most sacred ones? How does the authority of the Psalms stand with their opinions, except at best by a forced figurative interpretation? And our Lord's discourses in the Gospels, especially the Sermon on the Mount, are they not virtually considered as chiefly important to the persons immediately addressed, and of inferior instructiveness to us now that the Spirit (as it is profanely said) is come? In short, is not the rich and varied Revelation of our merciful Lord practically reduced to a few chapters of St. Paul's Epistles, whether rightly (as they maintain) or (as we should say) perversely understood? If then the Romanists have added to the Word of God, is it not undeniable that there is a school of religionists among us who have taken from it?

4. I would remark, that the immediate tendency of these opinions is to undervalue ordinances as well as doctrines. The same argument evidently applies; for, if the renewed state of heart is (as it is supposed) attained, what matter whether Sacraments have or have not been administered? The notion of invisible grace and invisible privileges is, on this supposition, altogether superseded; that of communion with Christ is limited to the mere exercise of the affections in prayer and

meditation,—to sensible effects; and he who considers he has already gained this one essential gift of grace (as he calls it), may plausibly inquire, after the fashion of the day, why he need wait upon ordinances which he has anticipated in his religious attainments,—which are but means to an end, which *he* has not to seek, even if they be not outward forms altogether,—and whether Christ will not accept at the last day all who believe, without inquiring if they were members of the Church, or were confirmed, or were baptized, or received the blessing of mere men who are "earthen vessels."

5. The foregoing remarks go to show the utterly unevangelical character of the system in question; unevangelic in the full sense of the word, whether by the Gospel be meant the inspired document of it, or the doctrines brought to light through it, or the Sacramental Institutions which are the gift of it, or the theology which interprets it, or the Covenant which is the basis of it. A few words shall now be added, to show the inherent mischief of the system as such; which I conceive to lie in its necessarily involving a continual self-contemplation and reference to self, in all departments of conduct. He who aims at attaining sound doctrine or right practice, more or less looks out of himself; whereas, in labouring after a certain frame of mind, there is an habitual reflex action of the mind upon itself. That this is really involved in the modern system, is evident from the very doctrine principally insisted on by it; for, as if it were not enough for a man to look up simply to Christ for salvation, it is declared to be necessary that he should be able to recognise this in himself, that he should define his own state of mind, confess he is justified by faith along, and explain what is mean by that confession. Now, the truest obedience is indisputably that which is done from love of God, without narrowly measuring the magnitude or nature of the sacrifice involved in it. He who has learned to give names to his thoughts and deeds, to appraise them as if for the market, to attach to each its due measure of commendation or usefulness, will soon involuntarily corrupt his motives by pride or selfishness. A sort of self-approbation will insinuate itself into his mind: so subtle as not at once to be recognised by himself,—an habitual quiet self-esteem, leading him to prefer his own views to those of others, and a secret, if not avowed persuasion, that he is in a different state from the generality of those around him. This is an incidental, though of course not a necessary evil of religious journals; nay, of such compositions as Ministerial duties involve. They lead those who write them, in some respect or other, to a contemplation of self. Moreover, as to religious journals, useful as they often are, at the same time I believe persons find great difficulty, while recording their feelings in banishing the thought that one day these good feelings will be known to the world, and are thus insensibly led to modify and prepare their

language as if for a representation. Seldom indeed is any one in the *practice* of contemplating his better thoughts or doings without proceeding to display them to others; and hence it is that it is so easy to discover a conceited man. When this is encouraged in the sacred province of religion, it produces a certain unnatural solemnity of manner, arising from a wish to be, nay, to appear spiritual, which is at once very painful to beholders, and surely quite at variance with our Saviour's rule of anointing our head and washing our face, even when we are most self-abased in heart. Another mischief arising from this self-contemplation is the peculiar kind of selfishness (if I may use so harsh a term) which it will be found to foster. They who make self instead of their Maker the great object of their contemplation will naturally exalt themselves. Without denying that the glory of God is the great end to which all things are to be referred, they will be led to connect indissolubly His glory with their own certainty of salvation; and this partly accounts for its being so common to find rigid predestinarian views, and the exclusive maintenance of justification by Faith in the same persons. And for the same reason, the Scripture doctrines relative to the Church and its offices will be unpalatable to such persons; no one thing being so irreconcilable with another, as the system which makes a man's thoughts centre in himself, with that which directs them to a fountain of grace and truth, on which God has made him dependent.

And as self-confidence and spiritual pride are the legitimate results of these opinions in one set of persons, so in another they lead to a feverish anxiety about their religious state and prospects, and fears lest they are under the reprobation of their All-merciful Saviour. It need scarcely be said that a contemplation of self is a frequent attendant, and a frequent precursor of a deranged state of the mental powers.

To conclude. It must not be supposed from the foregoing remarks that I am imputing all the consequences enumerated to every one who holds the main doctrine from which they legitimately follow. Many men zealously maintain principles which they never follow out in their own minds, or after a time silently discard, except as far as words go, but which are sure to receive a full development in the history of any school or party of men which adopts them. Considered thus, as the characteristics of a school, the principles in question are doubtless antichristian; for they destroy all positive doctrine, all ordinances, all good works; they foster pride, invite hypocrisy, discourage the weak, and deceive most fatally, while they profess to be the especial antidotes to self-deception. We have seen these effects of them two centuries since in the history of the English Branch of the Church; for what we know, a more fearful triumph is still in store for them. But, however that may be, let not the watchmen of Jerusalem fail to give timely warning of the approaching enemy, or to acquit themselves of all cowardice or compliance as regards it. Let them prefer the Old Commandment, as it has been from the beginning, to any novelties of man, recollecting Christ's words, "Blessed is he that watcheth, and keepeth his garments, lest he walk naked, and they see his shame."[1]

[vol. ii, no. 15]

The Danger of Accomplishments
(The Feast of St. Luke the Evangelist)

In the hearts of all that are wise hearted, I have put wisdom.

Exod. 31:6

St. Luke differed from his fellow-evangelists and fellow-disciples in having received the advantages of (what is called) a liberal education. In this respect he resembled St. Paul, who, with equal accomplishments, appears to have possessed even more learning. He is said to have been a native of Antioch, a city celebrated for the refined habits and cultivated intellect of its inhabitants; and his profession was that of a physician or surgeon, which of itself evidences him to have been in point of education something above the generality of men. This is confirmed by the character of his writings, which are superior in composition to any part of the New Testament, excepting some of St. Paul's Epistles.

There are persons who doubt whether what are called "accomplishments," whether in literature or in the fine arts, can be consistent with deep and practical seriousness of mind. They think that attention to these argues a lightness of mind, and, at least, takes up time which might be better employed; and, I confess, at first sight they seem to be able to say much in defence of their opinion. Yet, notwithstanding, St. Luke and St. Paul were accomplished men, and evidently took pleasure in their accomplishments.

I am not speaking of human *learning*; this also many men think inconsistent with simple uncorrupted faith. They suppose that learning must make a man proud. This is of course a great mistake; but of it I am not speaking, but of an over-jealousy of *accomplishments*, the elegant arts and studies, such as poetry, literary composition, painting, music, and the like; which are considered (not indeed to make a man *proud*, but) to make him *trifling*. Of this opinion, how far it is true, and how far not true, I am going to speak: being led to the consideration of it by the known fact, that St. Luke was a polished writer, and yet an Evangelist.

Now, that the accomplishments I speak of have a *tendency* to make us

[1] Rev. 16: 15.

trifling and unmanly, and therefore are to be viewed by each of us with suspicion as far as regards himself, I am ready to admit, and shall presently make clear. I allow, that in matter of fact, refinement and luxury, elegance and effeminacy, go together. Antioch, the most polished, was the most voluptuous city of Asia. But the *abuse* of good things is no argument against the things themselves; mental cultivation *may* be divine gift, though it is abused. All God's gifts are perverted by man; health, strength, intellectual power, are all turned by sinners to bad purposes, yet they are not evil in themselves: therefore an acquaintance with the elegant arts may be a gift and a good, and intended to be an instrument of God's glory, though numbers who have it are rendered thereby indolent, luxurious, and feeble-minded.

But the account of the building of the Tabernacle in the wilderness, from which the text is taken, is decisive on this point. It is too long to read to you, but a few verses will remind you of the nature of it. "Thou shalt speak unto all that are wise hearted, whom I have filled with the spirit of wisdom, that they may make Aaron's garments to consecrate him, that he may minister unto me in the priest's office." "See I have called by name Bezaleel . . . and have filled him with the Spirit of God, in wisdom and in understanding, and in knowledge, and in all manner of workmanship, to devise cunning works, to work in gold, and in silver, and in brass, and in cutting of stones, to set them, and in carving of timber, to work all manner of workmanship." "Take ye from among you an offering unto the lord; whosoever is of a willing heart let him bring it, an offering of the Lord, gold, and silver, and brass, and blue, and purple, and scarlet and fine linen, and goats' hair, and rams' skins dyed red, and badgers' skins, and shittim wood, and oil for the light, and spices for anointing oil, and for the sweet incense, and onyx stones, and stones to be set for the ephod, and for the breat-plate. And every wise hearted among you shall come and make all that the Lord hath commanded."[1]

How then is it, that what in itself is of so excellent, and (I may say) divine a nature, is yet so commonly perverted? I proceed to state what is the danger, as it appears to me, of being accomplished, with a view to answer this question.

Now the *danger* of an elegant and polite education is, that it separates feeling and acting; it teaches us to think, speak, and be affected aright, without forcing us to practise what is right. I will take an illustration of this, though somewhat a familiar one, from the effect produced upon the mind by reading what is commonly called a romance or novel, which comes under the description of polite literature, of which I am speaking. Such works contain many good sentiments (I am taking the better sort of

---

[1] Exod. 28: 3; 31: 2–5, 35: 5–10.

them): characters too are introduced, virtuous, noble, patient under suffering, and triumphing at length over misfortune. The great truths of religion are upheld, we will suppose, and enforced; and our affections excited and interested in what is good and true. But it is all fiction; it does not exist out of a book which contains the beginning and end of it. *We* have nothing *to do*; we read, are affected, softened or roused, and that is all; we cool again,—nothing comes of it. Now observe the effect of this. God has made us feel in order that we may *go on to act* in consequence of feeling; if then we allow our feelings to be excited without acting upon them, we do mischief to the moral system within us, just as we might spoil a watch, or other piece of mechanism, by playing with the wheels of it. We weaken its springs, and they cease to act truly. Accordingly, when we have got into the habit of amusing ourselves with these works of fiction, we come at length to feel the excitement without the slightest thought or tendency to act upon it; and since it is very difficult to begin any duty *without* some emotion or other (that is, to begin on mere principles of dry reasoning), a grave question arises, how, after destroying the connexion between feeling and acting, how shall we get ourselves to act when circumstances make it our duty to do so? For instance, we will say we have read again and again, of the heroism of facing danger, and we have glowed with the thought of its nobleness. We have felt how great it is to bear pain, and submit to indignities, rather than wound our conscience; and all this, again and again, when we had no opportunity of carrying our good feelings into practice. Now, suppose at length we actually come into trial, and let us say, our feelings become roused, as often before, at the thought of boldly resisting temptations to cowardice, shall we therefore do our duty, quitting ourselves like men? rather, we are likely to talk loudly, and then run from the danger. Why?—rather let us ask, why *not*? what is to keep us from yielding? Because we *feel* aright? nay, we have again and again felt aright, and thought aright, without accustoming ourselves to act aright, and, though there was an original connexion in our minds between feeling and acting, there is none now; the wires within us, as they may be called, are loosened and powerless.

And what is here instanced of fortitude, is true in all cases of duty. The refinement which literature gives, is that of thinking, feeling, knowing and speaking, right, not of acting right; and thus, while it makes the manners amiable, and the conversation decorous and agreeable, it has no tendency to make the conduct, the practice of the man *virtuous*.

Observe, I have supposed the works of fiction I speak of to inculcate right sentiments; though such works (play-books for example) are often vicious and immoral. But even at best, supposing them well principled, still after all, at best, they are, I say, dangerous, in themselves;—that is, if we allow refinement to stand in the place of hardy, rough-handed obedience. It follows, that I am much opposed to certain *religious* novels, which some persons think so useful: that they sometimes do good, I am far from denying;—but they do more harm than good. They do harm on the whole; they lead men to cultivate the religious affections separated from religious practice. And here I might speak of that entire religious system (miscalled religious) which makes Christian faith consist, not in the honest and plain practice of what is right, but in the luxury of excited religious feeling, in a mere meditating on our Blessed Lord, and dwelling as in a reverie on what He has done for us;—for such indolent contemplation will no more sanctify a man *in fact*, than reading a poem or listening to a chant or psalm-tune.

The case is the same with the arts last alluded to, poetry and music. These are especially likely to make us unmanly, if we are not on our guard, as exciting emotions without insuring correspondent practice, and so destroying the connexion between feeling and acting; for I here mean by unmanliness the inability to do with ourselves what we wish,— the saying fine things and yet lying slothfully on our couch, as if we could not get up, though we ever so much wished it.

And here I must notice something besides in elegant accomplish-ments, which goes to make us over-refined and fastidious, and falsely delicate. In books, everything is made beautiful in its way. Pictures are drawn of *complete* virtue; little is said about failures, and little or nothing of the drudgery of ordinary, every-day obedience, which is neither poetical nor interesting. True faith teaches us to do numberless disagreeable things for Christ's sake, to bear petty annoyances, which we find written down in no book. In most books Christian conduct is made grand, elevated, and splendid; so that any one, who only knows of true religion from books, and not from actual endeavours to be religious, is sure to be offended at religion when he actually comes upon it, from the roughness and humbleness of his duties, and his necessary deficiencies in doing them. It is beautiful in a picture to wash the disciples' feet; but the sands of the real desert have no lustre in them to compensate for the servile nature of the occupation.

And further still, it must be observed, that the art of composing, which is a chief accomplishment, has in itself a tendency to make us artificial and insincere. For to be ever attending to the fitness and propriety of our words, is (or at least there is the risk of its being) a kind of acting; and knowing what can be said on both sides of a subject, is a main step towards thinking the one side as good as the other. Hence men in ancient times, who cultivated polite literature, went by the name of "Sophists;" that is, men who wrote elegantly, and talked eloquently, on any subject whatever, right or wrong. St. Luke perchance might have been such a Sophist, had he not been a Christian.

Such are some of the dangers of elegant accomplishments; and they beset more or less all educated persons; and of these perhaps not the least such females as happen to have no very direct duties, and are above the drudgery of common life, and hence are apt to become fastidious and fine,—to love a luxurious ease, and to amuse themselves in mere elegant pursuits, the while they admire and profess what is religious and virtuous, and think that they really possess the character of mind which they esteem.

With these thoughts before us, it is necessary to look back to the Scripture instances which I began by adducing, to avoid the conclusion that accomplishments are positively dangerous, and unworthy a Christian. But St. Luke and St. Paul show us, that we may be sturdy workers in the Lord's service, and bear our cross manfully, though we be adorned with all the learning of the Egyptians; or rather, that the resources of literature, and the graces of a cultivated mind, may be made both a lawful source of enjoyment to the possessor, and a means of introducing and recommending the Truth to others; while the history of the Tabernacle shows that all the cunning arts and precious possessions of this world may be consecrated to a religious service, and be made to speak of the world to come.

I conclude then with the following cautions, to which the foregoing remarks lead. First, we must avoid giving too much time to lighter occupations; and next, we must never allow ourselves to read works of fiction or poetry, or to interest ourselves in the fine arts for the mere sake of the things themselves: but keep in mind all along that we are Christians and accountable beings, who have fixed principles of right and wrong, by which all things must be tried, and have religious habits to be matured within them, towards which all things are to be made subservient. Nothing is more common among accomplished people than the habit of reading books so entirely for reading's sake, as to praise and blame the actions and persons described in a random way, according to their fancy, not considering whether they are really good or bad according to the standard of moral truth. I would not be austere; but when this is done habitually, surely it is dangerous. Such too is the abuse of poetical talent, that sacred gift. Nothing is more common than to fall into the practice of uttering fine sentiments, particularly in letter writing, as a matter of course, or a kind of elegant display. Nothing more common in singing than to use words which have a light meaning, or a bad one. All these things are hurtful to seriousness of character. It is for this reason (to put aside others) that the profession of stage-players, and again of orators, is a dangerous one. They learn to say good things, and to excite in themselves vehement feelings, about nothing at all.

If we are in earnest we shall let nothing lightly pass by which may do us good, nor shall we dare to trifle with such sacred subjects as morality and religious duty. We shall apply all we read to ourselves; and this almost without intending to do so, from the mere sincerity and honesty of our desire to please God. We shall be suspicious of all such good thoughts and wishes, and we shall shrink from all such exhibitions of our principles, as fall short of action. We shall aim at doing right, and so glorifying our Father, and shall exhort and constrain others to do so also; but as for talking on the appropriate subjects of religious meditation, and *trying* to show piety, and to excite corresponding feelings in another, even though our nearest friend, far from doing this, we shall account it a snare and a mischief. Yet this is what many persons consider the highest part of religion, and call it spiritual conversation, the test of a spiritual mind; whereas, putting aside the incipient and occasional hypocrisy, and again the immodesty of it, I call all formal and intentional expression of religious emotions, all studied passionate discourse, *dissipation,*—dissipation the same in nature, though different in subject, as what is commonly so called; for it is a drain and a waste of our religious and moral strength, a general weakening of our spiritual powers (as I have already shown); and all for what?—for the pleasure of the immediate excitement. Who can deny that this religious disorder is a parallel case to that of the sensualist? Nay, precisely the same as theirs, from whom the religionists in question think themselves very far removed, of the fashionable world I mean, who read works of fiction, frequent the public shows, are ever on the watch for novelities, and affect a pride of manners and a "mincing"[1] deportment, and are ready with all kinds of good thoughts and keen emotions on all occasions. . . .

[vol. ii, no. 30]

Unreal Words
(Advent)

Thine eyes shall see the King in His beauty: they shall behold the land
that is very far off.

Isa. 33:17

The Prophet tells us, that under the Gospel covenant God's servants will have the privilege of seeing those heavenly sights which were but shadowed out in the Law. Before Christ came was the time of shadows but when He came, He brought truth as well as grace; and as He who is the Truth has come to us, so does He in return require that we should be true and sincere in our dealings with Him. To be true and sincere is really to see with our minds those great wonders which He has wrought in order that we might see them. When God opened the eyes of the ass on

[1] Isa. 3: 16.

which Balaam rode, she saw the Angel and acted upon the sight. When He opened the eyes of the young man, Elisha's servant, he too saw the chariots and horses of fire, and took comfort. And in like manner, Christians are now under the protection of a Divine Presence, and that more wonderful than any which was vouchsafed of old time. God revealed Himself visibly to Jacob, to Job, to Moses, to Joshua, and to Isaiah; to us He reveals Himself not visibly, but more wonderfully and truly; not without the co-operation of our own will, but upon our faith, and for that very reason more truly; for faith is the special means of gaining spiritual blessings. Hence St. Paul prays for the Ephesians "that Christ may dwell in their hearts by faith," and that "the eyes of their understanding may be enlighted." And St. John declares that "the Son of God hath given us an understanding that we may know Him that is true: and we are in Him that is true, even in His Son Jesus Christ."[1]

We are no longer then in the region of shadows: we have the true Saviour set before us, the true reward, and the true means of spiritual renewal. We know the true state of the soul by nature and by grace, the evil of sin, the consequences of sinning, the way of pleasing God, and the motives to act upon. God has revealed Himself clearly to us; He has "destroyed the face of the covering cast over all people, and the veil that is spread over all nations." "The darkness is past, and the True Light now shineth."[2] And therefore, I say, He calls upon us in turn to "walk in the light as He is in the light." The Pharisees might have this excuse in their hypocrisy, that the plain truth had not been revealed to them; we have not even this poor reason for insincerity. We have no opportunity of mistaking one thing for another: the promise is expressly made to us that "our teachers shall not be removed into a corner any more, but our eyes shall see our teachers;" that "the eyes of them that see shall not be dim;" that every thing shall be called by its right name; that "the vile person shall be no more called liberal, nor the churl said to be bountiful;"[3] in a word, as the text speaks, that "our eyes shall see the King in His beauty; we shall behold the land that is very far off." Our professions, our creeds, our prayers, our dealings, our conversation, our arguments, our teaching must henceforth be sincere, or, to use an expressive word, must be *real*. What St. Paul says of himself and his fellow-labourers, that they were true because Christ is true, applies to all Christians: "Our rejoicing is this, the testimony of our conscience, that in simplicity and godly sincerity, not with fleshly wisdom, but by the grace of God, we have had our conversation in the world, and more abundantly to you-ward. . . . The things that I purpose, do I purpose

---

[1] Eph. 3: 17: 1: 18: 1 John 5: 20. [2] Isa. 25: 7; 1 John 2: 8.
[3] Isa. 30: 20; 32: 3, 5.

according to the flesh, that with me there should be yea yea, and nay nay? But as God is true, our word toward you was not yea and nay. For the Son of God, Jesus Christ, . . . was not yea and nay, but in Him was yea. For all the promises of God in Him are yea, and in Him Amen, unto the glory of God by us."[1]

And yet it need scarcely be said, nothing is so rare as honesty and singleness of mind; so much so, that a person who is really honest, is already perfect. Insincerity was an evil which sprang up within the Church from the first; Ananias and Simon were not open opposers of the Apostles, but false brethren. And, as foreseeing what was to be, our Saviour is remarkable in His ministry for nothing more than the earnestness of the dissuasives which He addressed to those who came to Him, against taking up religion lightly, or making promises which they were likely to break.

Thus He, "the True Light, which lighteth every man that cometh into the world," "the Amen, the faithful and true Witness, the Beginning of the creation of God,"[2] said to the young Ruler, who lightly called Him "Good Master," "Why callest thou Me good?" as bidding him weigh his words; and then abruptly told him, "One thing thou lackest." When a certain man professed that he would follow Him whithersoever He went, He did not respond to him, but said, "The foxes have holes, and the birds of the air have nests, but the Son of Man hath not where to lay His head." When St. Peter said with all his heart in the name of himself and brethren, "To whom shall we go? Thou hast the words of eternal life," He answered pointedly, "Have not I chosen you twelve, and one of you is a devil?" as if He said, "Answer for thyself." When the two Apostles professed their desire to cast their lot with Him, He asked whether they could "drink of His cup, and be baptized with His baptism." And when "there went great multitudes with Him," He turned and said, that unless a man hated relations, friends, and self, he could not be His disciple. And then he proceeded to warn all men to "count the cost" ere they followed Him. Such is the merciful severity with which He repels us that He may gain us more truly. And what He thinks of those who, after coming to Him, relapse into a hollow and hypocritical profession, we learn from His language towards the Laodiceans: "I know they works, that thou art neither cold nor hot: I would thou wert cold or hot. So then, because thou art lukewarm, and neither cold nor hot, I will cast thee out of My mouth."[3]

We have a striking instance of the same conduct on the part of that ancient Saint who prefigured our Lord in name and office, Joshua, the

---

[1] Cor. 1: 12–20.   [2] John 1: 9; Rev. 3: 14.
[3] Mark 10: 17–21; Matt. 8: 20; John 6: 68–70; Matt. 20: 22; Luke 14: 25–8; Rev. 3: 15, 16.

captain of the chosen people in entering Canaan. When they had at length taken possession of that land which Moses and their fathers had seen "very far off," they said to him, "God forbid that we should forsake the Lord, and serve other gods. We will . . . serve the Lord, for He is our God." He made answer, "Ye cannot serve the Lord; for He is a holy God; He is a jealous God; He will not forgive your transgressions nor your sins."[1] Not as if he would hinder them from obeying, but to sober them in professing. How does his answer remind us of St. Paul's still more awful words, about the impossibility of renewal after utterly falling away!

And what is said of profession of *discipleship* applies undoubtedly in its degree to *all* profession. To make professions is to play with edged tools, unless we attend to what we are saying. Words have a meaning, whether we mean that meaning or not; and they are imputed to us in their real meaning, when our not meaning it is our own fault. He who takes God's Name in vain, is not counted guiltless because he means nothing by it,— he cannot frame a language for himself; and they who make professions, of whatever kind, are heard in the sense of those professions, and are not excused because they themselves attach no sense to them. "By thy words thou shalt be justified, and by thy words thou shalt be condemned."[2]

Now this consideration needs especially to be pressed upon Christians at this day; for this is especially a day of professions. You will answer in my own words, that all ages have been ages of profession. So they have been, in one way or other, but this day in its own especial sense;— because this is especially a day of individual profession. This is a day in which there is (rightly or wrongly) so much of private judgment, so much of separation and difference, so much of preaching and teaching, so much of authorship, that it involves individual profession, responsibility, and recompense in a way peculiarly its own. It will not then be out of place if, in connexion with the text, we consider some of the many ways in which persons, whether in this age or in another, make unreal professions, or seeing see not, and hearing hear not, and speak without mastering, or trying to master, their words. This I will attempt to do at some length, and in matters of detail, which are not the less important because they are minute.

Of course it is very common in all matters, not only in religion, to speak in an unreal way; viz., when we speak on a subject with which our minds are not familiar. If you were to hear a person who knew nothing about military matters, giving directions how soldiers on service should conduct themselves, or how their food and lodging, or their marching, was to be duly arranged, you would be sure that his mistakes would be

---

[1] Josh. 24: 16–19.   [2] Matt. 12: 37.

such as to excite the ridicule and contempt of men experienced in warfare. If a foreigner were to come to one of our cities, and without hesitation offer plans for the supply of our markets, or the management of our police, it is so certain that he would expose himself, that the very attempt would argue a great want of good sense and modesty. We should feel that he did not understand us, and that when he spoke about us, he would be using words without meaning. If a dim-sighted man were to attempt to decide questions of proportion and colour, or a man without ear to judge of musical compositions, we should feel that he spoke on and from general principles, on fancy, or by deduction and argument, not from a real apprehension of the matters which he discussed. His remarks would be theoretical and unreal.

This unsubstantial way of speaking is instanced in the case of persons who fall into any new company, among strange faces and amid novel occurrences. They sometimes form amiable judgments of men and things, sometimes the reverse,—but whatever their judgments be, they are to those who know the men and the things strangely unreal and distorted. They feel reverence where they should not; they discern slights where none were intended; they discover meaning in events which have none; they fancy motives; they misinterpret manner; they mistake character; and they form generalizations and combinations which exist only in their own minds.

Again persons who have not attended to the subject of morals, or to politics, or to matters ecclesiastical, or to theology, do not know the relative value of questions which they meet with in these departments of knowledge. They do not understand the difference between one point and another. The one and the other are the same to them. They look at them as infants gaze at the objects which meet their eyes, in a vague unapprehensive way, as if not knowing whether a thing is a hundred miles off or close at hand, whether great or small, hard or soft. They have no means of judging, no standard to measure by,—and they give judgments at random, saying yea or nay on very deep questions, according as their fancy is struck at the moment, or as some clever or specious argument happens to come across them. Consequently they are inconsistent; say one thing one day, another the next;—and if they must act, act in the dark; or if they can help acting, do not act; or if they act freely, act from some other reason not avowed. All this is to be unreal.

Again, there cannot be a more apposite specimen of unreality than the way in which judgments are commonly formed upon important questions by the mass of the community. Opinions are continually given in the world on matters, about which those who offer them are as little qualified to judge as blind men about colours, and that because they have never exercised their minds upon the points in question. This is a day in which all men are obliged to have an opinion on all questions, political, social, and religious, because they have in some way or other an influence upon the decision; yet the multitude are for the most part absolutely without capacity to take their part in it. In saying this, I am far from meaning that this need be so,—I am far from denying that there is such a thing as plain good sense, or (what is better) religious sense, which will see its way through very intricate matters, or that this is in fact sometimes exerted in the community at large on certain great questions; but at the same time this practical sense is so far from existing as regards the vast mass of questions which in this day come before the public, that (as all persons who attempt to gain the influence of the people on their side know well) their opinions must be purchased by interesting their prejudices or fears in their favour;—not by presenting a question in its real and true substance, but by adroitly colouring it, or selecting out of it some particular point which may be exaggerated, and dressed up, and be made the means of working on popular feelings. And thus government and the art of government becomes, as much as popular religion, hollow and unsound.

And hence it is that the popular voice is so changeable. One man or measure is the idol of the people to-day, another to-morrow. They have never got beyond accepting shadows for things.

What is instanced in the mass is instanced also in various ways in individuals, and in points of detail. For instance, some men are set perhaps on being eloquent speakers. They use great words and imitate the sentences of others; and they fancy that those whom they imitate had as little meaning as themselves, or they perhaps contrive to think that they themselves have a meaning adequate to their words.

Another sort of unreality, or voluntary profession of what is above us, is instanced in the conduct of those who suddenly come into power or place. They affect a manner such as they think the office requires, but which is beyond them, and therefore unbecoming. They wish to act with dignity, and they cease to be themselves.

And so again, to take a different case, many men, when they come near persons in distress and wish to show sympathy, often condole in a very unreal way. I am not altogether laying this to their fault; for it is very difficult to know what to do, when on the one hand we cannot realize to ourselves the sorrow, yet withal wish to be kind to those who feel it. A tone of grief seems necessary, yet (if so be) cannot under our circumstances be genuine. Yet even here surely there is a true way, if we could find it, by which pretence may be avoided, and yet respect and consideration shown.

And in like manner as regards religious emotions. Persons are aware from the mere force of the doctrines of which the Gospel consists, that they ought to be variously affected, and deeply and intensely too, in consequence of them. The doctrines of original and actual sin, of Christ's Divinity and Atonement, and of Holy Baptism, are so vast, that no one can realize them without very complicated and profound feelings. Natural reason tells a man this, and that if he simply and genuinely believes the doctrines, he must have these feelings; and he professes to believe the doctrines absolutely, and therefore he professes the correspondent feelings. But in truth he perhaps does *not* really believe them absolutely, because such absolute belief is the work of long time, and therefore his profession of feeling outruns the real inward existence of feeling, or he becomes unreal. Let us never lose sight of two truths,— that we ought to have our hearts penetrated with the love of Christ and full of self-renunciation; but that if they be not, professing that they are does not make them so.

Again, to take a more serious instance of the same fault, some persons pray, not as sinners addressing their God, not as the Publican smiting on his breast, and saying, "God be merciful to me a sinner," but in such a way as they conceive to be becoming *under* circumstances of guilt, in a way becoming such a strait. They are self-conscious, and reflect on what they are about, and instead of actually approaching (as it were) the mercy-seat, they are filled with the thought that God is great, and man His creature, God on high and man on earth, and that they are engaged in a high and solemn service, and that they ought to rise up to its sublime and momentous character.

Another still more common form of the same fault, yet without any definite pretence or effort, is the mode in which people speak of the shortness and vanity of life, the certainty of death, and the joys of heaven. They have commonplaces in their mouths, which they bring forth upon occasions for the good of others, or to console them, or as a proper and becoming mark of attention towards them. Thus they speak to clergymen in a professedly serious way, making remarks true and sound, and in themselves deep, yet unmeaning in their mouths, or they give advice to children or young men; or perhaps in low spirits or sickness they are led to speak in a religious strain as if it was spontaneous. Or when they fall into sin, they speak of man being frail, of the deceitfulness of the human heart, of God's mercy, and so on:—all these great words, heaven, hell, judgment, mercy, repentance, works, the world that now is, the world to come, being little more than "lifeless sounds, whether of pipe or harp," in their mouths and ears, as the "very lovely song of one that had a pleasant voice and can play well on an instrument,"—as the proprieties of conversation, or the civilities of good breeding.

I am speaking of the conduct of the world at large, called Christian;

but what has been said applies, and necessarily, to the case of a number of well-disposed or even religious men. I mean, that before men come to know the realities of human life, it is not wonderful that their view of religion should be unreal. Young people who have never known sorrow or anxiety, or the sacrifices which conscientiousness involves, want commonly that depth and seriousness of character, which sorrow only and anxiety and self-sacrifice can give. I do not notice this as a fault, but as a plain fact, which may often be seen, and which it is well to bear in mind. This is the legitimate use of this world, to make us seek for another. It does its part when it repels us and disgusts us and drives us elsewhere. Experience of it gives experience of that which is its antidote, in the case of religious minds; and we become real in our view of what is spiritual by the contact of things temporal and earthly. And much more are men unreal when they have some secret motive urging them a different way from religion, and when their professions therefore are forced into an unnatural course in order to subserve their secret motive. When men do not like the conclusions to which their principles lead, or the precepts which Scripture contains, they are not wanting in ingenuity to blunt their force. They can frame some theory, or dress up certain objections, to defend themselves withal; a theory, that is, or objections, which it is difficult to refute perhaps, but which any rightly-ordered mind, nay, any common bystander, perceives to be unnatural and insincere.

What has been here noticed of individuals, takes place even in the case of whole Churches, at times when love has waxed cold and faith failed. The whole system of the Church, its discipline and ritual, are all in their origin the spontaneous and exuberant fruit of the real principle of spiritual religion in the hearts of its members. The invisible Church has developed itself into the Church visible, and its outward rites and forms are nourished and animated by the living power which dwells within it. Thus every part of it is real, down to the minutest details. But when the seductions of the world and the lusts of the flesh have eaten out this divine inward life, what is the outward Church but a hollowness and a mockery, like the whited sepulchres of which our Lord speaks, a memorial of what was and is not? and though we trust that the Church is nowhere thus utterly deserted by the Spirit of truth, at least according to God's ordinary providence, yet may we not say that in proportion as it approaches to this state of deadness, the grace of its ordinances, though not forfeited, at least flows in but a scanty or uncertain stream?

And lastly, if this unreality may steal over the Church itself, which is in its very essence a practical institution, much more is it found in the philosophies and literature of men. Literature is almost in its essence unreal; for it is the exhibition of thought disjoined from practice. Its very home is supposed to be ease and retirement; and when it does more than speak or write, it is accused of transgressing its bounds. This indeed constitutes what is considered its true dignity and honour, viz. its abstraction from the actual affairs of life; its security from the world's currents and vicissitudes; its saying without doing. A man of literature is considered to preserve his dignity by doing nothing; and when he proceeds forward into action, he is thought to lose his position, as if he were degrading his calling by enthusiasm, and becoming a politician or a partisan. Hence mere literary men are able to say strong things against the opinions of their age, whether religious or political, without offence; because no one thinks they mean anything by them. They are not expected to go forward to act upon them, and mere words hurt no one.

Such are some of the more common or more extended specimens of profession without action, or of speaking without really seeing and feeling. In instancing which, let it be observed, I do not mean to say that such profession, as has been described, is always culpable and wrong; indeed I have implied the contrary throughout. It is often a misfortune. It takes a long time really to feel and understand things as they are; we learn to do so only gradually. Profession beyond our feelings is only a fault when we might help it;—when either we speak when we need not speak, or do not feel when we might have felt. Hard insensible hearts, ready and thoughtless talkers, these are they whose unreality, as I have termed it, is a sin; it is the sin of every one of us, in proportion as our hearts are cold, or our tongues excessive.

But the mere fact of our saying more than we feel is not necessarily sinful. St. Peter did not rise up to the full meaning of his confession, "Thou art the Christ," yet he was pronounced blessed. St. James and St. John said, "We are able," without clear apprehension, yet without offence. We ever promise things greater than we master, and we wait on God to enable us to perform them. Our promising involves a prayer for light and strength. And so again we all say the Creed, but who comprehends it fully? All we can hope is, that we are in the way to understand it; that we partly understand it; that we desire, pray, and strive to understand it more and more. Our Creed becomes a sort of prayer. Persons are culpably unreal in their way of speaking, not when they say more than they feel, but when they say things different from what they feel. A miser praising almsgiving, or a coward giving rules for courage, is unreal; but it is not unreal for the less to discourse about the greater, for the liberal to descant upon munificence, or the generous to praise the noble-minded, or the self-denying to use the language of the austere, or the confessor to exhort to martyrdom.

What I have been saying comes to this:—be in earnest, and you will speak of religion where, and when, and how you should; aim at things, and your words will be right without aiming. There are ten thousand ways of looking at this world, but only one right way. The man of pleasure has his way, the man of gain his, and the man of intellect his. Poor men and rich men, governors and governed, prosperous and discontented, learned and unlearned, each has his own way of looking at the things which come before him, and each has a wrong way. There is but one right way; it is the way in which God looks at the world. Aim at looking at it in God's way. Aim at seeing things as God sees them. Aim at forming judgments about persons, events, ranks fortunes, changes, objects, such as God forms. Aim at looking at this life as God looks at it. Aim at looking at the life to come, and the world unseen, as God does. Aim at "seeing the King in His beauty." All things that we see are but shadows to us and delusions, unless we enter into what they really mean.

It is not an easy thing to learn that new language which Christ has brought us. He has interpreted all things for us in a new way; He has brought us a religion which sheds a new light on all that happens. Try to learn this language. Do not get it by rote, or speak it as a thing of course. Try to understand what you say. Time is short, eternity is long; God is great, man is weak; he stands between heaven and hell; Christ is his Saviour; Christ has suffered for him. The Holy Ghost sanctifies him; repentance purifies him, faith justifies, works save. These are solemn truths, which need not be actually spoken, except in the way of creed or of teaching; but which must be laid up in the heart. That a thing is true, is no reason that it should be said, but that it should be done; that it should be acted upon; that it should be made our own inwardly.

Let us avoid talking, of whatever kind; whether mere empty talking, or censorious talking, or idle profession, or descanting upon Gospel doctrines, or the affectation of philosophy, or the pretence of eloquence. Let us guard against frivolity, love of display, love of being talked about, love of singularity, love of seeming original. Let us aim at meaning what we say, and saying what we mean; let us aim at knowing when we understand a truth, and when we do not. When we do not, let us take it on faith, and let us profess to do so. Let us receive the truth in reverence, and pray God to give us a good will, and divine light, and spiritual strength, that it may bear fruit within us.

[vol. v, no. 3]

The Thought of God, the Stay of the Soul
(Quinquagesima)

Ye have not received the spirit of bondage again to fear, but ye have received the Spirit of adoption, whereby we cry, Abba, Father.

Rom. 8: 15

When Adam fell, his soul lost its true strength; he forfeited the inward light of God's presence, and became the wayward, fretful, excitable, and miserable being which his history has shown him to be ever since; with alternate strength and feebleness, nobleness and meanness, energy in the beginning and failure in the end. Such was the state of his soul in itself, not to speak of the Divine wrath upon it, which followed, or was involved in the Divine withdrawal. It lost its spiritual life and health, which was necessary to complete its nature, and to enable it to fulfil the ends for which it was created,—which was necessary both for its moral integrity and its happiness; and as if faint, hungry, or sick, it could no longer stand upright, but sank on the ground. Such is the state in which every one of us lies as born into the world; and Christ has come to reverse this state, and restore us the great gift which Adam lost in the beginning. Adam fell from his Creator's favour to be a bond-servant; and Christ has come to set us free again, to impart to us the Spirit of adoption, whereby we become God's children, and again approach Him as our Father.

I say, by birth we are in a state of defect and want; we have not all that is necessary for the perfection of our nature. As the body is not complete in itself, but requires the soul to give it a meaning, so again the soul till God is present with it and manifested in it, has faculties and affections without a ruling principle, object, or purpose. Such it is by birth, and this Scripture signifies to us by many figures; sometimes calling human nature blind, sometimes hungry, sometimes unclothed, and calling the gift of the Spirit light, health, food, warmth, and raiment; all by way of teaching us what our first state is, and what our gratitude should be to Him who has brought us into a new state. For instance, "Because thou sayest, I am rich, and increased in goods, and have need of nothing; and knowest not that thou art wretched, and miserable, and poor, and blind, and naked: I counsel thee to buy of Me gold tried in the fire, that thou mayest be rich; and white raiment, that thou mayest be clothed, . . . and anoint thine eyes with eye-salve, that thou mayest see." Again, "God, who commanded the light to shine out of darkness, hath shined in our hearts, to give the light of the knowledge of the glory of God, in the face of Jesus Christ." Again, "Awake, thou that sleepest, and arise from the dead, and Christ shall give thee light." Again, "Whosoever drinketh of the water that I shall give him, shall never thirst; but the water that I shall give him shall be in him a well of water springing up into everlasting life." And in the Book of Psalms, "They shall be satisfied with the plenteousness of Thy house; and Thou shalt give them drink of Thy pleasures as out of the river. For with Thee is the well of life, and in Thy Light shall we see light." And in another Psalm, "My soul shall be satisfied, even as it were with marrow and fatness, when my mouth praiseth Thee with joyful lips." And so again, in the Prophet Jeremiah, "I will satiate the souls of the priests with fatness; and My people shall be satisfied with My goodness. . . . I have satiated the weary soul, and I have replenished every sorrowful soul."[1]

Now the doctrine which these passages contain is often truly expressed thus: that the soul of man is made for the contemplation of its Maker; and that nothing short of that high contemplation is its happiness; that, whatever it may possess besides, it is unsatisfied till it is vouchsafed God's presence, and lives in the light of it. There are many aspects in which the same solemn truth may be viewed; there are many ways in which it may be signified. I will now dwell upon it as I have been stating it.

I say, then, that the happiness of the soul consists in the exercise of the affections; not in sensual pleasures, not in activity, not in excitement, not in self-esteem, not in the consciousness of power, not in knowledge; in none of these things lies our happiness, but in our affections being elicited, employed, supplied. As hunger and thirst, as taste, sound, and smell, are the channels through which this bodily frame receives pleasure, so the affections are the instruments by which the soul has pleasure. When they are exercised duly, it is happy; when they are undeveloped, restrained, or thwarted, it is not happy. This is our real and true bliss, not to know, or to affect, or to pursue; but to love, to hope, to joy, to admire, to revere, to adore. Our real and true bliss lies in the possession of those objects on which our hearts may rest and be satisfied.

Now, if this be so, here is at once a reason for saying that the thought of God, and nothing short of it, is the happiness of man; for though there is much besides to serve as subject of knowledge, or motive for action, or means of excitement, yet the affections require a something more vast and more enduring than anything created. What is novel and sudden excites, but does not influence; what is pleasurable or useful raises no awe; self moves no reverence, and mere knowledge kindles no love. He alone is sufficient for the heart who made it. I do not say, of course, that nothing short of the Almighty Creator can awaken and answer to our love, reverence, and trust; man can do this for man. Man doubtless is an object to rouse his brother's love, and repays it in his measure. Nay, it is a great duty, one of the two chief duties of religion, thus to be minded towards our neighbour. But I am not speaking here of what we can do, or ought to do, but what it is our happiness to do: and surely it may be said that though the love of the brethren, the love of all men, be one half of our obedience, yet exercised by itself, were that possible, which it is not, it would be no part of our reward. And for this reason, if for no other, that our hearts require something more permanent and uniform than man can be. We gain much for a time from fellowship with each other. It is a relief to us, as fresh air to the fainting, or meat and drink to

[1] Rev. 3: 17, 18; 2 Cor. 4: 6; Eph. 5: 14; John 4: 14; Ps 36: 8, 9; 63: 5; Jer. 31: 14, 25.

the hungry, or a flood of tears to the heavy in mind. It is a soothing comfort to have those whom we may make our confidants; a comfort to have those to whom we may confess our faults; a comfort to have those to whom we may look for sympathy. Love of home and family in these and other ways is sufficient to make this life tolerable to the multitude of men, which otherwise it would not be; but still, after all, our affections exceed such exercise of them, and demand what is more stable. Do not all men die? are they not taken from us? are they not as uncertain as the grass of the field? We do not give our hearts to things irrational, because these have no permanence in them. We do not place our affections in sun, moon, and stars, or this rich and fair earth, because all things material come to nought, and vanish like day and night. Man, too, though he has an intelligence within him, yet in his best estate he is altogether vanity. If our happiness consists in our affections being employed and recompensed, "man that is born of a woman" cannot be our happiness; for how can he stay another, who "continueth not in one stay" himself?

But there is another reason why God alone is the happiness of our souls, to which I wish rather to direct attention:—the contemplation of Him, and nothing but it, is able fully to open and relieve the mind, to unlock, occupy, and fix our affections. We may indeed love things created with great intenseness, but such affection, when disjoined from the love of the Creator, is like a stream running in a narrow channel, impetuous, vehement, turbid. The heart runs out, as it were, only at one door; it is not an expanding of the whole man. Created natures cannot open us, or elicit the ten thousand mental senses which belong to us, and through which we really live. None but the presence of our Maker can enter us; for to none besides can the whole heart in all its thoughts and feelings be unlocked and subjected. "Behold," He says, "I stand at the door and knock; if any man hear My voice and open the door, I will come in to him, and will sup with him, and he with Me." "My Father will love him, and We will come unto him, and make Our abode with him." "God hath sent forth the Spirit of His Son into your hearts." "God is greater than our heart, and knoweth all things."[1] It is this feeling of simple and absolute confidence and communion, which soothes and satisfies those to whom it is vouchsafed. We know that even our nearest friends enter into us but partially, and hold intercourse with us only at times; whereas the consciousness of a perfect and enduring Presence, and it alone, keeps the heart open. Withdraw the Object on which it rests, and it will relapse again into its state of confinement and constraint; and in proportion as it is limited, either to certain seasons or to certain affections, the heart is straitened and distressed. If it be not

---

[1] Rev. 3: 20; John 14: 23; Gal. 4: 6; 1 John 3: 20.

over bold to say it, He who is infinite can alone be its measure; He alone can answer to the mysterious assemblage of feelings and thoughts which it has within it. "There is no creature that is not manifest in His sight, but all things are naked and opened unto the eyes of Him with whom we have to do."[1]

This is what is meant by the peace of a good conscience; it is the habitual consciousness that our hearts are open to God, with a desire that they should be open. It is a confidence in God, from a feeling that there is nothing in us which we need be ashamed or afraid of. You will say that no man on earth is in such a state; for we are all sinners, and that daily. It is so; certainly we are quite unfitted to endure God's all-searching Eye, to come into direct contact (if I may so speak) with His glorious Presence, without any medium of intercourse between Him and us. But, first, there may be degrees of this confidence in different men, though the perfection of it be in none. And again, God in his great mercy, as we all well know, has revealed to us that there is a Mediator between the sinful soul and Himself. And as His merits most wonderfully intervene between our sins and God's judgment, so the thought of those merits, when present with the Christian, enables him, in spite of his sins, to lift up his heart to God; and believing, as he does, that he is (to use Scripture language) in Christ, or, in other words, that he addresses Almighty God, not simply face to face, but in and through Christ, he can bear to submit and open his heart to God, and to wish it open. For while he is very conscious both of original and actual sin, yet still a feeling of his own sincerity and earnestness is possible; and in proportion as he gains as much as this, he will be able to walk unreservedly with Christ his God and Saviour, and desire His continual presence with him, though he be a sinner, and will wish to be allowed to make Him the one Object of his heart. Perhaps, under somewhat of this feeling, Hagar said, "Thou, God, seest me." It is under this feeling that holy David may be supposed to say, "Examine me, O Lord, and prove me; try out my reins and my heart." "Try me, O God, and seek the ground of my heart; prove me, and examine my thoughts. Look well, if there be any way of wickedness in me; and lead me in the way everlasting."[2] And especially is it instanced in St. Paul, who seems to delight in the continual laying open of his heart to God, and submitting it to His scrutiny, and waiting for His Presence upon it; or, in other words, in the joy of a good conscience. For instance, "I have lived in all good conscience before God until this day." "Herein do I exercise myself, to have always a conscience void of offence toward God, and toward men." "I say the truth in Christ, I lie not; my conscience also bearing me witness in the Holy Ghost." "Our rejoicing is

---

[1] Heb. 4: 12.　　　　[2] Ps. 26: v. 2; 139: vv. 23, 24.

this, the testimony of our conscience, that in simplicity and godly sincerity, not with fleshly wisdom, but by the grace of God, we have had our conversation in the world, and more abundantly to you-ward."[1] It is, I say, the characteristic of St. Paul, as manifested to us in his Epistles, to live in the sight of Him who "searcheth the reins and the heart," to love to place himself before Him, and, while contemplating God, to dwell on the thought of God's contemplating him.

And, it may be, this is something of the Apostle's meaning, when he speaks of the witness of the Spirit. Perhaps he is speaking of that satisfaction and rest which the soul experiences in proportion as it is able to surrender itself wholly to God, and to have no desire, no aim, but to please Him. When we are awake, we are conscious we are awake, in a sense in which we cannot fancy we are, when we are asleep. When we have discovered the solution of some difficult problem in science, we have a conviction about it which is distinct from that which accompanies fancied discoveries or guesses. When we realize a truth we have a feeling which they have not, who take words for things. And so, in like manner, if we are allowed to find that real and most sacred Object on which our heart may fix itself, a fulness of peace will follow, which nothing but it can give. In proportion as we have given up the love of the world, and are dead to the creature, and, on the other hand, are born of the Spirit unto love of our Maker and Lord, this love carries with it its own evidence whence it comes. Hence the Apostle says, "The Spirit itself beareth witness with our spirit, that we are the children of God." Again, he speaks of Him "who hath sealed us, and given the earnest of the Spirit in our hearts."[2]

I have been saying that our happiness consists in the contemplation of God;—(such a contemplation is alone capable of accompanying the mind always and everywhere, for God alone can be always and everywhere present;)—and that what is commonly said about the happiness of a good conscience, confirms this; for what is it to have a good conscience, when we examine the force of our words, but to be ever reminded of God by our own hearts, to have our hearts in such a state as to be led thereby to look up to Him, and to desire His eye to be upon us through the day? It is in the case of holy men the feeling attendant on the contemplation of Almighty God.

But, again, this sense of God's presence is not only the ground of the peace of a good conscience, but of the peace of repentance also. At first sight it might seem strange how repentance can have in it anything of comfort and peace. The Gospel, indeed, promises to turn all sorrow into joy. It makes us take pleasure in desolateness, weakness, and contempt.

---

[1] Acts 23: 1; 24: 16; Rom. 9: 1; 2 Cor. 1: 12. [2] Rom. 8: 16; 2 Cor. 1: 22.

"We glory in tribulations also," says the Apostle, "because the love of God is shed abroad in our hearts by the Holy Ghost which is given unto us." It destroys anxiety: "Take no thought for the morrow, for the morrow shall take thought for the things of itself." It bids us take comfort under bereavement: "I would not have you ignorant, brethren, concerning them which are asleep, that ye sorrow not, even as others which have no hope."[1] But if there be one sorrow, which might seem to be unmixed misery, if there be one misery left under the Gospel, the awakened sense of having abused the Gospel might have been considered that one. And, again, if there be a time when the presence of the Most High would at first sight seem to be intolerable, it would be then, when first the consciousness vividly bursts upon us that we have ungratefully rebelled against Him. Yet so it is that true repentance cannot be without the thought of God; it has the thought of God, for it seeks Him; and it seeks Him, because it is quickened with love; and even sorrow must have a sweetness, if love be in it. For what is to repent but to surrender ourselves to God for pardon or punishment; as loving His presence for its own sake, and accounting chastisement from Him better than rest and peace from the world? While the prodigal son remained among the swine, he had sorrow enough, but no repentance; remorse only; but repentance led him to rise and go to his Father, and to confess his sins. Thus he relieved his heart of its misery, which before was like some hard and fretful tumour weighing upon it. Or, again, consider St. Paul's account of the repentance of the Corinthians; there is sorrow in abundance, nay, anguish, but no gloom, no dryness of spirit, no sternness. The penitents afflict themselves, but it is from the fulness of their hearts, from love, gratitude, devotion, horror of the past, desire to escape from their present selves into some state holier and more heavenly. St. Paul speaks of their "earnest desire, their mourning, their fervent mind towards him." He rejoices, "not that they were made sorry, but that they sorrowed to repentance." "For ye were made sorry," he proceeds, "after a godly manner, that ye might receive damage by us in nothing." And he describes this "sorrowing after a godly sort," to consist in "carefulness, which it wrought to them," "clearing of themselves,"—"indignation,"—"fear,"—"vehement desire,"—"zeal," —"revenge,"[2]—feelings, all of them, which open the heart, yet without relaxing it, in that they terminate in acts or works.

On the other hand, remorse, or what the Apostle calls "the sorrow of the world," worketh death. Instead of coming to the Fount of Life, to the God of all consolation, remorseful men feed on their own thoughts, without any confidant of their sorrow. They disburden themselves to no

---

[1] Rom. 5: 3, 5; Matt. 6: 34; 1 Thess. 4: 13. [2] 2 Cor. 7: 7, 9, 11.

one: to God they will not, to the world they cannot confess. The world will not attend to their confession; it is a good associate, but it cannot be an intimate. It cannot approach us or stand by us in trouble; it is no Paraclete; it leaves all our feelings buried within us, either tumultuous, or, at best, dead: it leaves us gloomy or obdurate. Such is our state, while we live to the world, whether we be in sorrow or in joy. We are pent up within ourselves, and are therefore miserable. Perhaps we may not be able to analyse our misery, or even to realize it, as persons oftentimes who are in bodily sicknesses. We do not know, perhaps, what or where our pain is; we are so used to it that we do not call it pain. Still so it is; we need a relief to our hearts, that they may be dark and sullen no longer, or that they may not go on feeding upon themselves; we need to escape from ourselves to something beyond; and much as we may wish it otherwise, and may try to make idols to ourselves, nothing short of God's presence is our true refuge; everything else is either a mockery, or but an expedient useful for its season or in its measure.

How miserable then is he, who does not practically know this great truth! Year after year he will be a more unhappy man, or, at least, he will emerge into a maturity of misery at once, when he passes out of this world of shadows into that kingdom where all is real. He is at present attempting to satisfy his soul with that which is not bread; or he thinks the soul can thrive without nourishment. He fancies he can live without an object. He fancies that he is sufficient for himself; or he supposes that knowledge is sufficient for his happiness; or that exertion, or that the good opinion of others, or (what is called) fame, or that the comforts and luxuries of wealth, are sufficient for him. What a truly wretched state is that coldness and dryness of soul, in which so many live and die, high and low, learned and unlearned. Many a great man, many a peasant, many a busy man, lives and dies with closed heart, with affections undeveloped, unexercised. You see the poor man, passing day after day, Sunday after Sunday, year after year, without a thought in his mind, to appearance almost like a stone. You see the educated man, full of thought, full of intelligence, full of action, but still with a stone heart, as cold and dead as regards his affections, as if he were the poor ignorant countryman. You see others, with warm affections, perhaps, for their families, with benevolent feelings towards their fellow-men, yet stopping there; centring their hearts on what is sure to fail them, as being perishable. Life passes, riches fly away, popularity is fickle, the senses decay, the world changes, friends die. One alone is constant; One alone is true to us; One alone can be true; One alone can be all things to us; One alone can supply our needs; One alone can train us up to our full perfection; One alone can give a meaning to our complex and intricate nature; One alone can give us tune and harmony; One alone can form and possess us. Are we allowed to put ourselves under His guidance? this surely is the only question. Has He really made us His children, and taken possession of us by His Holy Spirit? Are we still in His kingdom of grace, in spite of our sins? The question is not whether we should go, but whether He will receive. And we trust, that, in spite of our sins, He will receive us still, every one of us, if we seek His face in love unfeigned, and holy fear. Let us then do our part, as He has done His, and much more. Let us say with the Psalmist, "Whom have I in heaven but Thee? and there is none upon earth I desire in comparison of Thee. My flesh and my heart faileth; but God is the strength of my heart, and my portion for ever."[1]

[vol. v, no. 22]

---

[1] Ps. 73: 25, 26.

4
THE THEOLOGIAN

Although Newman never regarded himself as a professional theologian and although he was clearly no systematic theologian, he is without doubt one of the great theological geniuses in the history of Christianity. His first book, *The Arians of the Fourth Century* (1833), may seem in the light of modern research to be less than impeccable in its historical scholarship, but its originality and penetration as a theological work are extremely impressive. In its carefully nuanced approach to the key issue of dogmatic formulations, it shows the kind of balance which is so distinguishing a feature of Newman's mature Catholic ecclesiology. And in its openness to the possibility of salvation outside Christianity and of revelation in non-Christian religions, the book strikes a very modern note, with obvious implications for Christian apologetics and missionary strategy.

Newman's general theological stance as an Anglican may be justly characterized as 'conservative'. His protest against religious liberalism and rationalism was as unremitting as his opposition to the emotional and untheological Evangelicalism which had exerted such influence on him in the years following his conversion at the age of 15, but which he came to regard as an incomplete and doctrinally unsatisfactory form of Christianity. His writings on both these subjects continue to have an obvious bearing and relevance today. *Lectures on the Doctrine of Justification* (1838), however, which contain a sustained and vigorous attack on the doctrine of justification by faith alone, are nevertheless a very significant milestone in the history of ecumenical theology. As part of Newman's effort to construct an Anglican theology of the 'Via Media' or 'middle way' between Protestantism and Roman Catholicism, it attempts to find a solution to the classic Catholic–Protestant controversy over justification by a fresh return to the sources in Scripture and the Fathers. The work is arguably Newman's most profound theological work.

It remains true, however, that Newman's most celebrated achievement as a theologian is *An Essay on the Development of Christian Doctrine* (1845), a work which is foreshadowed in the last of the *Oxford University Sermons* (1843). Newman completed it, or rather left it unfinished, as an Anglican and published it as a Catholic. It is one of the great seminal works of theology, a kind of theological counterpart to that other Victorian intellectual classic, Darwin's *Origin of Species*. If the problem of justification is no longer the contentious issue that it once was, the questions raised by the phenomenon of doctrinal development, which is now generally recognized as an inescapable historical fact by

Protestant as well as Catholic theologians, have become some of the most controversial and crucial in modern theology.

As a Catholic, Newman's theological position did not change in any radical way, but his context and situation were greatly altered. In place of a church uncertain about its beliefs and identity, he found himself in a highly authoritarian church where little heed was given to the rights either of the laity or of theologians. His own theological investigation and thought had led him to the conclusion that the modern Roman Catholic Church was indeed the same Church as that of the Fathers. The second half of his life saw the gradual, often very painful, development of a theology of the church (or ecclesiology), which achieves perhaps its finest expression in the last great chapter of the *Apologia pro Vita sua* (1864), where both the authority of the hierarchy and the freedom of theologians are simultaneously upheld, while the tension between the two is seen as not simply inevitable but as essentially creative for the life of the Church. A similar balance is to be found in Newman's *Letter to the Duke of Norfolk* (1875) where he both defends the dogma of papal infallibility, while insisting on its limited scope as well as on the responsibility of theologians in interpreting doctrinal statements, and also asserts the rights of the individual conscience not against but alongside the claims of authority. His final great contribution to ecclesiology is the lengthy 1877 preface to the *Via Media*, where he explains that the Church has three quite separate offices—the 'Prophetical', the 'Regal', and the 'Sacerdotal'—which are liable to conflict with one another. Once again the keynote is equipoise as opposed to encroachment: like his idea of the university, Newman's idea of the Church is of a wholeness and unity comprising a variety of elements and parts held together in fruitful tension, each sustained by mutual dependence rather than threatened by the collision of interaction.

THE ARIANS OF THE FOURTH CENTURY

Dogma

If I avow my belief, that freedom from symbols and articles is abstractedly the highest state of Christian communion, and the peculiar privilege of the primitive Church, it is not from any tenderness towards that proud impatience of control in which many exult, as in a virtue: but first, because technicality and formalism are, in their degree, inevitable results of public confessions of faith; and next, because when confessions do not exist, the mysteries of divine truth, instead of being exposed to the gaze of the profane and uninstructed, are kept hidden in the bosom of the Church, far more faithfully than is otherwise possible; and reserved by a private teaching, through the channel of her ministers, as rewards in due measure and season, for those who are prepared to profit by them; for those, that is, who are diligently passing through the successive stages of faith and obedience. And thus, while the Church is not committed to declarations, which, most true as they are, still are daily wrested by infidels to their ruin; on the other hand, much of that mischievous fanaticism is avoided, which at present abounds from the vanity of men, who think that they can explain the sublime doctrines and exuberant promises of the Gospel, before they have yet learned to know themselves and to discern the holiness of God, under the preparatory discipline of the Law and of Natural Religion. Influenced, as we may suppose, by these various considerations, from reverence for the free spirit of Christian faith, and still more for the sacred truths which are the objects of it, and again from tenderness both for the heathen and the neophyte, who were unequal to the reception of the strong meat of the full Gospel, the rulers of the Church were dilatory in applying a remedy, which nevertheless the circumstances of the times imperatively required. They were loth to confess, that the Church had grown too old to enjoy the free, unsuspicious teaching with which her childhood was blest; and that her disciples must, for the future, calculate and reason before they spoke and acted. . . .

Non-Christian Revelation

. . .St. Paul evidently connects the true religion with the existing systems which he laboured to supplant, in his speech to the Athenians in the Acts, and his example is a sufficient guide to missionaries now, . . . but are we able to account for his conduct, and ascertain the principle by which it was regulated? I think we can; and the exhibition of it will set before the reader another doctrine . . . which I shall call *the divinity of Traditionary Religion*.

We know well enough for practical purposes what is meant by Revealed Religion; viz. that it is the doctrine taught in the Mosaic and Christian dispensations, and contained in the Holy Scriptures, and is from God in a sense in which no other doctrine can be said to be from Him. Yet if we would speak correctly, we must confess, on the authority of the Bible itself, that all knowledge of religion is from Him, and not only that which the Bible has transmitted to us. There never was a time when God had not spoken to man, and told him to a certain extent his duty. His injunctions to Noah, the common father of all mankind, is the first recorded fact of the sacred history after the deluge. Accordingly, we are expressly told in the New Testament, that at no time He left Himself without witness in the world, and that in every nation He accepts those who fear and obey Him. It would seem, then, that there is someting true and divinely revealed, in every religion all over the earth, overloaded, as it may be, and at times even stifled by the impieties which the corrupt will and understanding of man have incorporated with it. Such are the doctrines of the power and presence of an invisible God, of His moral law and governance, of the obligation of duty, and the certainty of a just judgment, and of reward and punishment, as eventually dispensed to individuals; so that Revelation, properly speaking, is an universal, not a local gift; and the distinction between the state of Israelites formerly and Christians now, and that of the heathen, is, not that we can, and they cannot attain to future blessedness, but that the Church of God ever has had, and the rest of mankind never have had, authoritative documents of truth, and appointed channels of communication with Him. The word and the Sacraments are the characteristic of the elect people of God; but all men have had more or less the guidance of Tradition, in addition to those internal notions of right and wrong which the Spirit has put into the heart of each individual.

This vague and uncertain family of religious truths, originally from God, but sojourning without the sanction of miracle, or a definite home, as pilgrims up and down the world, and discernible and separable from the corrupt legends with which they are mixed, by the spiritual mind alone, may be called the *Dispensation of Paganism* ... And further, Scripture gives us reason to believe that the traditions, thus originally delivered to mankind at large, have been secretly re-animated and enforced by new communications from the unseen world; though these were not of such a nature as to be produced as evidence, or used as criteria and tests, and roused the attention rather than informed the understandings of the heathen. The book of Genesis contains a record of the Dispensation of Natural Religion, or Paganism, as well as of the patriarchal. The dreams of Pharaoh and Abimelech, as of Nebuchadnezzar afterwards, are instances of the dealings of God with those to whom He did not vouchsafe a written revelation. Or should it be said, that these particular cases merely come within the range of the Divine supernatural Governance which was in their neighbourhood,—an assertion which requires proof,—let the book of Job be taken as a less suspicious instance of the dealings of God with the heathen. Job was a pagan in the same sense in which the Eastern nations are Pagans in the present day. He lived among idolaters, yet he and his friends had cleared themselves from the superstitions with which the true creed was beset; and while one of them was divinely instructed by dreams, he himself at length heard the voice of God out of the whirlwind, in recompense for his long trial and his faithfulness under it. Why should not the book of Job be accepted by us, as a gracious intimation given us, who are God's sons, for our comfort, when we are anxious about our brethren who are still "scattered abroad" in an evil world; an intimation that the Sacrifice, which is the hope of Christians, has its power and its success, wherever men seek God with their whole heart?—If it be objected that Job lived in a less corrupted age than the times of ignorance which followed, Scripture, as if for our full satisfaction, draws back the curtain farther still in the history of Balaam. There a bad man and a heathen is made the oracle of true divine messages about doing justly, and loving mercy, and walking humbly; nay, even among the altars of superstition, the Spirit of God vouchsafes to utter prophecy. And so in the cave of Endor, even a saint was sent from the dead to join the company of an apostate king, and of the sorceress whose aid he was seeking. Accordingly, there is nothing unreasonable in the notion, that there may have been heathen poets and sages, or sibyls again, in a certain extent divinely illuminated, and organs through whom religious and moral truth was conveyed to their countrymen; though their knowledge of the Power from whom the gift came, nay, and their perception of the gift as existing in themselves, may have been very faint or defective. . . .

If this doctrine be scriptural, it is not difficult to determine the line of conduct which is to be observed by the Christian apologist and missionary. Believing God's hand to be in every system, so far forth as it is true (though Scripture alone is the depositary of His unadulterated and complete revelation), he will, after St. Paul's manner, seek some points in the existing superstitions as the basis of his own instructions, instead of indiscriminately condemning and discarding the whole assemblage of heathen opinions and practices; and he will address his hearers, not as men in a state of actual perdition, but as being in imminent danger of "the wrath to come," because they are in bondage and ignorance, and probably under God's displeasure, that is, the vast majority of them are so in fact; but not necessarily so, from the very circumstance of their being heathen. And while he strenuously opposes all that is idolatrous, immoral, and profane, in their creed, he will profess to be leading them on to perfection, and to be recovering and purifying, rather than reversing the essential principles of their belief.

[ch. 1]

Doctrinal Comprehensiveness

Thus the systematic doctrine of the Trinity may be considered as the shadow, projected for the contemplation of the intellect, of the Object of scripturally-informed piety: a representation, economical; necessarily imperfect, as being exhibited in a foreign medium, and therefore involving apparent inconsistencies or mysteries; given to the Church by tradition contemporaneously with those apostolic writings, which are addressed more directly to the heart; kept in the background in the infancy of Christianity, when faith and obedience were vigorous, and brought forward at a time when, reason being disproportionately developed, and aiming at sovereignty in the province of religion, its presence became necessary to expel an usurping idol from the house of God.

If this account of the connexion between the theological system and the Scripture implication of it be substantially correct, it will be seen how ineffectual all attempts ever will be to secure the doctrine by mere general language. It may be readily granted that the intellectual representation should ever be subordinate to the cultivation of the religious affections. And after all, it must be owned, so reluctant is a well-constituted mind to reflect on its own motive principles, that the correct intellectual image, from its hardness of outline, may startle and offend those who have all along been acting upon it. Doubtless there are portions of the ecclesiastical doctrine, presently to be exhibited, which may at first sight seem a refinement, merely because the object and bearings of them are not understood without reflection and experience. But what is left to the church but to speak out, in order to exclude error? Much as we may wish it, we cannot restrain the rovings of the intellect, or silence its clamorous demand for a formal statement concerning the Object of our worship. If, for instance, Scripture bids us adore God, and adore His Son, our reason at once asks, whether it does not follow that there are two Gods; and a system of doctrine becomes unavoidable; being framed, let it be observed, not with a view of explaining, but of arranging the inspired notices concerning the Supreme Being, of providing, not a consistent, but a connected statement. There the inquisitiveness of a pious mind rests, viz., when it has pursued the subject into the mystery which is its limit. But this is not all. The intellectual expression of theological truth not only excludes heresy, but directly assists the acts of religious worship and obedience; fixing and stimulating the Christian spirit in the same way as the knowledge of the One God relieves and illuminates the perplexed conscience of the religious heathen.—And thus much on the importance of Creeds to tranquillize the mind; the text of Scripture being addressed principally to the affections, and of a religious, not a philosophical character.

Nor, in the next place, is an assent to the text of Scripture sufficient for the purposes of Christian fellowship. As the sacred text was not intended to satisfy the intellect, neither was it given as a test of the religious temper which it forms, and of which it is an expression. Doubtless no combination of words will ascertain an unity of sentiment in those who adopt them; but one form is more adapted for the purpose than another. Scripture being unsystematic, and the faith which it propounds being scattered through it documents, and understood only when they are viewed as a whole, the Creeds aim at concentrating its general spirit, so as to give security to the Church, as far as may be, that its members take that definite view of that faith which alone is the true one. But, if this be the case, how idle is it to suppose that to demand assent to a form of words which happens to be scriptural, is on that account sufficient to effect an unanimity in thought and action! If the Church would be vigorous and influential, it must be decided and plain-spoken in its doctrine, and must regard its faith rather as a character of mind than as a notion. To attempt comprehensions of opinion, amiable as the motive frequently is, is to mistake arrangements of words, which have no existence except on paper, for habits which are realities; and ingenious generalizations of discordant sentiments for that practical agreement which alone can lead to cooperation. We may indeed artificially classify light and darkness under one term or formula; but nature has her own fixed courses, and unites mankind by the sympathy of moral character, not by those forced resemblances which the imagination singles out at pleasure even in the most promiscuous collection of materials. However plausible may be the veil thus thrown over heterogeneous doctrines, the flimsy artifice is discomposed so soon as the principles beneath it are called upon to move and act. Nor are these attempted comprehensions innocent; for, it being the interest of our enemies to weaken the Church, they have always gained a point, when they have put upon us words for things, and persuaded us to fraternize with those who, differing from us in essentials, nevertheless happen, in the excursive range of opinion, somewhere to intersect that path of faith, which centres in supreme and zealous devotion to the service of God.

Let it be granted, then, as indisputable, that there are no two opinions so contrary to each other, but some form of words may be found vague enough to comprehend them both. . . .

[ch. 2]

'ON THE INTRODUCTION OF RATIONALISTIC PRINCIPLES INTO REVEALED RELIGION'

Rationalism is a certain abuse of Reason; that is, a use of it for purposes for which it never was intended, and is unfitted. To rationalize in matters of Revelation is to make our reason the standard and measure of the doctrines revealed; to stipulate that those doctrines should be such as to carry with them their own justification; to reject them, if they come in collision with our existing opinions or habits of thought, or are with difficulty harmonized with our existing stock of knowledge. And thus a rationalistic spirit is the antagonist of Faith; for Faith is, in its very nature, the acceptance of what our reason cannot reach, simply and absolutely upon testimony.

There is, of course, a multitude of cases in which we allowably and rightly accept statements as true, partly on reason, and partly on testimony. We supplement the information of others by our own knowledge, by our own judgment of probabilities; and, if it be very strange or extravagant, we suspend our assent. This is undeniable; still, after all, there are truths which are incapable of reaching us except on testimony, and there is testimony, which by and in itself, has an imperative claim on our acceptance.

As regards Revealed Truth, it is not Rationalism to set about to ascertain, by the exercise of reason, what things are attainable by reason, and what are not; nor, in the absence of an express Revelation, to inquire into the truths of Religion, as they come to us by nature; nor to determine what proofs are necessary for the acceptance of a Revelation, if it be given; nor to reject a Revelation on the plea of insufficient proof; nor, after recognizing it as divine, to investigate the meaning of its declarations, and to interpret its language; nor to use its doctrines, as far as they can be fairly used, in inquiring into its divinity; nor to compare and connect them with our previous knowledge, with a view of making them parts of a whole; nor to bring them into dependence on each other, to trace their mutual relations, and to pursue them to their legitimate issues. This is not Rationalism; but it is Rationalism to accept the Revelation, and then to explain it away; to speak of it as the Word of God, and to treat it as the word of man; to refuse to let it speak for itself; to claim to be told the *why* and the *how* of God's dealings with us, as therein described, and to assign to Him a motive and a scope of our own; to stumble at the partial knowledge which He may give us of them; to put aside what is obscure, as if it had not been said at all; to accept one half of what has been told us, and not the other half; to assume that the contents of Revelation are also its proof; to frame some gratuitous hypothesis about them, and then to garble, gloss, and colour them, to trim, clip, pare away, and twist them, in order to bring them into conformity with the idea to which we have subjected them. . . .

. . . The Rationalist makes himself his own centre, not his Maker; he does not go to God, but he implies that God must come to him. And this, it is to be feared, is the spirit in which multitudes of us act at the present day. Instead of looking out of ourselves, and trying to catch glimpses of God's workings, from any quarter,—throwing ourselves forward upon Him and waiting on Him, we sit at home bringing everything to ourselves, enthroning ourselves in our own views, and refusing to believe anything that does not force itself upon us as true. Our private judgment is made everything to us,—is contemplated, recognized, and consulted as the arbiter of all questions, and as independent of everything external to us. Nothing is considered to have an existence except so far forth as our minds discern it. The notion of half views and partial knowledge, of guesses, surmises, hopes and fears, of truths faintly apprehended and not understood, of isolated facts in the great scheme of Providence, in a word, the idea of Mystery, is discarded.

Hence a distinction is drawn between what is called Objective and Subjective Truth, and Religion is said to consist in a reception of the latter. By Objective Truth is meant the Religious System considered as existing in itself, external to this or that particular mind: by Subjective, is meant that which each mind receives in particular, and considers to be such. To believe in Objective Truth is to throw ourselves forward upon that which we have but partially mastered or made subjective; to embrace, maintain, and use general propositions which are larger than our own capacity, of which we cannot see the bottom, which we cannot follow out into their multiform details; to come before and bow before the import of such propositions, as if we were contemplating what is real and independent of human judgment. Such a belief, implicit, and symbolized as it is in the use of creeds, seems to the Rationalist superstitious and unmeaning, and he consequently confines Faith to the province of Subjective Truth, or to the reception of doctrine, as, and so far as, it is met and apprehended by the mind, which will be differently, as he considers, in different persons, in the shape of orthodoxy in one, heterodoxy in another. That is, he professes to *believe* in that which he *opines*; and he avoids the obvious extravagance of such an avowal by maintaining that the moral trial involved in Faith does not lie in the submission of the reason to external realities partially disclosed, but in what he calls that candid pursuit of truth which ensures the eventual adoption of that opinion on the subject, which is best for us individually, which is most natural according to the constitution of our own minds, and, therefore, divinely intended for us. I repeat, he owns that Faith, viewed with reference to its objects is never more than an opinion, and is pleasing to God, not as an active principle apprehending definite doctrines, but as a result and fruit, and therefore an evidence of past diligence, independent inquiry, dispassionateness, and the like. Rationalism takes the words of Scripture as signs of Ideas; Faith, of Things or Realities. . . .

. . . Revelation, as a Manifestation, is a doctrine variously received by various minds, but nothing more to each than that which each mind comprehends it to be. Considered as a Mystery, it is a doctrine enunciated by inspiration, in human language, as the only possible medium of it, and suitably, according to the capacity of language; a doctrine *lying hid* in language, to be received in that language from the first by every mind, whatever be its separate power of understanding it; entered into more or less by this or that mind, as it may be; and admitting of being apprehended more and more perfectly according to the diligence of this mind and that. It is one and the same, independent and real, of depth unfathomable, and illimitable in its extent.

This is a fit place to make some remarks on the Scripture sense of the word Mystery. It may seem a contradiction in terms to call Revelation a Mystery; but is not the book of the Revelation of St. John as great a mystery from beginning to end as the most abstruse doctrine the mind ever imagined? yet it is even called a *Revelation*. How is this? The answer is simple. No revelation can be complete and systematic, from the weakness of the human intellect; *so far as* it is not such, it is mysterious. When nothing is revealed, nothing is known, and there is nothing to contemplate or marvel at; but when something is revealed, and only something, for all cannot be, there are forthwith difficulties and perplexities. A Revelation is religious doctrine viewed on its illuminated side; a Mystery is the selfsame doctrine viewed on the side unilluminated. Thus Religious Truth is neither light nor darkness, but both together; it is like the dim view of a country seen in the twilight, with forms half extricated from the darkness, with broken lines, and isolated masses. Revelation, in this way of considering it, is not a revealed *system*, but consists of a number of detached and incomplete truths belonging to a vast system unrevealed, of doctrines and injunctions mysteriously connected together; that is, connected by unknown media, and bearing upon unknown portions of the system. . . .

The practical inference to be drawn from this view is, first, that we should be very reverent in dealing with Revealed Truth; next, that we should avoid all rash theorizing and systematizing as relates to it, which is pretty much what looking into the Ark was under the Law: further, that we should be solicitous to hold it safely and entirely; moreover, that we should be zealous and pertinacious in guarding it; and lastly, which is implied in all these, that we should religiously adhere to the form of words and the ordinances under which it comes to us, through which it is revealed to us, and apart from which the Revelation does not exist, there being nothing else given us by which to ascertain or enter into it.

Striking indeed is the contrast presented to this view of the Gospel by the popular theology of the day! That theology is as follows: that the Atonement is the chief doctrine of the Gospel; again, that it is chiefly to be regarded, not as a wonder in heaven, and in its relation to the attributes of God and to the unseen world, but in its experienced effects on our minds, in the change it effects when it is believed. To this, as if to the point of sight in a picture, all the portions of the Gospel system are directed and made to converge; as if this doctrine were so fully understood, that it might fearlessly be used to regulate, adjust, correct, complete, everything else. Thus, the doctrine of the Incarnation is viewed as necessary and important to the Gospel, *because* it gives virtue to the Atonement; of the Trinity, *because* it includes the revelation, not only of the Redeemer, but also of the Sanctifier, by whose aid and influence the Gospel message is to be blessed to us. It follows that faith is nearly the whole of religious service, for through it the message or Manifestation is received; on the other hand, the scientific language of Catholicism, concerning the Trinity and Incarnation, is disparaged, as having no tendency to enforce the effect upon our minds of the doctrine of the Atonement, while the Sacraments are limited to the office of representing, and promising, and impressing on us the promise of divine influences, in no measure of conveying them. Thus the Dispensation, in its length, depth, and height, is practically identified with its Revelation, or rather its necessarily superficial Manifestation. Not that the reality of the Atonement, in itself, is formally denied, but it is cast in the background, except so far as it can be discovered to be influential, viz., to show God's hatred of sin, the love of Christ, and the like; and there is an evident tendency to consider it as a *mere* Manifestation of the love of Christ, to the denial of all real virtue in it as an expiation for sin: as if His death took place merely to show His love for us as a sign of God's infinite mercy, to calm and assure us, without any real connexion existing between it and God's forgiveness of our sins. And the Dispensation thus being hewn and chiselled into an intelligible human system, is represented, when thus mutilated, as affording a remarkable evidence of the truth of the Bible, an evidence level to the reason, and superseding the testimony of the Apostles. That is, according to the above observations, that Rationalism, or want of faith, which has in the first place invented a spurious gospel, next looks complacently on its own offspring, and pronounces it to be the very image of that notion of the Divine Providence, according to which it was originally modelled; a procedure, which, besides more serious objections, incurs the logical absurdity of arguing in a circle. . . .

[*Essays Critical and Historical*, vol. 1]

LECTURES ON THE DOCTRINE OF JUSTIFICATION

Love considered as the Formal Cause of Justification

. . . When, then, divines, however high in repute, come to me with their visionary system, an unreal righteousness and a real corruption, I answer that the Law is past, and that I will not be brought into bondage by shadows. . . . Reputed justification was the gift of the Law; but grace and truth came by Jesus Christ. Away then with this modern, this private, this arbitrary, this unscriptural system, which promising liberty conspires against it; which abolishes Christian Sacraments to introduce barren and dead ordinances; and for the real participation of the Son, and justification through the Spirit, would, at the very marriage feast, feed us on shells and husks, who hunger and thirst after righteousness. It is a new gospel, unless three hundred years stand for eighteen hundred; and if men are bent on seducing us from the ancient faith, let them provide a more specious error, a more alluring sophism, a more angelic tempter, than this. It is surely too bold an attempt to take from our hearts the power, the fulness, the mysterious presence of Christ's most holy death and resurrection, and to soothe us for our loss with the name of having it. . . .

[Lecture 2]

The Characteristics of the Gift of Righteousness

. . . It is the fashion of the day to sever these two from one another, which God has joined, the seal and the impression, justification and renewal. You hear men speak of glorying in the Cross of Christ, who are utter strangers to the notion of the Cross as actually applied to them in water and blood, in holiness and mortification. They think the Cross can be theirs *without* being applied,—without its coming near them,—while they keep at a distance from it, and only gaze at it. They think individuals are justified immediately by the great Atonement,—justified by Christ's death, and not, as St. Paul says, by means of His Ressurection,—justified by what they consider *looking* at His death. Because the Brazen Serpent in the wilderness healed by being looked at, they consider that Christ's Sacrifice saves by the mind's contemplating it. This is what they call casting themselves upon Christ,—coming before Him simply and without self-trust, and being saved by faith. Surely we ought so to *come* to Christ; surely we must believe; surely we must look; but the question is, in what form and manner He *gives* Himself to us: and it will be found that, when He enters into us, glorious as He is Himself, pain and self-denial are His attendants. Gazing on the Brazen Serpent did not heal; but God's invisible communication of the gift of health to those who gazed. So also justification is wholly the work of God; it comes from God to us; it is a power exerted on our souls by Him, as the healing of the Israelites was a power exerted on their bodies. The gift must be brought *near* to us; it is not like the Brazen Serpent, a mere external, material, local sign; it is a spiritual gift, and, as being such, admits of being applied to us individually. Christ's Cross does not justify by being looked at, but by being applied; not by as merely beheld by faith, but by being actually set up within us, and that not by our act, but by God's invisible grace. Men sit, and gaze, and speak of the great Atonement, and think this is appropriating it; not more truly than kneeling to the material cross itself is appropriating it. Men say that faith is an apprehending and applying; faith cannot really apply the Atonement; man cannot make the Saviour of the world his own; the Cross must be brought home to us, not in word, but in power, and this is the work of the Spirit. This is justification; but when imparted to the soul, it draws blood, it heals, it purifies, it glorifies.

[Lecture 7]

On Preaching the Gospel

... it may fairly be questioned whether religion does not necessarily imply the belief in such sensible tokens of God's favour, as the Sacraments are accounted by the Church. Religion is of a personal nature, and implies the acknowledgment of a particular Providence, of a God speaking, not merely to the world at large, but to this person or that, to me and not to another. The Sacred Volume is a common possession, and speaks to one man as much and as little as to his neighbour. Our nature requires something special; and if we refuse what has been actually given, we shall be sure to adopt what has not been given. We shall set up calves at Dan and Bethel, if we give up the true Temple and the Apostolic Ministry. This we see fulfilled before our eyes in many ways; those who will not receive Baptism as the token of God's election, have recourse to certain supposed experiences of it in their hearts. This is the idolatry of a refined age, in which the superstitions of barbarous times displease, in consequence of their grossness. Men congratulate themselves on their emancipation from forms and their enlightened worship, when they are but in the straight course to a worse captivity, and are exchanging dependence on the creature for dependence on self.

And thus we are led to the consideration of the opposite side of the question before us, that is, whether at this day it is not rather the accusing party itself than the Church that is accused, to which the charge of Judaism properly attaches. At first sight a suggestion of this kind will look like a refinement, or as only a sharp retort urged in controversy, and not to be seriously dwelt on. But I wish it dwelt on most seriously, and if rejected, rejected after being dwelt on. I observe, then, that what the Jews felt concerning their Law, is exactly what many upholders of the tenet of "faith only," feel concerning what they consider faith; that they substitute faith for Christ; that they so regard it, that instead of being the way to Him, it is in the way; that they make it a something to rest in; nay, that they alter the meaning of the word, as the Jews altered the meaning of the word Law; in short, that, under the pretence of light and liberty; they have brought into the Gospel the narrow, minute, technical, nay, I will say carnal and hollow system of the Pharisees. Let me explain what I mean.

I would say this then:—that a system of doctrine has risen up during the last three centuries, in which faith or spiritual-mindedness is contemplated and rested on as the end of religion instead of Christ. I do not mean to say that Christ is not mentioned as the Author of all good, but that stress is laid rather on the believing than on the Object of belief, on the comfort and persuasiveness of the doctrine rather than on the doctrine itself. And in this way religion is made to consist in contemplating ourselves instead of Christ; not simply in looking to Christ, but in ascertaining that we look to Christ, not in His Divinity and Atonement, but in our conversion and our faith in those truths. . . .

. . . The true preaching of the Gospel is to preach Christ. But the fashion of the day has been, instead of this, to preach conversion; to attempt to convert by insisting on conversion; to exhort men to undergo a change; to tell them to be sure they look at Christ, instead of simply holding up Christ to them; to tell them to have faith, rather than to supply its Object; to lead them to stir up and work up their minds, instead of impressing on them the thought of Him who can savingly work in them; to bid them take care that their faith is justifying, not dead, formal, self-righteous, and merely moral, whereas the image of Christ fully delineated of itself destroys deadness, formality, and self-righteousness; to rely on words, vehemence, eloquence, and the like, rather than to aim at conveying the one great evangelical idea whether in words or not. And thus faith and (what is called) spiritual-mindedness are dwelt on as *ends*, and obstruct the view of Christ, just as the Law was perverted by the Jews. . . .

. . . Poor miserable captives, to whom such doctrine is preached as the Gospel! What! is *this* the liberty wherewith Christ has made us free, and wherein we stand, the home of our own thoughts, the prison of our own sensations, the province of self, a monotonous confession of what we are by nature, not what Christ is in us, and a resting at best not on His love towards us, but in our faith towards Him! This is nothing but a specious idolatry; a man thus minded does not simply think of God when he prays to Him, but is observing whether he feels properly or not; does not believe and obey, but considers it enough to be conscious that he is what he calls warm and spiritual; does not contemplate the grace of the Blessed Eucharist, the Body and Blood of His Saviour Christ, except—O shameful and fearful error!—except as a quality of his own mind. . . .

The doctrine of justifying faith is a summary of the whole process of salvation from first to last; a sort of philosophical analysis of the Gospel, a contemplation of it as a whole, rather than a practical direction. If it must be taken as a practical direction, and in a certain sense it may, then we must word it, not, "justification through faith," but, "justification by Christ." Thus, interpreted, the rule it gives is, "*go* to Christ;" but taken in the letter, it seems to say merely, "Get *faith*; become spiritual; see that you are not mere moralists, mere formalists, see that you feel. If you do not feel, Christ will profit you nothing: you must have a spiritual taste; you must see yourself to be a sinner; you must accept, apprehend, appropriate the gift; you must understand and acknowledge that Christ is the 'pearl of great price;' you must be conscious of a change wrought in you, for the most part going through the successive stages of darkness, trouble, error, light, and comfort." Thus the poor and sorrowful soul, instead of being led at once to the source of all good, is taught to make much of the conflict of truth and falsehood within itself as the pledge of God's love, and to picture to itself faith, as a sort of passive quality which sits amid the ruins of human nature, and keeps up what may be called a silent protest, or indulges a pensive meditation over its misery. And, indeed, faith thus regarded cannot do more; for while it acts, not to lead the soul to Christ, but to detain it from Him, how can the soul but remain a prisoner, in that legal or natural state described by the Apostle in the seventh of Romans?—a passage of Scripture which the upholders of this doctrine confess, nay boast that they feel to be peculiarly their own. Such is their first error, and a second obviously follows. True faith is what may be called colourless, like air or water; it is but the medium through which the soul sees Christ; and the soul as little really rests upon it and contemplates it, as the eye can see the air. When, then, men are bent on holding it (as it were) in their hands, curiously inspecting, analyzing, and so aiming at it, they are obliged to colour and thicken it, that it may be seen and touched. That is, they substitute for it something or other, a feeling, notion, sentiment, conviction, or act of reason, which they may hang over, and doat upon. They rather aim at experiences (as they are called) within them, than at Him that is without them. They are led to enlarge upon the signs of conversion, the variations of their feelings, their aspirations and longings, and to tell all this to others;—to tell others how they fear, and hope, and sin, and rejoice, and renounce themselves, and rest in Christ only; how conscious they are that their best deeds are but "filthy rags," and all is of grace, till in fact they have little time left them to guard against what they are condemning, and to exercise what they think they are so full of. Now men in a battle are brief-spoken; they realize their situation and are intent upon it. And men who are acted upon by news good or bad, or sights beautiful or fearful, admire, rejoice, weep, or are pained, but are moved spontaneously, not with a direct consciousness of their emotion. Men of elevated minds are not their own historians and panegyrists. So it is with faith and other Christian graces. Bystanders see our minds; but our minds, if healthy, see but the objects which possess them. As God's grace elicits our faith, so His holiness stirs our fear, and His glory kindles our love. Others may say of us "here is faith," and "there is conscientiousness," and "there is love;" but we can only say, "this is God's grace," and "that is His holiness," and "that is His glory."

And this being the difference between true faith and self-contemplation, no wonder that where the thought of self obscures the thought of God, prayer and praise languish, and only preaching flourishes. Divine worship is simply contemplating our Maker, Redeemer, Sanctifier, and Judge; but discoursing, conversing, making speeches, arguing, reading, and writing about religion, tend to make us forget Him in ourselves. The Ancients worshipped; they went out of their own minds into the Infinite Temple which was around them. They saw Christ in the Gospels, in the Creed, in the Sacraments and other Rites; in the visible structure and ornaments of His House, in the Altar, and in the Cross; and, not content with giving the service of their eyes, they gave Him their voices, their bodies, and their time, gave up their rest by night and their leisure by day, all that could evidence the offering of their hearts to Him. Theirs was not a service once a week, or some one day, now and then, painfully, as if ambitiously and lavishly given to thanksgiving or humiliation; not some extraordinary address to the throne of grace, offered by one for many, when friends met, with much point and impressiveness, and as much like an exhortation, and as little like a prayer, as might be; but every day and every portion of the day was begun and sanctified with devotion. Consider those Seven Services of the Holy Church Catholic in her best ages, which, without encroaching upon her children's duties towards this world, secured them in their duties to the world unseen. Unwavering, unflagging, not urged by fits and starts, not heralding forth their feelings, but resolutely, simply, perseveringly, day after day, Sunday and week-day, fast-day and festival, week by week, season by season, year by year, in youth and in age, through a life, thirty years, forty years, fifty years, in prelude of the everlasting chant before the Throne,—so they went on, "continuing *instant* in prayer," after the pattern of Psalmists and Apostles, in the day with David, in the night with Paul and Silas, winter and summer, in heat and in cold, in peace and in danger, in a prison or in a cathedral, in the dark, in the day-break, at sun-rising, in the forenoon, at noon, in the afternoon, at eventide, and on going to rest, still they had Christ before them; His thought in their mind, His emblems in their eye, His name in their mouth, His service in their posture, magnifying Him, and calling on all that lives to magnify Him, joining with Angels in heaven and Saints in Paradise to bless and praise Him for ever and ever. O great and noble system, not of the Jews who rested in their rights and privileges, not of those Christians who are taken up with their own feelings, and who describe what they should exhibit, but of the true Saints of God, the undefiled and virgin souls who follow the Lamb whithersoever He goeth! Such is the difference between those whom Christ praises and those whom He condemns or warns. The Pharisee recounted the signs of God's mercy upon and in Him; the Publican simply looked to God. The young Ruler boasted of his correct life, but the penitent woman anointed Jesus' feet and kissed them. Nay, holy Martha herself spoke of her "much service;" while Mary waited on Him for the "one thing needful." The one thought of themselves; the others thought of Christ. To look to Christ is to be justified by faith; to think of being justified by faith is to look from Christ and to fall from grace. He who worships Christ and works for Him, is acting out that doctrine which another does but enunciate; his worship and his works are acts of faith, and avail to his salvation, because he does not do them *as* availing.

But I must end a train of thought, which, left to itself would run on into a whole work. And in doing so I make one remark, which is perhaps the great moral of the history of Protestantism. Luther found in the Church great moral corruptions countenanced by its highest authorities; he felt them; but instead of meeting them with divine weapons, he used one of his own. He adopted a doctrine original, specious, fascinating, persuasive, powerful against Rome, and wonderfully adapted, as if prophetically, to the genius of the times which were to follow. He found Christians in bondage to their works and observances; he released them by his doctrine of faith; and he left them in bondage to their feelings. He weaned them from seeking assurance of salvation in standing ordinances, at the cost of teaching them that a personal consciousness of it was promised to every one who believed. For outward signs of grace he substituted inward; for reverence towards the Church contemplation of self. And thus, whereas he himself held the proper efficacy of the Sacraments, he has led others to disbelieve it; whereas he preached against reliance on self, he introduced it in a more subtle shape; whereas he professed to make the written word all in all, he sacrificed it in its length and breadth to the doctrine which he had wrested from a few texts.

This is what comes of fighting God's battles in our own way, of extending truths beyond their measure, of anxiety after a teaching more compact, clear, and spiritual, than the Creed of the Apostles. Thus the Pharisees were more careful of their Law than God who gave it; thus Saul saved the cattle he was bid destroy, "to sacrifice to the Lord;" thus Judas was concerned at the waste of the ointment, which might have been given to the poor. In these cases bad men professed to be more zealous for God's honour, more devotional, or more charitable, than the servants of God; and in a parallel way Protestants would be more spiritual. Let us be sure things are going wrong with us, when we see doctrines more clearly, and carry them out more boldly, than they are taught us in Revelation.

[Lecture 13]

OXFORD UNIVERSITY SERMONS

The Theory of Developments in Religious Doctrine
(Preached on the Purification, 1843)

But Mary kept all these things, and pondered them in her heart

Luke 2: 19

Little is told us in Scripture concerning the Blessed Virgin, but there is one grace of which the Evangelists make her the pattern, in a few simple sentences,—of Faith. Zacharias questioned the Angel's message, but "Mary said, Behold the handmaid of the Lord; be it unto me according to thy word." Accordingly Elisabeth, speaking with an apparent allusion to the contrast thus exhibited between her own highly-favoured husband, righteous Zacharias, and the still more highly-favoured Mary, said, on receiving her salutation, "Blessed art thou among women, and blessed is the fruit of thy womb; Blessed is she that believed, for there shall be a performance of those things which were told her from the Lord."

2. But Mary's faith did not end in a mere acquiescence in Divine providences and revelations: as the text informs us, she "pondered" them. When the shepherds came, and told of the vision of Angels which they had seen at the time of the Nativity, and how one of them announced that the Infant in her arms was "the Saviour, which is Christ the Lord," while others did but wonder, "Mary kept all these things, and pondered them in her heart." Again, when her Son and Saviour had come to the age of twelve years, and had left her for awhile for His Father's service, and had been found, to her surprise, in the Temple, amid the doctors, both hearing them and asking them questions, and had, on her addressing Him, vouchsafed to justify His conduct, we are told, "His mother kept all these sayings in her heart." And accordingly, at the marriage-feast in Cana, her faith anticipated His first miracle, and she said to the servants, "Whatsoever He saith unto you, do it."

3. Thus St. Mary is our pattern of Faith, both in the reception and in the study of Divine Truth. She does not think it enough to accept, she dwells upon it; not enough to possess, she uses it; not enough to assent, she developes it; not enough to submit the Reason, she reasons upon it; not indeed reasoning first, and believing afterwards, with Zacharias, yet first believing without reasoning, next from love and reverence, reasoning after believing. And thus she symbolizes to us, not only the faith of the unlearned, but of the doctors of the Church also, who have to investigate, and weigh, and define, as well as to profess the Gospel; to draw the line between truth and heresy; to anticipate or remedy the various aberrations of wrong reason; to combat pride and recklessness with their own arms; and thus to triumph over the sophist and the innovator.

4. If, then, on a Day dedicated to such high contemplations as the Feast which we are now celebrating, it is allowable to occupy the thoughts with a subject not of a devotional or practical nature, it will be some relief of the omission to select one in which St. Mary at least will be our example,—the use of Reason in investigating the doctrines of Faith; a subject, indeed, far fitter for a volume than for the most extended notice which can here be given to it; but one which cannot be passed over altogether in silence, in any attempt at determining the relation of Faith to Reason.

5. The overthrow of the wisdom of the world was one of the earliest, as well as the noblest of the triumphs of the Church; after the pattern of her Divine Master, who took His place among the doctors before He preached His new Kingdom, or opposed Himself to the world's power. St. Paul, the learned Pharisee, was the first fruits of that gifted company,

in whom the pride of science is seen prostrated before the foolishness of preaching. From his day to this the Cross has enlisted under its banner all those great endowments of mind, which in former times had been expended on vanities, or dissipated in doubt and speculation. Nor was it long before the schools of heathenism took the alarm, and manifested an unavailing jealousy of the new doctrine, which was robbing them of their most hopeful disciples. They had hitherto taken for granted that the natural home of the Intellect was the Garden or the Porch; and it reversed their very first principles to be called on to confess, what yet they could not deny, that a Superstition, as they considered it, was attracting to itself all the energy, the keenness, the originality, and the eloquence of the age. But these aggressions upon heathenism were only the beginning of the Church's conquests; in the course of time the whole mind of the world, as I may say, was absorbed into the philosophy of the Cross, as the element in which it lived, and the form upon which it was moulded. And how many centuries did this endure, and what vast ruins still remain of its dominion! In the capitals of Christendom the high cathedral and the perpetual choir still witness to the victory of Faith over the world's power. To see its triumph over the world's wisdom, we must enter those solemn cemeteries in which are stored the relics and the monuments of ancient Faith—our libraries. Look along their shelves, and every name you read there is, in one sense or other, a trophy set up in record of the victories of Faith. How many long lives, what high aims, what single-minded devotion, what intense contemplation, what fervent prayer, what deep erudition, what untiring diligence, what toilsome conflicts has it taken to establish its supremacy! This has been the object which has given meaning to the life of Saints, and which is the subject-matter of their history. For this they have given up the comforts of earth and the charities of home, and surrendered themselves to an austere rule, nay, even to confessorship and persecution, if so be they could make some small offering, or do some casual service, or provide some additional safeguard towards the great work which was in progress. This has been the origin of controversies, long and various, yes, and the occasion of much infirmity, the test of much hidden perverseness, and the subject of much bitterness and tumult. The world has been moved in consequence of it, populations excited, leagues and alliances formed, kingdoms lost and won: and even zeal, when excessive, evinced a sense of its preciousness; nay, even rebellions in some sort did homage to it, as insurgents imply the actual sovereignty of the power which they are assailing. Meanwhile the work went on, and at length a large fabric of divinity was reared, irregular in its structure, and diverse in its style, as beseemed the slow growth of centuries; nay, anomalous in its details, from the peculiarities of individuals, or the interference of strangers, but still, on the whole, the development of an idea, and like itself, and unlike any thing else, its most widely-separated parts having relations with each other, and betokening a common origin.

6. Let us quit this survey of the general system, and descend to the history of the formation of any Catholic dogma. What a remarkable sight it is, as almost all unprejudiced persons will admit, to trace the course of the controversy, from its first disorders to its exact and determinate issue. Full of deep interest, to see how the great idea takes hold of a thousand minds by its living force, and will not be ruled or stinted, but is "like a burning fire," as the Prophet speaks, "shut up" within them, till they are "weary of forebearing, and cannot stay," and grows in them, and at length is born through them, perhaps in a long course of years, and even successive generations; so that the doctrine may rather be said to use the minds of Christians, than to be used by them. Wonderful it is to see with what effort, hesitation, suspense, interruption,—with how many swayings to the right and to the left— with how many reverses, yet with what certainty of advance, with what precision in its march, and with what ultimate completeness, it has been evolved; till the whole truth "self-balanced on its centre hung," part answering to part, one, absolute, integral, indissoluble, while the world lasts! Wonderful, to see how heresy has but thrown that idea into fresh forms, and drawn out from it farther developments, with an exuberance which exceeded all questioning, and a harmony which baffled all criticism, like Him, its Divine Author, who, when put on trial by the Evil One, was but fortified by the assault, and is ever justified in His sayings, and overcomes when He is judged.

7. And this world of thought is the expansion of a few words, uttered, as if casually, by the fishermen of Galilee. Here is another topic which belongs more especially to that part of the subject to which I propose to confine myself. Reason has not only submitted, it has ministered to Faith; it has illustrated it documents; it has raised illiterate peasants into philosophers and divines; it has elicited a meaning from their words which their immediate hearers little suspected. Stranger surely is it that St. John should be a theologian, than that St. Peter should be a prince. This is a phenomenon proper to the Gospel, and a note of divinity. Its half sentences, its overflowings of language, admit of development;[1] they have a life in them which shows itself in progress; a truth, which has the token of consistency; a reality, which is fruitful in resources; a depth, which extends into mystery: for they are representations of what is actual, and has a definite location and necessary bearings and a meaning in the great system of things, and a harmony in what it is, and a

---

[1] Vide Butler's Analogy, part ii. ch. iii.

compatibility in what it involves. What form of Paganism can furnish a parallel? What philosopher has left his words to posterity as a talent which could be put to usury, as a mine which could be wrought? Here, too, is the badge of heresy; its dogmas are unfruitful; it has no theology; so far forth as it is heresy, it has none. Deduct its remnant of Catholic theology, and what remains? Polemics, explanations, protests. It turns to Biblical Criticism, or to the Evidences of Religion, for want of a province. Its *formulæ* end in themselves, without development, because they are words; they are barren, because they are dead. If they had life, they would increase and multiply; or, if they do live and bear fruit, it is but as "sin, when it is finished, bringeth forth death." It developes into dissolution; but it creates nothing, it tends to no system, its resultant dogma is but the denial of all dogmas, any theology, under the Gospel. No wonder it denies what it cannot attain.

8. Heresy denies to the Church what is wanting in itself. Here, then, we are brought to the subject to which I wish to give attention. It need not surely formally be proved that this disparagement of doctrinal statements, and in particular of those relating to the Holy Trinity and Incarnation, is especially prevalent in our times. There is a suspicion widely abroad,—felt, too, perhaps, by many who are unwilling to confess it,—that the development of ideas and formation of dogmas is a mere abuse of Reason, which, when it attempted such sacred subjects, went beyond its powers, and could do nothing more than multiply words without meaning, and deductions which come to nothing. The conclusion follows, that such an attempt does but lead to mischievous controversy, from that discordance of doctrinal opinions, which is its immediate consequence; that there is, in truth, no necessary or proper connexion between inward religious belief and scientific expositions; and that charity, as well as good sense, is best consulted by reducing creeds to the number of private opinions, which, if individuals will hold for themselves, at least they have no right to impose upon others.

9. It is my purpose, then, in what follows, to investigate the connexion between faith and Dogmatic Confession, as far as relates to the sacred doctrines which were just now mentioned, and to show the office of the Reason in reference to it; and, in doing so, I shall make as little allusion as may be to erroneous views on the subject, which have been mentioned only for the sake of perspicuity; following rather the course which the discussion may take, and pursuing those issues on which it naturally opens. Nor am I here in any way concerned with the question, who is the legitimate framer and judge of these dogmatic inferences under the Gospel, or if there be any. Whether the Church is infallible, or the individual, or the first ages, or none of these, is not the point here, but the theory of developments itself.

10. Theological dogmas are propositions expressive of the judgments which the mind forms, or the impressions which it receives, of Revealed Truth. Revelation sets before it certain supernatural facts and actions, beings and principles; these make a certain impression or image upon it; and this impression spontaneously, or even necessarily, becomes the subject of reflection on the part of the mind itself, which proceeds to investigate it, and to draw it forth in successive and distinct sentences. Thus the Catholic doctrine of Original Sin, or of Sin after Baptism, or of the Eucharist, or of Justification, is but the expression of the inward belief of Catholics on these several points, formed upon an analysis of that belief. Such, too, are the high doctrines with which I am especially concerned.

11. Now, here I observe, first of all, that, naturally as the inward idea of divine truth, such as has been described, passes into explicit form by the activity of our reflective powers, still such an actual delineation is not essential to its genuineness and perfection. A peasant may have such a true impression, yet be unable to give any intelligible account of it, as will easily be understood. But what is remarkable at first sight is this, that there is good reason for saying that the impression made upon the mind need not even be recognized by the parties possessing it. It is no proof that persons are not possessed, because they are not conscious, of an idea. Nothing is of more frequent occurrence, whether in things sensible or intellectual, than the existence of such unperceived impressions. What do we mean when we say, that certain persons do not know themselves, but that they are ruled by views, feelings, prejudices, objects which they do not recognize? How common is it to be exhilarated or depressed, we do not recollect why, though we are aware that something has been told us, or has happened, good or bad, which accounts for our feeling, could we recall it! What is memory itself, but a vast magazine of such dormant, but present and excitable ideas? Or consider, when persons would trace the history of their own opinions in past years, how baffled they are in the attempt to fix the date of this or that conviction, their system of thought having been all the while in continual, gradual, tranquil expansion; so that it were as easy to follow the growth of the fruit of the earth, "first the blade, then the ear, after that the full corn in the ear," as to chronicle changes, which involved no abrupt revolution, or reaction, or fickleness of mind, but have been the birth of an idea, the development, in explicit form, of what was already latent within it. Or, again, critical disquisitions are often written about the idea which this or that poet might have in his mind in certain of his compositions and characters; and we call such analysis the philosophy of poetry, not implying thereby of necessity that the author wrote upon a theory in his actual delineation, or knew what he was doing; but that, in

matter of fact, he was possessed, ruled, guided by an unconscious idea. Moreover, it is a question whether that strange and painful feeling of unreality, which religious men experience from time to time, when nothing seems true, or good, or right, or profitable, when Faith seems a name, and duty a mockery, and all endeavours to do right, absurd and hopeless, and all things forlorn and dreary, as if religion were wiped out from the world, may not be the direct effect of the temporary obscuration of some master vision, which unconsciously supplies the mind with spiritual life and peace.

12. Or, to take another class of instances which are to the point so far as this, that at least they are real impressions, even though they be not influential. How common is what is called vacant vision, when objects meet the eye, without any effort of the judgment to measure or locate them; and that absence of mind, which recollects minutes afterwards the occurrence of some sound, the striking of the hour, or the question of a companion, which passed unheeded at the time it took place! How, again, happens it in dreams, that we suddenly pass from one state of feeling, or one assemblage of circumstances to another, without any surprise at the incongruity, except that, while we are impressed first in this way, then in that, we take no active cognizance of the impression? And this, perhaps, is the life of inferior animals, a sort of continuous dream, impressions without reflections; such, too, seems to be the first life of infants; nay, in heaven itself, such may be the high existence of some exalted orders of blessed spirits, as the Seraphim, who are said to be, not Knowledge, but all Love.

13. Now, it is important to insist on this circumstance, because it suggests the reality and permanence of inward knowledge, as distinct from explicit confession. The absence, or partial absence, or incompleteness of dogmatic statements is no proof of the absence of impressions or implicit judgments, in the mind of the Church. Even centuries might pass without the formal expression of a truth, which had been all along the secret life of millions of faithful souls. Thus, not till the thirteenth century was there any direct and distinct avowal, on the part of the Church, of the numerical Unity of the Divine Nature, which the language of some of the principal Greek fathers, *primâ facie*, though not really, denies. Again, the doctrine of the Double Procession was no Catholic dogma in the first ages, though it was more or less clearly stated by individual Fathers; yet, if it is now to be received, as surely it must be, as part of the Creed, it was really held every where from the beginning, and therefore, in a measure, held as a mere religious impression, and perhaps an unconscious one.

14. But, further, if the ideas may be latent in the Christian mind, by which it is animated and formed, it is less wonderful that they should be

difficult to elicit and define; and of this difficulty we have abundant proof in the History whether of the Church, or of individuals. Surely it is not at all wonderful, that, when individuals attempt to analyze their own belief, they should find the task arduous in the extreme, if not altogether beyond them; or, again, a work of many years; or, again, that they should shrink from the true developments, if offered to them, as foreign to their thoughts. This may be illustrated in a variety of ways.

15. It will often happen, perhaps from the nature of things, that it is impossible to master and express an idea in a short space of time. As to individuals, sometimes they find they cannot do so at all; at length, perhaps, they recognize, in some writer they meet, with the very account of their own thoughts, which they desiderate; and then they say, that "here is what they have felt all along, and wanted to say, but could not," or "what they have ever maintained, only better expressed." Again, how many men are burdened with an idea, which haunts them through a great part of their lives, and of which only at length, with much trouble, do they dispossess themselves? I suppose most of us have felt at times the irritation, and that for a long period, of thoughts and views which we felt, and felt to be true, only dimly showing themselves, or flitting before us; which at length we understood must not be forced, but must have their way, and would, if it were so ordered, come to light in their own time. The life of some men, and those not the least eminent among divines and philosophers, has centred in the development of one idea; nay, perhaps has been too short for the process. Again, how frequently it happens, that, on first hearing a doctrine propounded, a man hesitates, first acknowledges, then disowns it; then says that he has always held it, but finds fault with the mode in which it is presented to him, accusing it of paradox or over-refinement; that is, he cannot at the moment analyze his own opinions, and does not know whether he holds the doctrine or not, from the difficulty of mastering his thoughts.

16. Another characteristic, as I have said, of dogmatic statements, is the difficulty of recognizing them, even when attained, as the true representation of our meaning. This happens for many reasons; sometimes, from the faint hold we have of the impression itself, whether its nature be good or bad, so that we shrink from principles in substance, which we acknowledge in influence. Many a man, for instance, is acting on utilitarian principles, who is shocked at them in set treatises, and disowns them. Again, in sacred subjects, the very circumstance that a dogma professes to be a direct contemplation, and, if so be, a definition of what is infinite and eternal, is painful to serious minds. Moreover, from the hypothesis, it is the representation of an idea in a medium not native to it, not as originally conceived, but, as it were, in projection; no wonder, then, that, though there be an intimate correspondence, part

by part, between the impression and the dogma, yet there should be an harshness in the outline of the latter; as, for instance, a want of harmonious proportion; and yet this is unavoidable, from the infirmities of our intellectual powers.

17. Again, another similar peculiarity in developments in general, is the great remoteness of the separate results of a common idea, or rather at first sight the absence of any connexion. Thus it often happens that party spirit is imputed to persons, merely because they agree with one another in certain points of opinion and conduct, which are thought too minute, distant, and various, in the large field of religious doctrine and discipline, to proceed from any but an external influence and a positive rule; whereas an insight into the wonderfully expansive power and penetrating virtue of theological or philosophical ideas would have shown, that what is apparently arbitrary in rival or in kindred schools of thought, is after all rigidly determined by the original hypothesis. The remark has been made, for instance, that rarely have persons maintained the sleep of the soul before the Resurrection, without falling into more grievous errors; again, those who deny the Lutheran doctrine of Justification, commonly have tendencies towards a ceremonial religion; again, it is a serious fact that Protestantism has at various times unexpectedly developed into an allowance or vindication of polygamy; and heretics in general, however opposed in tenets, are found to have an inexplicable sympathy for each other, and never wake up from their ordinary torpor, but to exchange courtesies and meditate coalitions. One other remark is in point here, and relates to the length to which statements run, though, before we attempted them, we fancied our idea could be expressed in one or two sentences. Explanations grow under our hands, in spite of our effort at compression. Such, too, is the contrast between conversation and epistolary correspondence. We speak our meaning with little trouble; our voice, manner, and half words completing it for us; but in writing, when details must be drawn out, and misapprehensions anticipated, we seem never to be rid of the responsibility of our task. This being the case, it is surprising that the Creeds are so short, not surprising that they need a comment.

18. The difficulty, then, and hazard of developing doctrines implicitly received, must be fully allowed; and this is often made a ground for inferring that they have no proper developments at all; that there is no natural connexion between certain dogmas and certain impressions; and that theological science is a matter of time, and place, and accident, though inward belief is ever and every where one and the same. But surely the instinct of every Christian revolts from such a position; for the very first impulse of his faith is to try to express itself about the "great sight" which is vouchsafed to it; and this seems to argue

that a science there is, whether the mind is equal to its discovery or no. And, indeed, what science is open to every chance inquirer? which is not recondite in its principles? which requires not special gifts of mind for its just formation? All subject-matters admit of true theories and false, and the false are no prejudice to the true. Why should this class of ideas be different from all other? Principles of philosophy, physics, ethics, politics, taste, admit both of implicit reception and explicit statement; why should not the ideas, which are the secret life of the Christian, be recognized also as fixed and definite in themselves, and as capable of scientific analysis? Why should not there be real connexion between science and its subject-matter in religion, which exists in other departments of thought? No one would deny that the philosophy of Zeno or Pythagoras was the exponent of a certain mode of viewing things; or would affirm that Platonist and Epicurean acted on one and the same idea of nature, life, and duty, and meant the same thing, though they verbally differed, merely because a Plato or an Epicurus was needed to detect the abstruse elements of thought, out of which each philosophy was eventually constructed. A man surely may be a Peripatetic or an Academic in his feelings, views, aims, and acts, who never heard the names. Granting, then, extreme cases, when individuals who would analyze their views of religion are thrown entirely upon their own reason, and find that reason unequal to the task, this will be no argument against a general, natural, and ordinary correspondence between the dogma and the inward idea. Surely, if Almighty God is ever one and the same, and is revealed to us as one and the same, the true inward impression of Him, made on the recipient of the revelation, must be one and the same; and, since human nature proceeds upon fixed laws, the statement of that impression must be one and the same, so that we may as well say that there are two Gods as two Creeds. And considering the strong feelings and energetic acts and severe sufferings which age after age have been involved in the maintenance of the Catholic dogmas, it is surely a very shallow philosophy to account such maintenance a mere contest about words, and a very abject philosophy to attribute it to mere party spirit, or to personal rivalry, or to ambition, or to covetousness.

19. Reasonable, however, as is this view of doctrinal developments in general, it cannot be denied that those which relate to the Objects of Faith, of which I am particularly speaking, have a character of their own, and must be considered separately. Let us, then, consider how the case stands, as regards the sacred doctrines of the Trinity and the Incarnation.

20. The Apostle said to the Athenians, "Whom ye ignorantly worship, Him declare I unto you;" and the mind which is habituated to

the thought of God, of Christ, of the Holy Spirit, naturally turns, as I have said, with a devout curiosity to the contemplation of the Object of its adoration, and begins to form statements concerning Him before it knows whither, or how far, it will be carried. One proposition necessarily leads to another, and a second to a third; then some limitation is required; and the combination of these opposites occasions some fresh evolutions from the original idea, which indeed can never be said to be entirely exhausted. This process is its development, and results in a series, or rather body of dogmatic statements, till what was at first an impression on the Imagination has become a system or creed in the Reason.

21. Now such impressions are obviously individual and complete above other theorlogical ideas, *because* they are the impressions of Objects. Ideas and their developments are commonly not identical, the development being but the carrying out of the idea into its consequences. Thus the doctrine of Penance may be called a development of the doctrine of Baptism, yet still is a distinct doctrine; whereas the developments in the doctrines of the Holy Trinity and the Incarnation are mere portions of the original impression, and modes of representing it. As God is one, so the impression which He gives us of Himself is one; it is not a thing of parts; it is not a system; nor is it any thing imperfect, and needing a counterpart. It is the vision of an object. When we pray, we pray, not to an assemblage of notions, or to a creed, but to One Individual Being; and when we speak of Him we speak of a Person, not of a Law or a Manifestation. This being the case, all our attempts to delineate our impression of Him go to bring out one idea, not two or three or four; not a philosophy, but an individual idea in its separate aspects.

22. This may be fitly compared to the impressions made on us through the senses. Material objects are whole, and individual; and the impressions which they make on the mind, by means of the senses, are of a corresponding nature, complex and manifold in their relations and bearings, but considered in themselves integral and one. And in like manner the ideas which we are granted of Divine Objects under the Gospel, from the nature of the case and because they are ideas, answer to the Originals so far as this, that they are whole, indivisible, substantial, and may be called real, as being images of what is real. Objects which are conveyed to us through the senses, stand out in our minds, as I may say, with dimensions and aspects and influences various, and all of these consistent with one another, and many of them beyond our memory or even knowledge, while we contemplate the objects themselves; thus forcing on us a persuasion of their reality from the spontaneous congruity and coincidence of these accompaniments, as if they could not

be creations of our minds, but were the images of external and independent beings. This of course will take place in the case of the sacred ideas which are the objects of our faith. Religious men, according to their measure, have an idea or vision of the Blessed Trinity in Unity, of the Son Incarnate and of His Presence, not as a number of qualities, attributes, and actions, not as the subject of a number of propositions, but as one, and individual, and independent of words, as an impression conveyed through the senses.

23. Particular propositions, then, which are used to express portions of the great idea vouchsafed to us, can never really be confused with the idea itself, which all such propositions taken together can but reach, and cannot exceed. As definitions are not intended to go beyond their subject, but to be adequate to it, so the dogmatic statements of the Divine Nature used in our confessions, however multiplied, cannot say more than is implied in the original idea, considered in its completeness, without the risk of heresy. Creeds and dogmas live in the one idea which they are designed to express, and which alone is substantive; and are necessary only because the human mind cannot reflect upon it, except piecemeal, cannot use it in its oneness and entireness, nor without resolving it into a series of aspects and relations. And in matter of fact these expressions are never equivalent to it; we are able, indeed, to define the creations of our own minds, for they are what we make them and nothing else; but it were as easy to create what is real as to define it; and thus the Catholic dogmas are, after all, but symbols of a Divine fact, which, far from being compassed by those very propositions, would not be exhausted, nor fathomed, by a thousand.

24. Now of such sacred ideas and their attendant expressions, I observe:—

(1.) First, that an impression of this intimate kind seems to be what Scripture means by "knowledge." "This is life eternal," says our Saviour, "that they might know Thee the only True God, and Jesus Christ whom Thou hast sent." In like manner St. Paul speaks of willingly losing all things, "for the excellency of the knowledge of Christ Jesus;" and St. Peter of "the knowledge of Him who hath called us to glory and virtue."[1] Knowledge is the possession of those living ideas of sacred things, from which alone change of heart or conduct can proceed. This awful vision is what Scripture seems to designate by the phrases "Christ in us," "Christ dwelling in us by faith," "Christ formed in us," and "Christ manifesting Himself unto us." And though it is faint and doubtful in some minds, and distinct in others, as some remote object in the twilight or in the day, this arises from the circumstances of the

---

[1] John 17:3; Phil. 3:8; 2 Pet. 1:3.

particular mind, and does not interfere with the perfection of the gift itself.

25. (2.) This leads me next, however, to observe, that these religious impressions differ from those of material objects, in the mode in which they are made. The senses are direct, immediate, and ordinary informants, and act spontaneously without any will or effort on our part; but no such faculties have been given us, as far as we know, for realizing the Objects of Faith. It is true that inspiration may be a gift of this kind to those who have been favoured with it; nor would it be safe to deny to the illuminating grace of Baptism a power, at least of putting the mind into a capacity for receiving impressions; but the former of these is not ordinary, and both are supernatural. The secondary and intelligible means by which we receive the impression of Divine Verities, are, for instance, the habitual and devout perusal of Scripture, which gradually acts upon the mind; again, the gradual influence of intercourse with those who are in themselves in possession of the sacred ideas; again, the study of Dogmatic Theology, which is our present subject; again, a continual round of devotion; or again, sometimes, in minds both fitly disposed and apprehensive, the almost instantaneous operation of a keen faith. This obvious distinction follows between sensible and religious ideas, that we put the latter into language in order to fix, teach, and transmit them, but not the former. No one defines a material object by way of conveying to us what we know so much better by the senses, but we form creeds as a chief mode of perpetuating the impression.

26. (3.) Further, I observe, that though the Christian mind reasons out a series of dogmatic statements, one from another, this it has ever done, and always must do, not from those statements taken in themselves, as logical propositions, but as being itself enlightened and (as if) inhabited by that sacred impression which is prior to them, which acts as a regulating principle, ever present, upon the reasoning, and without which no one has any warrant to reason at all. Such sentences as "the Word was God," or "the Only-begotten Son who is in the bosom of the Father," or "the Word was made flesh," or "The Holy Ghost which proceedeth from the Father," are not a mere letter which we may handle by the rules of art at our own will, but august tokens of most simple, ineffable, adorable facts, embraced, enshrined according to its measure in the believing mind. For though the development of an idea is a deduction of proposition from proposition, these propositions are ever formed in and round the idea itself (so to speak), and are in fact one and all only aspects of it. Moreover, this will account both for the mode of arguing from particular texts or single words of Scripture, practised by the early Fathers, and for their fearless decision in practising it; for the great Object of Faith on which they lived both enabled them to

appropriate to itself particular passages of Scripture, and became to them a safeguard against heretical deductions from them. Also, it will account for the charge of weak reasoning, commonly brought against those Fathers; for never do we seem so illogical to others as when we are arguing under the continual influence of impressions to which they are insensible.

27. (4.) Again, it must of course be remembered, as I have just implied, (though as being an historical matter it hardly concerns us here), that Revelation itself has provided in Scripture the main outlines and also large details of the dogmatic system. Inspiration has superseded the exercise of human Reason in great measure, and left it but the comparatively easy task of finishing the sacred work. The question, indeed, at first sight occurs, why such inspired statements are not enough without further developments; but in truth, when Reason has once been put on the investigation, it cannot stop till it has finished it; one dogma creates another, by the same right by which it was itself created; the Scripture statements are sanctions as well as informants in the inquiry; they begin and they do not exhaust.

28. (5.) Scripture, I say, begins a series of developments which it does not finish; that is to say, in other words, it is a mistake to look for every separate proposition of the Catholic doctrine in Scripture. This is plain from what has gone before. For instance, the Athanasian Creed professes to lay down the right faith, which we must hold on its most sacred subjects, in order to be saved. This must mean that there is one view concerning the Holy Trinity, or concerning the Incarnation, which is true, and distinct from all others; one definite, consistent, entire view, which cannot be mistaken, not contained in any certain number of propositions, but held as a view by the believing mind, and not held, but denied by Arians, Sabellians, Tritheists, Nestorians, Monophysites, Socinians, and other heretics. That idea is not enlarged, if propositions are added, nor impaired if they are withdrawn: if they are added, this is with a view of conveying that one integral view, not of amplifying it. That view does not depend on such propositions: it does not consist in them; they are but specimens and indications of it. And they may be multiplied without limit. They are necessary, but not needful to it, being but portions or aspects of that previous impression which has at length come under the cognizance of Reason and the terminology of science. The question, then, is not whether this or that proposition of the Catholic doctrine is *in terminis* in Scripture, unless we would be slaves to the letter, but whether that one view of the Mystery, of which all such are the exponents, be not there; a view which would be some other view, and not itself, if any one of such propositions, if any one of a number of similar propositions, were not true. Those propositions imply each

other, as being parts of one whole; so that to deny one is to deny all, and to invalidate one is to deface and destroy the view itself. One thing alone has to be impressed on us by Scripture, the Catholic idea, and in it they all are included. To object, then, to the number of propositions, upon which an anathema is placed, is altogether to mistake their use; for their multiplication is not intended to enforce many things, but to express one,—to form within us that one impression concerning Almighty God, as the ruling principle of our minds, and that, whether we can fully recognize our own possession of it or no. And surely it is no paradox to say that such ruling ideas may exert a most powerful influence, at least in their various aspects, on our moral character, and on the whole man: as no one would deny in the case of belief or disbelief of a Supreme Being.

29. (6.) And here we see the ordinary mistake of doctrinal innovators, viz. to go away with this or that proposition of the Creed, instead of embracing that one idea which all of them together are meant to convey; it being almost a definition of heresy, that it fastens on some one statement as if the whole truth, to the denial of all others, and as the basis of a new faith; erring rather in what it rejects, than in what it maintains: though, in truth, if the mind deliberately rejects any portion of the doctrine, this is a proof that it does not really hold even that very statement for the sake of which it rejects the others. Realizing is the very life of true developments; it is peculiar to the Church, and the justificaiton of her definitions.

30. Enough has now been said on the distinction, yet connexion, between the implicit knowledge and the explicit confession of the Divine Objects of Faith, as they are revealed to us under the Gospel. An objection, however, remains, which cannot be satisfactorily treated in a few words. And what is worse than prolixity, the discussion may bear with it some appearance of unnecessary or even wanton refinement; unless, indeed, it is thrown into the form of controversy, a worse evil. Let it suffice to say, that my wish is, not to discover difficulties in any subject, but to solve them.

31. It may be asked, then, whether the mistake of words and names for things is not incurred by orthodox as well as heretics, in dogmatizing at all about the "secret things which belong unto the Lord our God," inasmuch as the idea of a supernatural object must itself be supernatural, and since no such ideas are claimed by ordinary Christians, no knowledge of Divine Verities is possible to them. How should any thing of this world convey ideas which are beyond and above this world? How can teaching and intercourse, how can human words, how can earthly images, convey to the mind an idea of the Invisible? They cannot rise above themselves. They can suggest no idea but what is resolvable into ideas natural and earthly. The words "Person," "Substance," "Consub-

stantial," "Generation," "Procession," "Incarnation," "Taking of the manhood into God," and the like, have either a very abject and human meaning, or none at all. In other words, there is no such inward view of these doctrines, distinct from the dogmatic language used to express them, as was just now supposed. The metaphors by which they are signified are not mere symbols of ideas which exist independently of them, but their meaning is coincident and identical with the ideas. When indeed, we have knowledge of a thing from other sources, then the metaphors we may apply to it are but accidental appendages to that knowledge; whereas our ideas of Divine things are just coextensive with the figures by which we express them, neither more nor less, and without them are not; and when we draw inferences from those figures, we are not illustrating one existing idea, but drawing mere logical inferences. We speak, indeed, of material objects freely, because our senses reveal them to us apart from our words; but as to these ideas about heavenly things, we learn them from words, yet (it seems) we are to say what we, without words, conceive of them, as if words could convey what they do not contain. It follows that our anathemas, our controversies, our struggles, our sufferings, are merely about the poor ideas conveyed to us in certain figures of speech.

32. Some obvious remarks suggest themselves in answer to this representation. First, it is difficult to determine what divine grace may not do for us, if not in immediately implanting new ideas, yet in refining and elevating those which we gain through natural informants. If, as we all acknowledge, grace renews our moral feelings, yet through outward means, if it opens upon us new ideas about virtue and goodness and heroism and heavenly peace, it does not appear why, in a certain sense, it may not impart ideas concerning the nature of God. Again, the various terms and figures which are used in the doctrine of the Holy Trinity or of the Incarnation, surely may by their combination create ideas which will be altogether new, though they are still of an earthly character. And further, when it is said that such figures convey no knowledge of the Divine Nature itself, beyond these figures, whatever they are, it should be considered whether our senses can be proved to suggest any real idea of matter. All that we know, strictly speaking, is the existence of the impressions our senses make on us; and yet we scruple not to speak as if they conveyed to us the knowledge of material substances. Let, then, the Catholic dogmas, as such, be freely admitted to convey no true idea of Almighty God, but only an earthly one, gained from earthly figures, provided it be allowed, on the other hand, that the senses do not convey to us any true idea of matter, but only an idea commensurate with sensible impressions.

33. Nor is there any reason why this should not be fully granted. Still

there may be a certain correspondence between the idea, though earthly, and its heavenly archetype, such, that that idea belongs to the archetype, in a sense in which no other earthly idea belongs to it, as being the nearest approach to it which our present state allows. Indeed Scripture itself intimates the earthly nature of our present ideas of Sacred Objects, when it speaks of our now "seeing in a glass *darkly*, ἐν αἰνίγματι, but then face to face;" and it has ever been the doctrine of divines that the Beatific Vision, or true sight of Almighty God, is reserved for the world to come. Meanwhile we are allowed such an approximation to the truth as earthly images and figures may supply to us.

34. It must not be supposed that this is the only case in which we are obliged to receive information needful to us, through the medium of our existing ideas, and consequently with but a vague apprehension of its subject-matter. Children, who are made our pattern in Scripture, are taught, by an accommodation, on the part of their teachers, to their immature faculties and their scanty vocabulary. To answer their questions in the language which we should use towards grown men, would be simply to mislead them, if they could construe it at all. We must dispense and "divide" the word of truth, if we would not have it changed, as far as they are concerned, into a word of falsehood; for what is short of truth in the letter may be to them the most perfect truth, that is, the nearest approach to truth, compatible with their condition. The case is the same as regards those who have any natural defect or deprivation which cuts them off from the circle of ideas common to mankind in general. To speak to a blind man of light and colours, in terms proper to those phenomena, would be to mock him; we must use other media of information accommodated to his circumstances, according to the well-known instance in which his own account of scarlet was to liken it to the sound of a trumpet. And so again, as regards savages, or the ignorant, or weak, or narrow-minded, our representations and arguments must take a certain form, if they are to gain admission into their minds at all, and to reach them. Again, what impediments do the diversities of language place in the way of communicating ideas! Language is a sort of analysis of thought; and, since ideas are infinite, and infinitely combined, and infinitely modified, whereas language is a method definite and limited, and confined to an arbitrary selection of a certain number of these innumerable materials, it were idle to expect that the courses of thought marked out in one language should, except in their great outlines and main centres, correspond to those of another. Multitudes of ideas expressed in the one do not even enter into the other, and can only be conveyed by some economy or accommodation, by circumlocutions, phrases, limiting

words, figures, or some bold and happy expedient. And sometimes, from
the continual demand, foreign words become naturalized. Again, the
difficulty is extreme, as all persons know, of leading certain individuals
(to use a familiar phrase) to understand one another; their habits of
thought turning apparently on points of mutual repulsion. Now this is
always in a measure traceable to moral diversities between the parties;
still, in many cases, it arises mainly from difference in the principle on
which they have divided and subdivided that world of ideas, which
comes before them both. They seem ever to be dodging each other, and
need a common measure or economy to mediate between them.

35. Fables, again, are economies or accommodations, being truths
and principles cast into that form in which they will be most vividly
recognized; as in the well-known instance attributed to Menenius
Agrippa. Again, mythical representations, at least in their better form,
may be considered facts or narratives, untrue, but like the truth,
intended to bring out the action of some principle, point of character,
and the like. For instance, the tradition that St. Ignatius was the child
whom our Lord took in His arms, may be unfounded; but it realizes to us
his special relation to Christ and His Apostles, with a keenness peculiar
to itself. The same remark may be made upon certain narratives of
martyrdoms, or of the details of such narratives, or of certain alleged
miracles, or heroic acts, or speeches, all which are the spontaneous
produce of religious feeling under imperfect knowledge. If the alleged
facts did not occur, they ought to have occurred (if I may so speak); they
are such as might have occurred, and would have occurred, under
circumstances; and they belong to the parties to whom they are
attributed, potentially, if not actually; or the like of them did occur; or
occurred to others similarly circumstanced, though not to those very
persons. Many a theory or view of things, on which an institution is
founded, or a party held together, is of the same kind. Many an
argument, used by zealous and earnest men, has this economical
character, being not the very ground on which they act, (for they
continue in the same course, though it be refuted,) yet, in a certain sense,
a representation of it, a proximate description of their feelings in the
shape of argument, on which they can rest, to which they can recur
when perplexed, and appeal when questioned. Now, in this reference to
accommodation or economy in human affairs, I do not meddle with the
question of casuistry, viz. which of such artifices, as they may be called,
are innocent, or where the line is to be drawn. That some are immoral,
common sense tells us; but it is enough for my purpose, if some are
necessary, as the same common sense will allow; and then the very
necessity of the use will account for the abuse and perversion.

36. Even between man and man, then, constituted, as men are, alike,

various distinct instruments, keys, or *calculi* of thought obtain, on which their ideas and arguments shape themselves respectively, and which we must use, if we would reach them. The cogitative method, as it may be called, of one man is notoriously very different from that of another; of the lawyer from that of the soldier, of the rich from that of the poor. The territory of thought is portioned out in a hundred different ways. Abstractions, generalizations, definitions, propositions, all are framed on distinct standards; and if this is found in matters of this world between man and man, surely much more must it exist between the ideas of men, and the thoughts, ways, and works of God.

37. One of the obvious instances of this contrariety is seen in the classifications we make of the subjects of the animal or vegetable kingdoms. Here a very intelligible order has been observed by the Creator Himself; still one of which we have not, after all, the key. We are obliged to frame one of our own; and when we apply it, we find that it will not exactly answer the Divine idea of arrangement, as it discovers itself to us; there being phenomena which we cannot locate, or which, upon our system of division, are anomalies in the general harmony of the Creation.

38. Mathematical science will afford us a more extended illustration of this distinction between supernatural and eternal laws, and our attempts to represent them, that is, our economies. Various methods or *calculi* have been adopted to embody those immutable principles and dispositions of which the science treats, which are really independent of any, yet cannot be contemplated or pursued without one or other of them. The first of these instruments of investigation employs the medium of extension; the second, that of number; the third, that of motion; the fourth proceeds on a more subtle hypothesis, that of increase. These methods are very distinct from each other, at least the geometrical and the differential; yet they are, one and all, analyses, more or less perfect, of those same necessary truths, for which we have not a name, of which we have no idea, except in the terms of such economical representations. They are all developments of one and the same range of ideas; they are all instruments of discovery as to those ideas. They stand for real things, and we can reason with them, though they be but symbols, as if they were the things themselves, for which they stand. Yet none of them carries out the lines of truth to their limits; first, one stops in the analysis, then another; like some calculating tables which answer for a thousand times, and miss in the thousand and first. While they answer, we can use them just as if they were the realities which they represent, and without thinking of those realities; but at length our instrument of discovery issues in some great impossibility or contradiction, or what we call in religion, a mystery. It has run its

length; and by its failure shows that all along it has been but an expedient for practical purposes, not a true analysis or adequate image of those recondite laws which are investigated by means of it. It has never fathomed their depth, because it now fails to measure their course. At the same time, no one, because it cannot do every thing, would refuse to use it within the range in which it will act; no one would say that it was a system of empty symbols, though it be but a shadow of the unseen. Though we use it with caution, still we use it, as being the nearest approximation to the truth which our condition admits.

39. Let us take another instance, of an outward and earthly form, or economy, under which great wonders unknown seem to be typified; I mean musical sounds, as they are exhibited most perfectly in instrumental harmony. There are seven notes in the scale; make them fourteen; yet what a slender outfit for so vast an enterprise! What science brings so much out of so little? Out of what poor elements does some great master in it create his new world! Shall we say that all this exuberant inventiveness is a mere ingenuity or trick of art, like some game or fashion of the day, without reality, without meaning? We may do so; and then, perhaps, we shall also account the science of theology to be a matter of words; yet, as there is a divinity in the theology of the Church, which those who feel cannot communicate, so is there also in the wonderful creation of sublimity and beauty of which I am speaking. To many men the very names which the science employs are utterly incomprehensible. To speak of an idea or a subject seems to be fanciful or trifling, to speak of the views which it opens upon us to be childish extravagance; yet it is possible that that inexhaustible evolution and disposition of notes, so rich yet so simple, so intricate yet so regulated, so various yet so majestic, should be a mere sound, which is gone and perishes? Can it be that those mysterious stirrings of heart, and keen emotions, and strange yearnings after we know not what, and awful impressions from we know not whence, should be wrought in us by what is unsubstantial, and comes and goes, and begins and ends in itself? It is not so; it cannot be. No; they have escaped from some higher sphere; they are the outpourings of eternal harmony in the medium of created sound; they are echoes from our Home; they are the voice of Angels, or the Magnificat of Saints, or the living laws of Divine Governance, or the Divine Attributes; something are they besides themselves, which we cannot compass, which we cannot utter,—though mortal man, and he perhaps not otherwise distinguished above his fellows, has the gift of eliciting them.

40. So much on the subject of musical sound; but what if the whole series of impressions, made on us through the senses, be, as I have already hinted, but a Divine economy suited to our need, and the token

of realities distinct from themselves, and such as might be revealed to us, nay, more perfectly, by other senses, different from our existing ones as they from each other? What if the properties of matter, as we conceive of them, are merely relative to us, so that facts and events, which seem impossible when predicated concerning it in terms of those impressions, are impossible only in those terms, not in themselves,—impossible only because of the imperfection of the idea, which, in consequence of those impressions, we have conceived of material substances? If so, it would follow that the laws of physics, as we consider them, are themselves but generalizations of economical exhibitions, inferences from figure and shadow, and not more real than the phenomena from which they are drawn. Scripture, for instance, says that the sun moves and the earth is stationary; and science, that the earth moves, and the sun is comparatively at rest. How can we determine which of these opposite statements is the very truth, till we know what motion is? If our idea of motion be but an accidental result of our present senses, neither proposition is true, and both are true; neither true philosophically, both true for certain practical purposes in the system in which they are respectively found; and physical science will have no better meaning when it says that the earth moves, than plane astronomy when it says that the earth is still.

41. And should any one fear lest thoughts such as these should tend to a dreary and hopeless scepticism, let him take into account the Being and Providence of God, the Merciful and True; and he will at once be relieved of his anxiety. All is dreary till we believe, what our hearts tell us, that we are subjects of His Governance; nothing is dreary, all inspires hope and trust, directly we understand that we are under His hand, and that whatever comes to us is from Him, as a method of discipline and guidance. What is it to us whether the knowledge He gives us be greater or less, if it be He who gives it? What is it to us whether it be exact or vague, if He bids us trust it? What have we to care whether we are or are not given to divide substance from shadow, if He is training us heavenwards by means of either? Why should we vex ourselves to find whether our deductions are philosophical or no, provided they are religious? If our senses supply the media by which we are put on trial, by which we are all brought together, and hold intercourse with each other, and are disciplined and are taught, and enabled to benefit others, it is enough. We have an instinct within us, impelling us, we have external necessity forcing us, to trust our senses, and we may leave the question of their substantial truth for another world, "till the day break, and the shadows flee away." And what is true of reliance on our senses, is true of all the information which it has pleased God to vouchsafe to us, whether in nature or in grace.

42. Instances, then, such as these, will be found both to sober and to encourage us in our theological studies,—to impress us with a profound sense of our ignorance of Divine Verities, when we know most; yet to hinder us from relinquishing their contemplation, though we know so little. On the one hand, it would appear that even the most subtle questions of the schools may have a real meaning, as the most intricate *formulæ* in analytics; and, since we cannot tell how far our instrument of thought reaches in the process of investigation, and at what point it fails us, no questions may safely be despised. "Whether God was any where before creation?" "whether He knows all creatures in Himself?" "whether the blessed see all things possible and future in Him?" "whether relation is the form of the Divine Persons?" "in what sense the Holy Spirit is Divine Love?" these, and a multitude of others, far more minute and remote, are all sacred from their subject.

43. On the other hand, it must be recollected that not even the Catholic reasonings and conclusions, as contained in Confessions, and most thoroughly received by us, are worthy of the Divine Verities which they represent, but are the truth only in as full a measure as our minds can admit it; the truth as far as they go, and under the conditions of thought which human feebleness imposes. It is true that God is without beginning, if eternity may worthily be considered to imply succession; in every place, if He who is a Spirit can have relations with space. It is right to speak of His Being and Attributes, if He be not rather super-essential; it is true to say that He is wise or powerful, if we may consider Him as other than the most simple Unity. He is truly Three, if He is truly One; He is truly One, if the idea of Him falls under earthly number. He has a triple Personality, in the sense in which the Infinite can be understood to have Personality at all. If we know any thing of Him,—if we may speak of Him in any way,—if we may emerge from Atheism or Pantheism into religious faith,—if we would have any saving hope, any life of truth and holiness within us,—this only do we know, with this only confession, we must begin and end our worship—that the Father is the One God, the Son the One God, and the Holy Ghost the One God; and that the Father is not the Son, the Son not the Holy Ghost, and the Holy Ghost not the Father.

44. The fault, then, which we must guard against in receiving such Divine intimations, is the ambition of being wiser than what is written; of employing the Reason, not in carrying out what is told us, but in impugning it; not in support, but in prejudice of Faith. Brilliant as are such exhibitions of its powers, they bear no fruit. Reason can but ascertain the profound difficulties of our condition, it cannot remove them; it has no work, it makes no beginning, it does but continually fall back, till it is content to be a little child, and to follow where Faith guides it.

45. What remains, then, but to make our prayer to the Gracious and Merciful God, the Father of Lights, that in all our exercises of Reason, His gift, we may thus use it,—as He would have us, in the obedience of Faith, with a view to His glory, with an aim at His Truth, in dutiful submission to His will, for the comfort of His elect, for the edification of Holy Jerusalem, His Church, and in recollection of His own solemn warning, "Every idle word that men shall speak, thy shall give account thereof in the day of judgment; for by thy words thou shalt be justified, and by thy words thou shalt be condemned."

[Sermon 15]

AN ESSAY ON THE DEVELOPMENT OF CHRISTIAN DOCTRINE

The following Essay is directed towards a solution of the difficulty which has been stated,—the difficulty, as far as it exists, which lies in the way of our using in controversy the testimony of our most natural informant concerning the doctrine and worship of Christianity, viz. the history of eighteen hundred years. The view on which it is written has at all times, perhaps, been implicitly adopted by theologians, and, I believe, has recently been illustrated by several distinguished writers of the continent, such as De Maistre and Möhler: viz. that the increase and expansion of the Christian Creed and Ritual, and the variations which have attended the process in the case of individual writers and Churches, are the necessary attendants on any philosophy or polity which takes possession of the intellect and heart, and has had any wide or extended dominion; that, from the nature of the human mind, time is necessary for the full comprehension and perfection of great ideas; and that the highest and most wonderful truths, though communicated to the world once for all by inspired teachers, could not be comprehended all at once by the recipients, but, as being received and transmitted by minds not inspired and through media which were human, have required only the longer time and deeper thought for their full elucidation. This may be called the *Theory of Development of Doctrine*; and, before proceeding to treat of it, one remark may be in place.

It is undoubtedly an hypothesis to account for a difficulty; but such too are the various explanations given by astronomers from Ptolemy to Newton of the apparent motions of the heavenly bodies, and it is as unphilosophical on that account to object to the one as to object to the other. Nor is it more reasonable to express surprise, that at this time of day a theory is necessary, granting for argument's sake that the theory is novel, than to have directed a similar wonder in disparagement of the theory of gravitation, or the Plutonian theory in geology. Doubtless, the theory of the Secret and the theory of doctrinal Developments are expedients, and so is the dictum of Vincentius; so is the art of grammar or the use of the quadrant; it is an expedient to enable us to solve what has now become a necessary and an anxious problem. For three hundred years the documents and the facts of Christianity have been exposed to a jealous scrutiny; works have been judged spurious which once were received without a question; facts have been discarded or modified which were once first principles in argument; new facts and new principles have been brought to light; philosophical views and polemical discussions of various tendencies have been maintained with more or less success. Not only has the relative situation of controversies and theologies altered, but infidelity itself is in a different,—I am obliged to say in a more hopeful position,—as regards Christianity. The facts of Revealed Religion, though in their substance unaltered, present a less compact and orderly front to the attacks of its enemies now than formerly, and allow of the introduction of new inquiries and theories concerning its sources and its rise. The state of things is not as it was, when an appeal lay to the supposed works of the Areopagite, or to the primitive Decretals, or to St. Dionysius's answers to Paul, or to the Cœna Domini of St. Cyprian. The assailants of dogmatic truth have got the start of its adherents of whatever Creed; philosophy is completing what criticism has begun; and apprehensions are not unreasonably excited lest we should have a new world to conquer before we have weapons for the warfare. Already infidelity has its views and conjectures, on which it arranges the facts of ecclesiastical history; and it is sure to consider the absence of any antagonist theory as an evidence of the reality of its own. That the hypothesis, here to be adopted, accounts not only for the Athanasian Creed, but for the Creed of Pope Pius, is no fault of those who adopt it. No one has power over the issues of his principles; we cannot manage our argument, and have as much of it as we please and no more. An argument is needed, unless Christianity is to abandon the province of argument; and those who find fault with the explanation here offered of its historical phenomena will find it their duty to provide one for themselves.

And as no special aim at Roman Catholic doctrine need be supposed to have given a direction to the inquiry, so neither can a reception of that doctrine be immediately based on its results. It would be the work of a life to apply the Theory of Developments so carefully to the writings of the Fathers, and to the history of controversies and councils, as thereby to vindicate the reasonableness of every decision of Rome; much less can such an undertaking be imagined by one who, in the middle of his days, is beginning life again. Thus much, however, might be gained even from an Essay like the present, an explanation of so many of the reputed corruptions, doctrinal and practical, of Rome, as might serve as a fair ground for trusting her in parallel cases where the investigation had not been pursued.

[Introduction]

On the Development of Ideas

The idea which represents an object or supposed object is commensurate with the sum total of its possible aspects, however they may vary in the separate consciousness of individuals; and in proportion to the variety of aspects under which it presents itself to various minds is its force and depth, and the arguments for its reality. Ordinarily an idea is not brought home to the intellect as objective except through this variety; like bodily substances, which are not apprehended except under the clothing of their properties and results, and which admit of being walked round, and surveyed on opposite sides, and in different perspectives, and in contrary lights, in evidence of their reality. And, as views of a material object may be taken from points so remote or so opposed, that they seem at first sight incompatible, and especially as their shadows will be disproportionate, or even monstrous, and yet all these anomalies will disappear and all these contrarieties be adjusted, on ascertaining the point of vision or the surface of projection in each case; so also all the aspects of an idea are capable of coalition, and of a resolution into the object to which it belongs and the *primâ facie* dissimilitude of its aspects becomes, when explained, an argument for its substantiveness and integrity, and their multiplicity for its originality and power.

There is no one aspect deep enough to exhaust the contents of a real idea, no one term or proposition which will serve to define it; though of course one representation of it is more just and exact than another, and though when an idea is very complex, it is allowable, for the sake of convenience, to consider its distinct aspects as if separate ideas. Thus, with all our intimate knowledge of animal life and of the structure of particular animals, we have not arrived at a true definition of any one of them, but are forced to enumerate properties and accidents by way of description. Nor can we inclose in a formula that intellectual fact, or system of thought, which we call the Platonic philosophy, or that historical phenomenon of doctrine and conduct, which we call the heresy of Montanus or of Manes. Again, if Protestantism were said to lie in its theory of private judgment, and Lutheranism in its doctrine of justification, this indeed would be an approximation to the truth; but it is plain that to argue or to act as if the one or the other aspect were a sufficient account of those forms of religion severally, would be a serious mistake. Sometimes an attempt is made to determine the "leading idea," as it has been called, of Christianity, an ambitious essay as employed on a supernatural work, when, even as regards the visible creation and the inventions of man, such a task is beyond us. Thus its one idea has been said by some to be the restoration of our fallen race, by others philanthropy, by others the tidings of immortality, or the spirituality of true religious service, or the salvation of the elect, or mental liberty, or the union of the soul with God. If, indeed, it is only thereby meant to use one or other of these as a central idea for convenience, in order to group others around it, no fault can be found with such a proceeding: and in this sense I should myself call the Incarnation the central aspect of Christianity, out of which the three main aspects of its teaching take their rise, the sacramental, the hierarchical, and the ascetic. But one aspect of Revelation must not be allowed to exclude or to obscure another; and Christianity is dogmatical, devotional, practical all at once; it is esoteric and exoteric; it is indulgent and strict; it is light and dark; it is love, and it is fear.

When an idea, whether real or not, is of a nature to arrest and possess the mind, it may be said to have life, that is, to live in the mind which is its recipient. Thus mathematical ideas, real as they are, can hardly properly be called living, at least ordinarily. But, when some great enunciation, whether true or false, about human nature, or present good, or government, or duty, or religion, is carried forward into the public throng of men and draws attention, then it is not merely received passively in this or that form into many minds, but it becomes an active principle within them, leading them to an ever-new contemplation of itself, to an application of it in various directions, and a propagation of it on every side. Such is the doctrine of the divine right of kings, or of the rights of man, or of the anti-social bearings of a priesthood, or utilitarianism, or free trade, or the duty of benevolent enterprises, or the philosophy of Zeno or Epicurus, doctrines which are of a nature to attract and influence, and have so far a *primâ facie* reality, that they may be looked at on many sides and strike various minds very variously. Let one such idea get possession of the popular mind, or the mind of any portion of the community, and it is not difficult to understand what will be the result. At first men will not fully realize what it is that moves them, and will express and explain themselves inadequately. There will be a general agitation of thought, and an action of mind upon mind. There will be a time of confusion, when conceptions and misconceptions are in conflict, and it is uncertain whether anything is to come of the idea at all, or which view of it is to get the start of the others. New lights will be brought to bear upon the original statements of the doctrine put forward; judgments and aspects will accumulate. After a while some definite teaching emerges; and, as time proceeds, one view will be modified or expanded by another, and then combined with a third; till the idea to which these various aspects belong, will be to each mind separately what at first it was only to all together. It will be surveyed too in its relation to other doctrines or facts, to other natural laws or established customs, to the varying circumstances of times and places, to other religions, polities, philosophies, as the case may be. How it stands affected towards other systems, how it affects them, how far it may be made to combine with them, how far it tolerates them, when it interferes with them, will be gradually wrought out. It will be interrogated and criticized by enemies, and defended by well-wishers. The multitude of opinions formed concerning it in these respects and many others will be collected, compared, sorted, sifted, selected, rejected, gradually attached to it, separated from it, in the minds of individuals and of the community. It will, in proportion to its native vigour and subtlety, introduce itself into the framework and details of social life, changing public opinion, and strengthening or undermining the foundations of established order. Thus in time it will have grown into an ethical code, or into a system of government, or into a theology, or into a ritual, according to its capabilities: and this body of thought, thus laboriously gained, will after all be little more than the proper representative of one idea, being in substance what that idea meant from the first, its complete image as seen in a combination of diversified aspects, with the suggestions and corrections of many minds, and the illustration of many experiences.

This process, whether it be longer or shorter in point of time, by which the aspects of an idea are brought into consistency and form, I call its development, being the germination and maturation of some truth or apparent truth on a large mental field. On the other hand this process will not be a development, unless the assemblage of aspects, which constitute its ultimate shape, really belongs to the idea from which they start. A republic, for instance, is not a development from a pure monarchy, though it may follow upon it; whereas the Greek "tyrant" may be considered as included in the idea of a democracy. Moreover a development will have this characteristic, that, its action being in the busy scene of human life, it cannot progress at all without cutting across, and thereby destroying or modifying and incorporating with itself existing modes of thinking and operating. The development then of an idea is not like an investigation worked out on paper, in which each successive advance is a pure evolution from a foregoing, but it is carried on through and by means of communities of men and their leaders and guides; and it employs their minds as its instruments, and depends upon them, while it uses them. And so, as regards existing opinions, principles, measures, and institutions of the community which it has invaded; it developes by establishing relations between itself and them; it employs itself, in giving them a new meaning and direction, in creating what may be called a jurisdiction over them, in throwing off whatever in them it cannot assimilate. It grows when it incorporates, and its identity is found, not in isolation, but in continuity and sovereignty. This it is that imparts to the history both of states and of religions, its specially turbulent and polemical character. Such is the explanation of the wranglings, whether of schools or of parliaments. It is the warfare of ideas under their various aspects striving for the mastery, each of them enterprising, engrossing imperious, more or less incompatible with the rest, and rallying followers or rousing foes, according as it acts upon the faith, the prejudices, or the interest of parties or classes.

Moreover, an idea not only modifies, but is modified, or at least influenced, by the state of things in which it is carried out, and is dependent in various ways on the circumstances which surround it. Its development proceeds quickly or slowly, as it may be; the order of succession in its separate stages is variable; it shows differently in a small sphere of action and in an extended; it may be interrupted, retarded, mutilated, distorted, by external violence; it may be enfeebled by the effort of ridding itself of domestic foes; it may be impeded and swayed or even absorbed by counter energetic ideas; it may be coloured by the received tone of thought into which it comes, or depraved by the intrusion of foreign principles, or at length shattered by the development of some original fault within it.

But whatever be the risk of corruption from intercourse with the world around, such a risk must be encountered if a great idea is duly to be understood, and much more if it is to be fully exhibited. It is elicited and expanded by trial, and battles into perfection and supremacy. Nor does it escape the collision of opinion even in its earlier years, nor does it remain truer to itself, and with a better claim to be considered one and the same, though externally protected from vicissitude and change. It is indeed sometimes said that the stream is clearest near the spring. Whatever use may fairly be made of this image, it does not apply to the history of a philosophy or belief, which on the contrary is more equable, and purer, and stronger, when its bed has become deep, and broad, and full. It necessarily rises out of an existing state of things, and for a time savours of the soil. Its vital element needs disengaging from what is foreign and temporary, and is employed in efforts after freedom which become more vigorous and hopeful as its years increase. Its beginnings are no measure of its capabilities, nor of its scope. At first no one knows what it is, or what it is worth. It remains perhaps for a time quiescent; it tries, as it were, its limbs, and proves the ground under it, and feels its way. From time to time it makes essays which fail, and are in consequence abandoned. It seems in suspense which way to go; it wavers, and at length strikes out in one definite direction. In time it enters upon strange territory; points of controversy alter their bearing; parties rise and fall around it; dangers and hopes appear in new relations; and old principles reappear under new forms. It changes with them in order to remain the same. In a higher world it is otherwise, but here below to live is to change, and to be perfect is to have changed often.

[ch. 1]

An Infallible Developing Authority to be Expected

Moreover, it must be borne in mind that, as the essence of all religion is authority and obedience, so the distinction between natural religion and revealed lies in this, that the one has a subjective authority, and the other an objective. Revelation consists in the manifestation of the Invisible Divine Power, or in the substitution of the voice of a Lawgiver for the voice of conscience. The supremacy of conscience is the essence of natural religion; the supremacy of Apostle, or Pope, or Church, or Bishop, is the essence of revealed; and when such external authority is taken away, the mind falls back again of necessity upon that inward guide which it possessed even before Revelation was vouchsafed. Thus, what conscience is in the system of nature, such is the voice of Scripture, or of the Church, or of the Holy See, as we may determine it, in the system of Revelation. It may be objected, indeed, that conscience is not infallible; it is true, but still it is ever to be obeyed. And this is just the prerogative which controversialists assign to the See of St. Peter; it is not in all cases infallible, it may err beyond its special province, but it has in all cases a claim on our obedience. . . . And as obedience to conscience, even supposing conscience ill-informed, tends to the improvement of our moral nature, and ultimately of our knowledge, so obedience to our ecclesiastical superior may subserve our growth in illumination and sanctity, even though he should command what is extreme or inexpedient, or teach what is external to his legitimate province.

The common sense of mankind does but support a conclusion thus forced upon us by analogical considerations. It feels that the very idea of revelation implies a present informant and guide, and that an infallible one; not a mere abstract declaration of Truths unknown before to man, or a record of history, or the result of an antiquarian research, but a message and a lesson speaking to this man and that. This is shown by the popular notion which has prevailed among us since the Reformation, that the Bible itself is such a guide; and which succeeded in overthrowing the supremacy of Church and Pope, for the very reason that it was a rival authority, not resisting merely, but supplanting it. In proportion, then, as we find, in matter of fact, that the inspired Volume is not adapted or intended to subserve that purpose, are we forced to revert to that living and present Guide, who, at the era of our rejection of her, had been so long recognized as the dispenser of Scripture, according to times and circumstances, and the arbiter of all true doctrine and holy practice to her children. We feel a need, and she alone of all things under heaven supplies it. We are told that God has spoken. Where? In a book? We have tried it and it disappoints; it disappoints us, that most holy and blessed gift, not from fault of its own, but because it is used for a purpose for which it was not given. The Ethiopian's reply, when St. Philip asked him if he understood what he was reading, is the voice of nature: "How can I, unless some man shall guide me?" The Church undertakes that office; she does what none else can do, and this is the secret of her power. . . . These sentences, whatever be the errors of their wording, surely express a great truth. The most obvious answer, then, to the question, why we yield to the authority of the Church in the questions and developments of faith, is, that some authority there must be if there is a revelation given, and other authority there is none but she. A revelation is not given, if there be no authority to decide what it is that is given. In the words of St. Peter to her Divine Master and Lord, "To whom shall we go?" . . .

And if the very claim to infallible arbitration in religious disputes is of so weighty importance and interest in all ages of the world, much more is it welcome at a time like the present, when the human intellect is so busy, and thought so fertile, and opinion so manifold. The absolute need of a spiritual supremacy is at present the strongest of arguments in favour of the fact of its supply. Surely, either an objective revelation has not been given, or it has been provided with means for impressing its objectiveness on the world. If Christianity be a social religion, as it certainly is, and if it be based on certain ideas acknowledged as divine, or a creed, (which shall here be assumed,) and if these ideas have various aspects, and make distinct impressions on different minds, and issue in consequence in a multiplicity of developments, true, or false, or mixed, as has been shown, what power will suffice to meet and to do justice to these conflicting conditions, but a supreme authority ruling and reconciling individual judgments by a divine right and a recognized wisdom? In barbarous times the will is reached through the senses; but in an age in which reason, as it is called, is the standard of truth and right, it is abundantly evident to any one, who mixes ever so little with the world, that, if things are left to themselves, every individual will have his own view of them,

and take his own course; that two or three will agree today to part
company to-morrow; that Scripture will be read in contrary ways, and
history, according to the apologue, will have to different comers its silver
shield and its golden; that philosophy, taste, prejudice, passion, party,
caprice, will find no common measure, unless there be some supreme
power to control the mind and to compel agreement.

There can be no combination on the basis of truth without an organ of
truth. As cultivation brings out the colours of flowers, and domestication
changes the character of animals, so does education of necessity develope
differences of opinion; and while it is impossible to lay down first
principles in which all will unite, it is utterly unreasonable to expect that
this man should yield to that, or all to one. I do not say there are no
eternal truths . . . which all acknowledge in private, but that there are
none sufficiently commanding to be the basis of public union and action.
The only general persuasive in matters of conduct is authority; that is,
(when truth is in question,) a judgment which we feel to be superior to
our own. If Christianity is both social and dogmatic, and intended for all
ages, it must humanly speaking have an infallible expounder. Else you
will secure unity of form at the loss of unity of doctrine, or unity of
doctrine at the loss of unity of form; you will have to choose between a
comprehension of opinions and a resolution into parties, between
latitudinarian and sectarian error. You may be tolerant or intolerant of
contrarieties of thought, but contrarieties you will have. By the Church
of England a hollow uniformity is preferred to an infallible chair; and by
the sects of England, an interminable division. Germany and Geneva
began with persecution, and have ended in scepticism. The doctrine of
infallibility is a less violent hypothesis than this sacrifice either of faith or
of charity. It secures the object, while it gives definiteness and force to
the matter, of the Revelation. . . .

The Existing Developments of Doctrine.

The Fulfilment of That Expectation

And this general testimony to the oneness of Catholicism extends to its
past teaching relatively to its present, as well as to the portions of its
present teaching one with another. No one doubts, with such exception
as has just been allowed, that the Roman Catholic communion of this
day is the successor and representative of the Medieval Church, or that
the Medieval Church is the legitimate heir of the Nicene; even allowing
that it is a question whether a line cannot be drawn between the Nicene
Church and the Church which preceded it. On the whole, all parties will
agree that, of all existing systems, the present communion of Rome is the

nearest approximation in fact to the Church of the Fathers, possible though some may think it, to be nearer still to that Church on paper. Did St. Athanasius or St. Ambrose come suddenly to life, it cannot be doubted what communion he would take to be his own. All surely will agree that these Fathers, with whatever opinions of their own, whatever protests, if we will, would find themselves more at home with such men as St. Bernard or St. Ignatius Loyola, or with the lonely priest in his lodging, or the holy sisterhood of mercy, or the unlettered crowd before the altar, than with the teachers or with the members of any other creed. And may we not add, that were those same Saints, who once sojourned, one in exile, one on embassy, at Treves, to come more northward still, and to travel until they reached another fair city, seated among groves, green meadows, and calm streams, the holy brothers would turn from many a high aisle and solemn cloister which they found there, and ask the way to some small chapel where mass was said in the populous alley or forlorn suburb? And, on the other hand, can any one who has not heard his name, and cursorily read his history, doubt for one instant how, in turn, the people of England, "we, our princes, our priests, and our prophets," Lords and Commons, Universities, Ecclesiastical Courts, marts of commerce, great towns, country parishes, would deal with Athanasius,—Athanasius, who spent his long years in fighting against sovereigns for a theological term?

[ch. 2]

Genuine Developments Contrasted with Corruptions

It may be said in answer to me that it is not enough that a certain large system of doctrine, such as that which goes by the name of Catholic, should admit of being referred to beliefs, opinions, and usages which prevailed among the first Christians, in order to my having a logical right to include a reception of the later teaching in the reception of the earlier; that an intellectual development may be in one sense natural, and yet untrue to its original, as diseases come of nature, yet are the destruction, or rather the negation of health; that the causes which stimulate the growth of ideas may also disturb and deform them; and that Christianity might indeed have been intended by its Divine Author for a wide expansion of the ideas proper to it, and yet this great benefit hindered by the evil birth of cognate errors which acted as its counterfeit; in a word, that what I have called developments in the Roman Church are nothing more or less than what used to be called her corruptions; and that new names do not destroy old grievances.

This is what may be said, and I acknowledge its force: it becomes necessary in consequence to assign certain characteristics of faithful developments, which none but faithful developments have, and the presence of which serves as a test to discriminate between them and corruptions. This I at once proceed to do, and I shall begin by determining what a corruption is, and why it cannot rightly be called, and how it differs from, a development.

To find then what a corruption or perversion of the truth is, let us inquire what the word means, when used literally of material substances. Now it is plain, first of all, that a corruption is a word attaching to organized matters only; a stone may be crushed to powder, but it cannot be corrupted. Corruption, on the contrary, is the breaking up of life, preparatory to its termination. This resolution of a body into its component parts is the stage before its dissolution; it begins when life has reached its perfection, and it is the sequel, or rather the continuation, of that process towards perfection, being at the same time the reversal and undoing of what went before. Till this point of regression is reached, the body has a function of its own, and a direction and aim in its action, and a nature with laws; these it is now losing, and the traits and tokens of former years; and with them its vigour and powers of nutrition, of assimilation, and of self-reparation.

Taking this analogy as a guide, I venture to set down seven Notes of varying cogency, independence and applicability, to discriminate healthy developments of an idea from its state of corruption and decay, as follows:—There is no corruption if it retains one and the same type, the same principles, the same organization; if its beginnings anticipate its subsequent phases, and its later phenomena protect and subserve its earlier; if it has a power of assimilation and revival, and a vigorous action from first to last. On these tests I shall now enlarge, nearly in the order in which I have enumerated them.

First Note of a Genuine Development.
Preservation of Type.

This is readily suggested by the analogy of physical growth, which is such that the parts and proportions of the developed form, however altered, correspond to those which belong to its rudiments. The adult animal has the same make, as it had on its birth; young birds do not grow into fishes, nor does the child degenerate into the brute, wild or domestic, of which he is by inheritance lord. . . .

However . . . this unity of type, characteristic as it is of faithful developments, must not be pressed to the extent of denying all variation, nay, considerable alteration of proportion and relation, as time goes on, in the parts or aspects of an idea. Great changes in outward appearance

and internal harmony occur in the instance of the animal creation itself. The fledged bird differs much from its rudimental form in the egg. The butterfly is the development, but not in any sense the image, of the grub.
. . .

More subtle still and mysterious are the variations which are consistent or not inconsistent with identity in political and religious developments. . . .

The same man may run through various philosophies or beliefs, which are in themselves irreconcilable, without inconsistency, since in him they may be nothing more than accidental instruments or expressions of what he is inwardly from first to last. The political doctrines of the modern Tory resemble those of the primitive Whig; yet few will deny that the Whig and Tory characters have each a discriminating type. Calvinism has changed into Unitarianism: yet this need not be called a corruption, even if it be not, strictly speaking, a development . . .

The history of national character supplies an analogy, rather than an instance strictly in point; yet there is so close a connexion between the development of minds and of ideas that it is allowable to refer to it here. Thus we find England of old the most loyal supporter, and England of late the most jealous enemy, of the Holy See. As great a change is exhibited in France, once the eldest born of the Church and the flower of her knighthood, now democratic and lately infidel. Yet, in neither nation, can these great changes be well called corruptions. . . .

And, in like manner, ideas may remain, when the expression of them is indefinitely varied; and we cannot determine whether a professed development is truly such or not, without some further knowledge than an experience of the mere fact of this variation. Nor will our instinctive feelings serve as a criterion. It must have been an extreme shock to St. Peter to be told he must slay and eat beasts, unclean as well as clean, though such a command was implied already in that faith which he held and taught; a shock, which a single effort, or a short period, or the force of reason would not suffice to overcome. Nay, it may happen that a representation which varies from its original may be felt as more true and faithful than one which has more pretensions to be exact. So it is with many a portrait which is not striking: at first look, of course, it disappoints us; but when we are familiar with it, we see in it what we could not see at first, and prefer it, not to a perfect likeness, but to many a sketch which is so precise as to be a caricature.

On the other hand, real perversions and corruptions are often not so unlike externally to the doctrine from which they come, as are changes which are consistent with it and true developments. When Rome changed from a Republic to an Empire, it was a real alteration of polity, or what may be called a corruption; yet in appearance the change was small. . . .

Nay, one cause of corruption in religion is the refusal to follow the course of doctrine as it moves on, and an obstinacy in the notions of the past. Certainly: as we see conspicuously in the history of the chosen race. The Samaritans who refused to add the Prophets to the Law, and the Sadducees who denied a truth which was covertly taught in the Book of Exodus, were in appearance only faithful adherents to the primitive doctrine. Our Lord found His people precisians in their obedience to the letter; He condemned them for not being led on to its spirit, that is, to its developments. The Gospel is the development of the Law; yet what difference seems wider than that which separates the unbending rule of Moses from the "grace and truth" which "came by Jesus Christ?" . . .

An idea then does not always bear about it the same external image; this circumstance, however, has no force to weaken the argument for its substantial identity, as drawn from its external sameness, when such sameness remains. On the contrary, for that very reason, *unity of type* becomes so much the surer guarantee of the healthiness and soundness of developments, when it is persistently preserved in spite of their number or importance.

Second Note. Continuity of Principles.

As in mathematical creations figures are formed on distinct formulæ, which are the laws under which they are developed, so it is in ethical and political subjects. Doctrines expand variously according to the mind, individual or social, into which they are received; and the peculiarities of the recipient are the regulating power, the law, the organization, or, as it may be called, the form of the development. The life of doctrines may be said to consist in the law or principle which they embody.

Principles are abstract and general, doctrines relate to facts; doctrines develope, and principles at first sight do not; doctrines grow and are enlarged, principles are permanent; doctrines are intellectual, and principles are more immediately ethical and practical. Systems live in principles and represent doctrines. Personal responsibility is a principle, the Being of a God is a doctrine; from that doctrine all theology has come in due course, whereas that principle is not clearer under the Gospel than in paradise, and depends, not on belief in an Almighty Governor, but on conscience.

Yet the difference between the two sometimes merely exists in our mode of viewing them; and what is a doctrine in one philosophy is a principle in another. Personal responsibility may be made a doctrinal basis, and develope into Arminianism or Pelagianism. Again, it may be discussed whether infallibility is a principle or a doctrine of the Church of Rome, and dogmatism a principle or doctrine of Christianity. Again, consideration for the poor is a doctrine of the Church considered as a religious body, and a principle when she is viewed as a political power.

Doctrines stand to principles, as the definitions to the axioms and postulates of mathematics. Thus the 15th and 17th propositions of Euclid's book I. are developments, not of the three first axioms, which are required in the proof, but of the definition of a right angle. Perhaps the perplexity, which arises in the mind of a beginner, on learning the early propositions of the second book, arises from these being more prominently exemplifications of axioms than developments of definitions. He looks for developments from the definition of the rectangle, and finds but various particular cases of the general truth, that "the whole is equal to its parts."

It might be expected that the Catholic principles would be later in development that the Catholic doctrines, inasmuch as they lie deeper in the mind, and are assumptions rather than objective professions. This has been the case. The Protestant controversy has mainly turned, or is turning, on one or other of the principles of Catholicity; and to this day the rule of Scripture Interpretation, the doctrine of Inspiration, the relation of Faith to Reason, moral responsibility, private judgment, inherent grace, the seat of infallibility, remain, I suppose, more or less undeveloped or, at least, undefined, by the Church.

Doctrines stand to principles, if it may be said without fancifulness, as fecundity viewed relatively to generation, though this analogy must not be strained. Doctrines are developed by the operation of principles, and develope variously according to those principles. Thus a belief in the transitiveness of worldly goods leads the Epicurean to enjoyment, and the ascetic to mortification; and, from their common doctrine of the sinfulness of matter, the Alexandrian Gnostics became sensualists, and the Syrian devotees. The same philosophical elements, received into a certain sensibility or insensibility to sin and its consequences, leads one mind to the Church of Rome; another to what, for want of a better word, may be called Germanism.

Again, religious investigation sometimes is conducted on the principle that it is a duty "to follow and speak the truth," which really means that it is no duty to fear error, or to consider what is safest, or to shrink from scattering doubts, or to regard the responsibility of misleading; and thus it terminates in heresy or infidelity, without any blame to religious investigation in itself.

Again, to take a different subject, what constitutes a chief interest of dramatic compositions and tales, is to use external circumstances, which may be considered their law of development, as a means of bringing out into different shapes, and showing under new aspects, the personal peculiarities of character, according as either those circumstances or those peculiarities vary in the case of the personages introduced.

Principles are popularly said to develope when they are but exemplified; thus the various sects of Protestantism, unconnected as they are with each other, are called developments of the principle of Private Judgment, of which really they are but applications and results.

A development, to be faithful, must retain both the doctrine and the principle with which it started. Doctrine without its correspondent principle remains barren, if not lifeless, of which the Greek Church seems an instance; or it forms those hollow professions which are familiarly called "shams," as a zeal for an established Church and its creed on merely conservative or temporal motives. Such, too, was the Roman Constitution between the reigns of Augustus and Dioclesian.

On the other hand, principle without its corresponding doctrine may be considered as the state of religious minds in the heathen world, viewed relatively to Revelation; that is, of the "children of God who are scattered abroad."

Pagans may have, heretics cannot have, the same principles as Catholics; if the latter have the same, they are not real heretics, but in ignorance. Principle is a better test of heresy than doctrine. Heretics are true to their principles, but change to and fro, backwards and forwards, in opinion; for very opposite doctrines may be exemplifications of the same principle. Thus the Antiochenes and other heretics sometimes were Arians, sometimes Sabellians, sometimes Nestorians, sometimes Monophysites, as if at random, from fidelity to their common principle, that there is no mystery in theology. Thus Calvinists become Unitarians from the principle of private judgment. The doctrines of heresy are accidents and soon run to an end; its principles are everlasting.

This, too, is often the solution of the paradox "Extremes meet," and of the startling reactions which take place in individuals; viz., the presence of some one principle or condition, which is dominant in their minds from first to last. If one of two contradictory alternatives be necessarily true on a certain hypothesis, then the denial of the one leads, by mere logical consistency and without direct reasons, to a reception of the other. Thus the question between the Church of Rome and Protestantism falls in some minds into the proposition, "Rome is either the pillar and ground of the Truth or she is Antichrist;" in proportion, then, as they revolt from considering her the latter are they compelled to receive her as the former. Hence, too, men may pass from infidelity to Rome, and from Rome to infidelity, from a conviction in both courses that there is no tangible intellectual position between the two.

Protestantism, viewed in its more Catholic aspect, is doctrine without active principle; viewed in its heretical, it is active principle without doctrine. Many of its speakers, for instance, use eloquent and glowing language about the Church and its characteristics: some of them do not realize what they say, but use high words and general statements about "the faith," and "primitive truth," and "schism," and "heresy," to which they attach no definite meaning; while others speak of "unity," "universality," and "Catholicity," and use the words in their own sense and for their own ideas. . . .

Third Note. Power of Assimilation.

In the physical world, whatever has life is characterized by growth, so that in no respect to grow is to cease to live. It grows by taking into its own substance external materials; and this absorption or assimilation is completed when the materials appropriated come to belong to it or enter into its unity. Two things cannot become one, except there be a power of assimilation in one or the other. Sometimes assimilation is effected only with an effort; it is possible to die of repletion, and there are animals who lie torpid for a time under the contest between the foreign substance and the assimilating power. And different food is proper for different recipients.

This analogy may be taken to illustrate certain peculiarities in the growth or development in ideas, which were noticed in the first Chapter. It is otherwise with mathetical and other abstract creations, which, like the soul itself, are solitary and self-dependent; but doctrines and views which relate to man are not placed in a void, but in the crowded world, and make way for themselves by interpenetration, and develope by absorption. Facts and opinions, which have hitherto been regarded in other relations and grouped round other centres, henceforth are gradually attracted to a new influence and subjected to a new sovereign. They are modified, laid down afresh, thrust aside, as the case may be. A new element of order and composition has come among them; and its life is proved by this capacity of expansion, without disarrangement or dissolution. An eclectic, conservative, assimilating, healing, moulding process, a unitive power, is of the essence, and a third test, of a faithful development.

Thus, a power of development is a proof of life, not only in its essay, but especially in its success; for a mere formula either does not expand or is shattered in expanding. A living idea becomes many, yet remains one.

The attempt at development shows the presence of a principle, and its success to the presence of an idea. Principles stimulate thought, and an idea concentrates it.

The idea never was that throve and lasted, yet, like mathematical truth, incorporated nothing from external sources. So far from the fact of such incorporation implying corruption, as is sometimes supposed, development is a process of incorporation. . . .

The stronger and more living is an idea, that is, the more powerful hold it exercises on the minds of men, the more able it is to dispense with safeguards, and trust to itself against the danger of corruption. As strong frames exult in their agility, and healthy constitutions throw off ailments, so parties or schools that live can afford to be rash, and will sometimes be betrayed into extravagances, yet are brought right by their inherent vigour. On the other hand, unreal systems are commonly decent externally. Forms, subscriptions, or Articles of religion are indispensable when the principle of life is weakly. Thus Presbyterianism has maintained its original theology in Scotland where legal subscriptions are enforced, whil it has run into Arianism or Unitarianism where that protection is away. We have yet to see whether the Free Kirk can keep its present theological ground. The Church of Rome can consult expedience more freely than other bodies, as trusting to her living tradition, and is sometimes thought to disregard principle and scruple, when she is but dispensing with forms. Thus Saints are often characterized by acts which are no pattern for others; and the most gifted men are, by reason of their very gifts, sometimes led into fatal inadvertences. Hence vows are the wise defence of unstable virtue, and general rules the refuge of feeble authority.

And so much may suffice on the *unitive power* of faithful developments, which constitutes their third characteristic.

Fourth Note. Logical Sequence.

Logic is the organization of thought, and, as being such, is a security for the faithfulness of intellectual developments; and the necessity of using it is undeniable as far as this, that its rules must not be transgressed. That it is not brought into exercise in every instance of doctrinal development is owing to the varieties of mental constitution, whether in communities or in individuals, with whom great truths or seeming truths are lodged. The question indeed may be asked whether a development can be other in any case than a logical operation; but, if by this is meant a conscious reasoning from premises to conclusion, of course the answer must be in the negative. An idea under one or other of its aspects grows in the mind by remaining there; it becomes familiar and distinct, and is viewed in its relations; it leads to other aspects, and these again to others, subtle, recondite, original, according to the character, intellectual and moral, of the recipient; and thus a body of thought is gradually formed without his recognizing what is going on within him. And all this while, or at least from time to time, external circumstances elicit into formal statement the thoughts which are coming into being in the depths of his mind; and soon he has to begin to defend them; and then again a further process must take place, of analyzing his statements and ascertaining their dependence one on another. And thus he is led to regard as consequences, and to trace to principles, what hitherto he has discerned by a moral perception, and adopted on sympathy; and logic is brought in to arrange and inculcate what no science was employed in gaining.

And so in the same way, such intellectual processes, as are carried on silently and spontaneously in the mind of a party or school, of necessity come to light at a later date, and are recognized, and their issues are scientifically arranged. And then logic has the further function of propagation; analogy, the nature of the case, antecedent probability, application of principles, congruity, expedience, being some of the methods of proof by which the development is continued from mind to mind and established in the faith of the community.

Yet even then the analysis is not made on a principle, or with any view to its whole course and finished results. Each argument is brought for an immediate purpose; minds develope step by step, without looking behind them or anticipating their goal, and without either intention or promise of forming a system. Afterwards, however, this logical character which the whole wears becomes a test that the process has been a true developement, not a perversion or corruption, from its evident naturalness; and in some cases from the gravity, distinctness, precision, and majesty of its advance, and the harmony of its proportions, like the tall growth, and graceful branching, and rich foliage, of some vegetable production. . . .

At the same time it may be granted that the spontaneous process which goes on within the mind itself is higher and choicer than that which is logical; for the latter, being scientific, is common property, and can be taken and made use of by minds who are personally strangers, in any true sense, both to the ideas in question and to their development.

Thus, the holy Apostles would without words know all the truths concerning the high doctrines of theology, which controversialists after them have piously and charitably reduced to formulæ, and developed through argument. Thus, St. Justin or St. Irenæus might be without any digested ideas of Purgatory or Original Sin, yet have an intense feeling, which they had not defined or located, both of the fault of our first nature and the responsibilities of our nature regenerate. Thus St. Antony said to the philosophers who came to mock him, "He whose mind is in health does not need letters;" and St. Ignatius Loyola, while yet an unlearned neophyte, was favoured with transcendent perceptions of the Holy Trinity during his penance at Manresa. Thus St. Athanasius himself is more powerful in statement and exposition than in proof; while in Bellarmine we find the whole series of doctrines carefully drawn out, duly adjusted with one another, and exactly analyzed one by one. . . .

Fifth Note. Anticipation of its Future.

Since, when an idea is living, that is, influential and effective, it is sure to develope according to its own nature, and the tendencies, which are carried out on the long run, may under favourable circumstances show themselves early as well as late, and logic is the same in all ages, instances of a development which is to come, though vague and isolated, may occur from the very first, though a lapse of time be necessary to bring them to perfection. And since developments are in great measure only aspects of the idea from which they proceed, and all of them are natural consequences of it, it is often a matter of accident in what order they are carried out in individual minds; and it is in no wise strange that here and there definite specimens of advanced teaching should very early occur, which in the historical course are not found till a late day. The fact, then, of such early or recurring intimations of tendencies which afterwards are fully realized, is a sort of evidence that those later and more systematic fulfilments are only in accordance with the original idea. . . .

Sixth Note. Conservative Action upon its Past.

As developments which are preceded by definite indications have a fair presumption in their favour, so those which do but contradict and reverse the course of doctrine which has been developed before them, and out of which they spring, are certainly corrupt; for a corruption is a development in that very stage in which it ceases to illustrate, and begins to disturb, the acquisitions gained in its previous history. . . .

A true development, then, may be described as one which is conservative of the course of antecedent developments being really those antecedents and something besides them: it is an addition which illustrates, not obscures, corroborates, not corrects, the body of thought from which it proceeds; and this is its characteristic as contrasted with a corruption.

For instance, a gradual conversion from a false to a true religion, plainly, has much of the character of a continuous process, or a development, in the mind itself, even when the two religions, which are the limits of its course, are antagonists. Now let it be observed, that such a change consists in addition and increase chiefly, not in destruction. . . .

When Roman Catholics are accused of substituting another Gospel for the primitive Creed, they answer that they hold, and can show that they hold, the doctrines of the Incarnation and Atonement, as firmly as any Protestant can state them. To this it is replied that they do certainly profess them, but that they obscure and virtually annul them by their additions; that the *cultus* of St. Mary and the Saints is no development of the truth, but a corruption and a religious mischief to those doctrines of which it is the corruption, because it draws away the mind and heart from Christ. But they answer that, so far from this, it subserves, illustrates, protects the doctrine of our Lord's loving kindness and mediation. Thus the parties in controversy join issue on the common ground, that a developed doctrine which reverses the course of development which has preceded it, is no true development but a corruption; also, that what is corrupt acts as an element of unhealthiness towards what is sound. . . .

Seventh Note. Chronic Vigour.

Since the corruption of an idea, as far as the appearance goes, is a sort of accident or affection of its development, being the end of a course, and a transition-state leading to a crisis, it is, as has been observed above, a brief and rapid process. While ideas live in men's minds, they are ever enlarging into fuller development: they will not be stationary in their corruption any more than before it; and dissolution is that further state to which corruption tends. Corruption cannot, therefore, be of long standing; and thus *duration* is another test of a faithful development. . . .

The course of heresies is always short; it is an intermediate state between life and death, or what is like death; or, if it does not result in death, it is resolved into some new, perhaps opposite, course of error, which lays no claim to be connected with it. And in this way indeed, but in this way only, an heretical principle will continue in life many years, first running one way, then another. . . .

It is true that decay, which is one form of corruption, is slow; but decay is a state in which there is no violent or vigorous action at all, whether of a conservative or a destructive character, the hostile influence being powerful enough to enfeeble the functions of life, but not to quicken its own process. And thus we see opinions, usages, and systems, which are of venerable and imposing aspect, but which have no soundness within them, and keep together from a habit of consistence, or from dependence on political institutions; or they become almost peculiarities of a country, or the habits of a race, or the fashions of society. And then, at length, perhaps, they go off suddenly and die out under the first rough influence from without. Such are the superstitions which pervade a population, like some ingrained dye or inveterate odour, and which at length come to an end, because nothing lasts for ever, but which run no course, and have no history; such was the established paganism of classical times, which was the fit subject of persecution, for its first breath made it crumble and disappear. Such apparently is the state of the Nestorian and Monophysite communions; such might have been the condition of Christianity had it been absorbed by the feudalism of the middle ages; such too is that Protestantism, or (as it sometimes calls itself) attachment to the Establishment, which is not unfrequently the boast of the respectable and wealthy among ourselves.

Whether Mahometanism external to Christendom, and the Greek Church within it, fall under this description is yet to be seen. Circumstances can be imagined which would even now rouse the fanaticism of the Moslem; and the Russian despotism does not meddle with the usages, though it may domineer over the priesthood, of the national religion.

Thus, while a corruption is distinguished from decay by its energetic action, it is distinguished from a development by its *transitory character*.

[ch. 5]

The Assimilating Power of Dogmatic Truth

That there is a truth then; that there is one truth; that religious error is in itself of an immoral nature; that its maintainers, unless involuntarily such, are guilty in maintaining it; that it is to be dreaded; that the search for truth is not the gratification of curiosity; that its attainment has nothing of the excitement of a discovery; that the mind is below truth, not above it, and is bound, not to descant upon it, but to venerate it; that truth and falsehood are set before us for the trial of our hearts; that our choice is an awful giving forth of lots on which salvation or rejection is inscribed; that "before all things it is necessary to hold the Catholic faith;" that "he that would be saved must thus think," and not otherwise; that, "if thou criest after knowledge, and liftest up thy voice for understanding, if thou seekest her as silver, and searchest for her as for hid treasure, then shalt thou understand the fear of the Lord, and find the knowledge of God,"—this is the dogmatical principle, which has strength.

That truth and falsehood in religion are but matter of opinion; that one doctrine is as good as another; that the Governor of the world does not intend that we should gain the truth; that there is no truth; that we are not more acceptable to God by believing this than by believing that; that no one is answerable for his opinions; that they are a matter of necessity or accident; that it is enough if we sincerely hold what we profess; that our merit lies in seeking, not in possessing; that it is a duty to follow what seems to us true, without a fear lest it should not be true; that it may be a gain to succeed, and can be no harm to fail; that we may take up and lay down opinions at pleasure; that belief belongs to the mere intellect, not to the heart also; that we may safely trust to ourselves in matters of Faith, and need no other guide,—this is the principle of philosophies and heresies, which is very weakness.

Two opinions encounter; each may be abstractedly true; or again, each may be a subtle, comprehensive doctrine, vigorous, elastic, expansive, various; one is held as a matter of indifference, the other as a matter of life and death; one is held by the intellect only, the other also by the heart: it is plain which of the two must succumb to the other. Such was the conflict of Christianity with the old established Paganism, which was almost dead before Christianity appeared; with the Oriental Mysteries, flitting wildly to and fro like spectres; with the Gnostics, who made Knowledge all in all, despised the many, and called Catholics mere children in the Truth; with the Neo-platonists, men of literature, pedants, visionaries, or courtiers; with the Manichees, who professed to seek Truth by Reason, not by Faith; with the fluctuating teachers of the school of Antioch, the time-serving Eusebians, and the reckless versatile Arians; with the fanatic Montanists and harsh Novatians, who shrank from the Catholic doctrine, without power to propagate their own. These sects had no stay or consistence, yet they contained elements of truth amid their error, and had Christianity been as they, it might have resolved into them; but it had that hold of the truth which gave its teaching a gravity, a directness, a consistency, a sternness, and a force, to which its rivals for the most part were strangers. It could not call evil good, or good evil, because it discerned the difference between them; it could not make light of what was so solemn, or desert what was so solid. Hence, in the collision, it broke in pieces its antagonists, and divided the spoils.

This was but another form of the spirit that made martyrs. Dogmatism was in teaching, what confession was in act. Each was the same strong principle of life in a different aspect, distinguishing the faith which was displayed in it from the world's philosophies on the one side, and the world's religions on the other. The heathen sects and the heresies of Christian history were dissolved by the breath of opinion which made them; paganism shuddered and died at the very sight of the sword of persecution, which it had itself unsheathed. Intellect and force were applied as tests both upon the divine and upon the human work; they prevailed with the human, they did but become instruments of the Divine. . . . Thus Christianity grew in its proportions, gaining aliment and medicine from all that it came near, yet preserving its original type, from its perception and its love of what had been revealed once for all and was no private imagination.

There are writers who refer to the first centuries of the Church as a time when opinion was free, and the conscience exempt from the obligation or temptation to take on trust what it had not proved; and that, apparently on the mere ground that the series of great theological decisions did not commence till the fourth. . . . The principle indeed of Dogmatism developes into Councils in the course of time; but it was active, nay sovereign from the first, in every part of Christendom. A conviction that truth was one: that it was a gift from without, a sacred trust, an inestimable blessing; that it was to be reverenced, guarded, defended, transmitted; that its absence was a grievous want, and its loss an unutterable calamity; and again, the stern words and acts of St. John, of Polycarp, Ignatius, Irenæus, Clement, Tertullian, and Origen;—all this is quite consistent with perplexity or mistake as to what was truth in particular cases, in what way doubtful questions were to be decided, or what were the limits of the Revelation. Councils and Popes are the guardians and instruments of the dogmatic principle: they are not that principle themselves; they presuppose the principle; they are summoned into action at the call of the principle, and the principle might act even before they had their legitimate place, and exercised a recognized power, in the movements of the Christian body.

The instance of Conscience, which has already served us in illustration, may assist us here. What Conscience is in the history of an individual mind, such was the dogmatic principle in the history of Christianity. Both in the one case and the other, there is the gradual formation of a directing power out of a principle. The natural voice of Conscience is far more imperative in testifying and enforcing a rule of duty, than successful in determining that duty in particular cases. It acts as a messenger from above, and says that there is a right and a wrong, and that the right must be followed; but it is variously, and therefore erroneously, trained in the instance of various persons. It mistakes error for truth; and yet we believe that on the whole, and even in those cases where it is ill-instructed, if its voice be diligently obeyed, it will gradually be cleared, simplified, and perfected, so that minds, starting differently will, if honest, in course of time converge to one and the same truth. I do not hereby imply that there is indistinctness so great as this in the theology of the first centuries; but so far is plain, that the early Church and Fathers exercised far more a ruler's than a doctor's office: it was the age of Martyrs, of acting not of thinking. Doctors succeeded Martyrs, light and peace of conscience follow upon obedience to it; yet, even before the Church had grown into the full measure of its doctrines, it was rooted in its principles.

So far, however, may be granted . . . that even principles were not so well understood and so carefully handled at first, as they were afterwards. In the early period, we see traces of a conflict, as well as of a variety, in theological elements, which were in course of combination, but which required adjustment and management before they could be used with precision as one. In a thousand instances of a minor character, the statements of the early Fathers are but tokens of the multiplicity of openings which the mind of the Church was making into the treasure-house of Truth; real openings, but incomplete or irregular. Nay, the doctrines even of the heretical bodies are indices and anticipations of the mind of the Church. As the first step in settling a question of doctrine is to raise and debate it, so heresies in every age may be taken as the measure of the existing state of thought in the Church, and of the movement of her theology; they determine in what way the current is setting, and the rate at which it flows.

[ch. 8]

Chronic Vigour

We have arrived at length at the seventh and last test, which was laid down when we started, for distinguishing the true development of an idea from its corruptions and perversions: it is this. A corruption, if vigorous, is of brief duration, runs itself out quickly, and ends in death; on the other hand, if it lasts, it fails in vigour and passes into a decay. This general law gives us additional assistance in determining the character of the developments of Christianity commonly called Catholic.

When we consider the succession of ages during which the Catholic system has endured, the severity of the trials it has undergone, the sudden and wonderful changes without and within which have befallen it, the incessant mental activity and the intellectual gifts of its maintainers, the enthusiasm which it has kindled, the fury of the controversies which have been carried on among its professors, the impetuosity of the assaults made upon it, the ever-increasing responsibilities to which it has been committed by the continuous development of its dogmas, it is quite inconceivable that it should not have been broken up and lost, were it a corruption of Christianity. Yet it is still living, if there be a living religion or philosophy in the world; vigorous, energetic, persuasive, progressive; *vires acquirit eundo*; it grows and is not overgrown; it spreads out, yet is not enfeebled; it is ever germinating, yet ever consistent with itself. Corruptions indeed are to be found which sleep and are suspended; and these, as I have said, are usually called "decays:" such is not the case with Catholicity; it does not sleep, it is not stationary even now; and that its long series of developments should be corruptions would be an instance of sustained error, so novel, so unaccountable, so preternatural, as to be little short of a miracle, and to rival those manifestations of Divine Power which constitute the evidence of Christianity. We sometimes view with surprise and awe the degree of pain and disarrangement which the human frame can undergo without succumbing; yet at length there comes an end. Fevers have their crisis, fatal or favourable; but this corruption of a thousand years, if corruption it be, has ever been growing nearer death, yet never reaching it, and has been strengthened, not debilitated, by its excesses.

For instance: when the Empire was converted, multitudes, as is very plain, came into the Church on but partially religious motives, and with habits and opinions infected with the false worships which they had professedly abandoned. History shows us what anxiety and effort it cost her rulers to keep Paganism out of her pale. To this tendency must be added the hazard which attended on the development of the Catholic ritual, such as the honours publicly assigned to Saints and Martyrs, the formal veneration of their relics, and the usages and observances which followed. What was to hinder the rise of a sort of refined Pantheism, and the overthrow of dogmatism *pari passu* with the multiplication of heavenly intercessors and patrons? If what is called in reproach "Saint-worship" resembled the polytheism which it supplanted, or was a corruption, how did Dogmatism survive? Dogmatism is a religion's profession of its own reality as contrasted with other systems; but polytheists are liberals, and hold that one religion is as good as another. Yet the theological system was developing and strengthening, as well as the monastic rule, which is intensely anti-pantheistic, all the while the ritual was assimilating itself, as Protestants say, to the Paganism of former ages.

Nor was the development of dogmatic theology, which was then taking place, a silent and spontaneous process. It was wrought out and carried through under the fiercest controversies, and amid the most fearful risks. The Catholic faith was placed in a succession of perils, and rocked to and fro like a vessel at sea. Large portions of Christendom were, one after another, in heresy or in schism; the leading Churches and the most authoritative schools fell from time to time into serious error; three Popes, Liberius, Vigilius, Honorius, have left to posterity the burden of their defence: but these disorders were no interruption to the sustained and steady march of the sacred science from implicit belief to formal statement. The series of ecclesiastical decisions, in which its progress was ever and anon signified, alternate between the one and the other side of the theological dogma especially in question, as if fashioning it into shape by opposite strokes. The controversy began in

Apollinaris, who confused or denied the Two Natures in Christ, and was condemned by Pope Damasus. A reaction followed, and Theodore of Mopsuestia suggested by his teaching the doctrine of Two Persons. After Nestorius had brought that heresy into public view, and had incurred in consequence the anathema of the Third Ecumenical Council, the current of controversy again shifted its direction; for Eutyches appeared, maintained the One Nature, and was condemned at Chalcedon. Something however was still wanting to the overthrow of the Nestorian doctrine of Two Persons, and the Fifth Council was formally directed against the writings of Theodore and his party. Then followed the Monothelite heresy, which was a revival of the Eutychian or Monophysite, and was condemned in the Sixth. Lastly, Nestorianism once more showed itself in the Adoptionists of Spain, and gave occasion to the great Council of Frankfort. Any one false step would have thrown the whole theory of the doctrine into irretrievable confusion; but it was as if some one individual and perspicacious intellect, to speak humanly, ruled the theological discussion from first to last. That in the long course of centuries, and in spite of the failure, in points of detail, of the most gifted Fathers and Saints, the Church thus wrought out the one and only consistent theory which can be taken on the great doctrine in dispute, proves how clear, simple, and exact her vision of that doctrine was. But it proves more than this. Is it not utterly incredible, that with this thorough comprehension of so great a mystery, as far as the human mind can know it, she should be at that very time in the commission of the grossest errors in religious worship, and should be hiding the God and Mediator, whose Incarnation she contemplated with so clear an intellect, behind a crowd of idols? . . .

It is true, there have been seasons when, from the operation of external or internal causes, the Church has been thrown into what was almost a state of *deliquium*; but her wonderful revivals, while the world was triumphing over her, is a further evidence of the absence of corruption in the system of doctrine and worship into which she has developed. If corruption be an incipient disorganization, surely an abrupt and absolute recurrence to the former state of vigour, after an interval, is even less conceivable than a corruption that is permanent. Now this is the case with the revivals I speak of. After violent exertion men are exhausted and fall asleep; they awake the same as before, refreshed by the temporary cessation of their activity; and such has been the slumber and such the restoration of the Church. She pauses in her course, and almost suspends her functions; she rises again, and she is herself once more; all things are in their place and ready for action. Doctrine is where it was, and usage, and precedence, and principle, and policy; there may be changes, but they are consolidations or adaptations; all is unequivocal and determinate, with an identity which there is no disputing. Indeed it is one of the most popular charges against the Catholic Church at this very time, that she is "incorrigible;"—change she cannot, if we listen to St. Athanasius or St. Leo; change she never will, if we believe the controversialist or alarmist of the present day.

Such were the thoughts concerning the "Blessed Vision of Peace," of one whose long-continued petition had been that the Most Merciful would not despise the work of His own Hands, nor leave him to himself:—while yet his eyes were dim, and his breast laden, and he could but employ Reason in the things of Faith. And now, dear Reader, time is short, eternity is long. Put not from you what you have here found; regard it not as mere matter of present controversy; set not out resolved to refute it, and looking about for the best way of doing so; seduce not yourself with the imagination that it comes of disappointment, or disgust, or restlessness, or wounded feeling, or undue sensibility, or other weakness. Wrap not yourself round in the associations of years past, nor determine that to be truth which you wish to be so, nor make an idol of cherished anticipations. Time is short, eternity is long.

[ch. 12]

'A CHARACTERISTIC OF THE POPES. ST. GREGORY THE GREAT'

Detachment, as we know from spiritual books, is a rare and high Christian virtue; a great Saint, St. Philip Neri, said that, if he had a dozen really detached men, he should be able to convert the world. To be detached is to be loosened from every tie which binds the soul to the earth, to be dependent on nothing sublunary, to lean on nothing temporal; it is to care simply nothing what other men choose to think or say of us, or do to us; to go about our own work, because it is our duty, as soldiers go to battle, without a care for the consequences; to account credit, honour, name, easy circumstances, comfort, human affections, just nothing at all, when any religious obligation involves the sacrifice of them. It is to be as reckless of all these goods of life on such occasions, as under ordinary circumstances we are lavish and wanton, if I must take an example, in our use of water,—or as we make a present of our words without grudging to friend or stranger,—or as we get rid of wasps or flies or gnats, which trouble us, without any sort of compunction, without hesitation before the act, and without a second thought after it.

Now this "detachment" is one of the special ecclesiastical virtues of the Popes. They are of all men most exposed to the temptation of secular connections; and, as history tells us, they have been of all men least subject to it. By their very office they are brought across every form of earthly power; for they have a mission to high as well as low, and it is on the high, and not the low, that their maintenance ordinarily depends. Cæsar ministers to Christ; the framework of society, itself a divine ordinance, receives such important aid from the sanction of religion, that it is its interest in turn to uphold religion, and to enrich it with temporal gifts and honours. Ordinarily speaking, then, the Roman Pontiffs owe their exaltation to the secular power, and have a great stake in its stability and prosperity. Under such circumstances, any men but they would have had a strong leaning towards what is called "Conservatism;" and they have been, and are, of course Conservatives in the right sense of the word; that is, they cannot bear anarchy, they think revolution an evil, they pray for the peace of the world and the prosperity of all Christian States, and they effectively support the cause of order and good government. The name of Religion is but another name for law on the one hand, freedom on the other; and at this very time, who are its professed enemies, but Socialists, Red Republicans, Anarchists, and Rebels? But a Conservative, in the political sense of the word, commonly signifies something else, which the Pope never is, and cannot be. It means a man who is at the top of the tree, and knows it, and means never to come down, whatever it may cost him to keep his place there. It means a man who upholds government and society and the existing state of things,—not because it exists,—not because it is good and desirable, because it is established, because it is a benefit to the population, because it is full of promise for the future,—but rather because he himself is well off in consequence of it, and because to take care of number one is his main political principle. It means a man who defends religion, not for religion's sake, but for the sake of its accidents and externals; and in this sense Conservative a Pope can never be, without a simple betrayal of the dispensation committed to him. Hence at this very moment the extreme violence against the Holy See, of the British legislature and constituency and their newspapers and other organs, mainly because it will not identify the cause of civil government with its own, because, while it ever benefits this world, it ever contemplates the world unseen.

So much, however, is intelligible enough; but there is a more subtle form of Conservatism, by which ecclesiastical persons are much more likely to be tempted and overcome, and to which also the Popes are shown in history to be superior. Temporal possessions and natural gifts may rightly be dedicated to the service of religion; however, since they do not lose their old nature by being invested by a new mission or quality, they still possess the pabulum of temptation, and may be fatal to ecclesiastical "detachment." To prefer the establishment of religion to its purity, is Conservatism, though in a plausible garb. It was once of no uncommon occurrence for saintly Bishops, in the time of famine or war, to break up the Church plate and sell it, in order to relieve the hungry or to redeem the captives by the sums which it brought them. Now this proceeding was not infrequently urged against them in their day as some great offence; but the Church has always justified them. Here we see, as in a typical instance, both the wrong Conservatism, of which I am speaking, and its righteous repudiation. This fault is an over-attachment to the ecclesiastical establishment, as such;—to the seats of its power, to its holy places, its sanctuaries, churches, and palaces,—to its various national hierarchies, with their several prescriptions, privileges, and possessions,—to traditional lines of policy, precedent, and discipline,— to rules and customs of long standing. But a great Pontiff must be detached from everything save the deposit of faith, the tradition of the Apostles, and the vital principles of the divine polity. He may use, he may uphold, he may and will be very slow to part with, a hundred things which have grown up, or taken shelter, or are stored, under the shadow of the Church; but, at bottom, and after all, he will be simply detached from pomp and etiquette, secular rank, secular learning, schools and libraries, Basilicas and Gothic cathedrals, old ways, old alliances, and old friends. He will be rightly jealous of their loss, but still he will "know nothing but" Him whose Vicar he is; he will not stake his fortunes, he will not rest his cause, upon any one else:—this is what he will do, and what he will not do, as in fact the great Popes of history have shown, in their own particular instances, on so many and various occasions.

Take the early Martyr-Popes, or the Gregories and the Leos; whether they were rich or poor, in power or in persecution, they were simply detached from every earthly thing save the Rock of Peter. This was their adamantine foundation, their starting-point in every enterprise, their refuge in every calamity, the point of leverage by which they moved the world. Secure in this, they have let other things come and go, as they would; or have deliberately made light of what they had, in order that they might gain what they had not. They have known, in the fulness of an heroic faith, that, while they were true to themselves and to their divinely appointed position, they could not but "inherit the earth," and that, if they lost ground here, it was only to make progress elsewhere. Old men usually get fond of old habits; they cannot imagine, understand, relish anything to which they are not accustomed. The Popes have been old men; but, wonderful to say, they have never been slow to venture out upon a new line, when it was necessary, and had even been looking about, sounding, exploring, taking observations, reconnoitring, attempting, even when there was no immediate reason why they should not let well alone, as the world would say, or even when they were hampered with difficulties at their door so great, that you would think that they had no time or thought to spare for anything in the distance. It is but a few years ago that a man of eighty, of humble origin, the most Conservative of Popes, as he was considered, with disaffection and sedition upheaving his throne, was found to be planning missions for the interior of Africa, and, when a moment's opportunity was given him, made the most autocratical of Emperors, the very hope of Conservatives, the very terror of Catholics, quail beneath his glance. And, thus independent of times and places, the Popes have never found any difficulty, when the proper moment came, of following out a new and daring line of policy (as their astonished foes have called it), of leaving the old world to shift for itself and to disappear from the scene in its due season, and of fastening on and establishing themselves in the new. . . .

[*Historical Sketches*, vol. iii]

APOLOGIA PRO VITA SUA

Liberalism

Liberty of thought is in itself a good; but it gives an opening to false liberty. Now by Liberalism I mean false liberty of thought, or the exercise of thought upon matters, in which, from the constitution of the human mind, thought cannot be brought to any successful issue, and therefore is out of place. Among such matters are first principles of whatever kind; and of these the most sacred and momentous are especially to be reckoned the truths of Revelation. Liberalism then is the mistake of subjecting to human judgment those revealed doctrines which are in their nature beyond the independent of it, and of claiming to determine on intrinsic grounds the truth and value of propositions which rest for their reception simply on the external authority of the Divine Word.

[note A]

Infallibility of the Church

Starting . . . with the being of a God, (which . . . is as certain to me as the certainty of my own existence, though when I try to put the grounds of that certainty into logical shape I find a difficulty in doing so in mood and figure to my satisfaction,) I look out of myself into the world of men, and there I see a sight which fills me with unspeakable distress. The world seems simply to give the lie to that great truth, of which my whole being is so full; and the effect upon me is, in consequence, as a matter of necessity, as confusing as if it denied that I am in existence myself. If I looked into a mirror, and did not see my face, I should have the sort of feeling which actually comes upon me, when I look into this living busy world, and see no reflexion of its Creator. This is, to me, one of those great difficulties of this absolute primary truth, to which I referred just now. Were it not for this voice, speaking so clearly in my conscience and my heart, I should be an atheist, or a pantheist, or a polytheist when I looked into the world. I am speaking for myself only; and I am far from denying the real force of the arguments in proof of a God, drawn from the general facts of human society and the course of history, but these do not warm me or enlighten me; they do not take away the winter of my desolation, or make the buds unfold and the leaves grow within me, and my moral being rejoice. The sight of the world is nothing else than the prophet's scroll, full of "lamentations, and mourning, and woe."

To consider the world in its length and breadth, its various history, the many races of man, their starts, their fortunes, their mutual alienation, their conflicts; and then their ways, habits, governments, forms of worship; their enterprises, their aimless courses, their random achievements and acquirements, the impotent conclusion of long-standing facts, the tokens so faint and broken of a superintending design, the blind evolution of what turn out to be great powers or truths, the progress of things, as if from unreasoning elements, not towards final causes, the greatness and littleness of man, his far-reaching aims, his short duration, the curtain hung over his futurity, the disappointments of life, the defeat of good, the success of evil, physical pain, mental anguish, the prevalence and intensity of sin, the pervading idolatries, the corruptions, the dreary hopeless irreligion, that condition of the whole race, so fearfully yet exactly described in the Apostle's words, "having no hope and without God in the world,"—all this is a vision to dizzy and appal; and inflicts upon the mind the sense of a profound mystery, which is absolutely beyond human solution.

What shall be said to this heart-piercing, reason-bewildering fact? I can only answer, that either there is no Creator, or this living society of men is in a true sense discarded from His presence. Did I see a boy of good make and mind, with the tokens on him of a refined nature, cast upon the world without provision, unable to say whence he came, his birth-place or his family connexions, I should conclude that there was some mystery connected with his history, and that he was one, of whom, from one cause or other, his parents were ashamed. Thus only should I be able to account for the contrast between the promise and the condition of his being. And so I argue about the world;—*If* there be a God, *since* there is a God, the human race is implicated in some terrible aboriginal calamity. It is out of joint with the purposes of its Creator. This is a fact, a fact as true as the fact of its existence; and thus the doctrine of what is theologically called original sin becomes to me almost as certain as that the world exists, and as the existence of God.

And now, supposing it were the blessed and loving will of the Creator to interfere in this anarchical condition of things, what are we to suppose would be the methods which might be necessarily or naturally involved in His purpose of mercy? Since the world is in so abnormal a state, surely it would be no surprise to me, if the interposition were of necessity equally extraordinary—or what is called miraculous. But that subject does not directly come into the scope of my present remarks. Miracles as evidence, involve a process of reason, or an argument; and of course I am thinking of some mode of interference which does not immediately run into argument. I am rather asking what must be the face-to-face antagonist, by which to withstand and baffle the fierce energy of passion and the all-corroding, all-dissolving scepticism of the intellect in religious inquiries? I have no intention at all of denying, that truth is the real object of our reason, and that, if it does not attain to truth, either the premiss or the process is in fault; but I am not speaking here of right reason, but of reason as it acts in fact and concretely in fallen man. I know that even the unaided reason, when correctly exercised, leads to a belief in God, in the immortality of the soul, and in a future retribution; but I am considering the faculty of reason actually and historically; and in this point of view, I do not think I am wrong in saying that its tendency is towards a simple unbelief in matters of religion. No truth, however sacred, can stand against it, in the long run; and hence it is that in the pagan world, when our Lord came, the last traces of the religious knowledge of former times were all but disappearing from those portions of the world in which the intellect had been active and had had a career.

And in these latter days, in like manner, outside the Catholic Church things are tending,—with far greater rapidity than in that old time from the circumstance of the age,—to atheism in one shape or other. What a scene, what a prospect, does the whole of Europe present at this day! and not only Europe, but every government and every civilization through the world, which is under the influence of the European mind! Especially, for it most concerns us, how sorrowful, in the view of religion, even taken in its most elementary, most attenuated form, is the spectacle presented to us by the educated intellect of England, France, and Germany! Lovers of their country and of their race, religious men, external to the Catholic Church, have attempted various expedients to arrest fierce wilful human nature in its onward course, and to bring it into subjection. The necessity of some form of religion for the interests of humanity, has been generally acknowledged: but where was the concrete representative of things invisible, which would have the force and the toughness necessary to be a breakwater against the deluge? Three centuries ago the establishment of religion, material, legal, and social, was generally adopted as the best expedient for the purpose, in those countries which separated from the Catholic Church; and for a long time it was successful; but now the crevices of those establishments are admitting the enemy. Thirty years ago, education was relied upon: ten years ago there was a hope that wars would cease for ever, under the influence of commercial enterprise and the reign of the useful and fine arts; but will any one venture to say that there is any thing any where on this earth, which will afford a fulcrum for us, whereby to keep the earth from moving onwards?

The judgment, which experience passes whether on establishments or on education, as a means of maintaining religious truth in this anarchical world, must be extended even to Scripture, though Scripture be divine. Experience proves surely that the Bible does not answer a purpose for which it was never intended. It may be accidentally the means of the conversion of individuals; but a book, after all, cannot make a stand against the wild living intellect of man, and in this day it begins to testify, as regards it own structure and contents, to the power of that universal solvent, which is so successfully acting upon religious establishments.

Supposing then it to be the Will of the Creator to interfere in human affairs, and to make provisions for retaining in the world a knowledge of Himself, so definite and distinct as to be proof against the energy of human scepticism, in such a case,—I am far from saying that there was no other way,—but there is nothing to surprise the mind, if He should think fit to introduce a power into the world, invested with the prerogative of infallibility in religious matters. Such a provision would be a direct, immediate, active, and prompt means of withstanding the difficulty; it would be an instrument suited to the need; and when I find that this is the very claim of the Catholic Church, not only do I feel no difficulty in admitting the idea, but there is a fitness in it, which recommends it to my mind. And thus I am brought to speak of the Church's infallibility, as a provision, adapted by the mercy of the Creator, to preserve religion in the world, and to restrain that freedom of thought, which of course in itself is one of the greatest of our natural gifts, and to rescue it from its own suicidal excesses. And let it be observed that, neither here nor in what follows, shall I have occasion to speak directly of Revelation in its subject-matter, but in reference to the sanction which it gives to truths which may be known independently of it,—as it bears upon the defence of natural religion. I say, that a power, possessed of infallibility in religious teaching, is happily adapted to be a working instrument, in the course of human affairs, for smiting hard and throwing back the immense energy of the aggressive, capricious, untrustworthy intellect . . .

And first, the initial doctrine of the infallible teacher must be an emphatic protest against the existing state of mankind. Man had rebelled against his Maker. It was this that caused the divine interposition: and to proclaim it must be the first act of the divinely-accredited messenger. The Church must denounce rebellion as of all possible evils the greatest. She must have no terms with it; if she would be true to her Master, she must ban and anathematize it. . . .

In like manner she has ever put forth, with most energetic distinctness, those other great elementary truths, which either are an explanation of her mission or give a character to her work. She does not teach that human nature is irreclaimable, else wherefore should she be sent? not, that it is to be shattered and reversed, but to be extricated, purified, and restored; not, that it is a mere mass of hopeless evil, but that it has the promise upon it of great things, and even now, in its present state of disorder and excess, has a virtue and a praise proper to itself. But in the next place she knows and she preaches that such a restoration, as she aims at effecting in it, must be brought about, not simply through certain outward provisions of preaching and teaching, even though they be her own, but from an inward spiritual power or grace imparted directly from above, and of which she is the channel. She has it in charge to rescue human nature from its misery, but not simply by restoring it on its own level, but by lifting it up to a higher level than its own. She recognizes in it real moral excellence though degraded, but she cannot set it free from earth except by exalting it towards heaven. It was for this end that a renovating grace was put into her hands; and therefore from the nature of the gift, as well as from the reasonableness of the case, she goes on, as a further point, to insist, that all true conversion must begin with the first springs of thought, and to teach that each individual man must be in his own person one whole and perfect temple of God, while he is also one of the living stones which build up a visible religious community. And thus the distinctions between nature and grace, and between outward and inward religion, become two further articles in what I have called the preamble of her divine commission.

Such truths as these she vigorously reiterates, and pertinaciously inflicts upon mankind; as to such she observes no half-measures, no economical reserve, no delicacy or prudence. "Ye must be born again," is the simple, direct form of words which she uses after her Divine Master: "your whole nature must be re-born; your passions, and your affections, and your aims, and your conscience, and your will, must all be bathed in a new element, and reconsecrated to your Maker,—and, the last not the least, your intellect." . . .

Passing now from what I have called the preamble of that grant of power, which is made to the Church, to that power itself, Infallibility, I premise two brief remarks:—1. on the one hand, I am not here determining any thing about the essential seat of that power, because that is a question doctrinal, not historical and practical; 2. nor, on the other hand, am I extending the direct subject-matter, over which that power of Infallibility has jurisdiction, beyond religious opinion:—and now as to the power itself.

This power, viewed in its fulness, is as tremendous as the giant evil which has called for it. It claims, when brought into exercise but in the legitimate manner, for otherwise of course it is but quiescent, to know for certain the very meaning of every portion of that Divine Message in detail, which was committed by our Lord to His Apostles. It claims to know its own limits, and to decide what it can determine absolutely and what it cannot. It claims, moreover, to have a hold upon statements not directly religious, so far as this,—to determine whether they indirectly relate to religion, and, according to its own definitive judgment, to pronounce whether or not, in a particular case, they are simply consistent with revealed truth. It claims to decide magisterially, whether as within its own province or not, that such and such statements are or are not prejudicial to the *Depositum* of faith, in their spirit or in their consequences, and to allow them, or condemn and forbid them, accordingly. It claims to impose silence at will on any matters, or controversies, of doctrine, which on its own *ipse dixit*, it pronounces to be dangerous, or inexpedient, or inopportune. It claims that, whatever may be the judgment of Catholics upon such acts, these acts should be received by them with those outward marks of reverence, submission, and loyalty, which Englishmen, for instance, pay to the presence of their sovereign, without expressing any criticism on them on the ground that in their matter they are inexpedient, or in their manner violent or harsh. And lastly, it claims to have the right of inflicting spiritual punishment, of cutting off from the ordinary channels of the divine life, and of simply excommunicating, those who refuse to submit themselves to its formal declarations. Such is the infallibility lodged in the Catholic Church, viewed in the concrete, as clothed and surrounded by the appendages of its high sovereignty: it is, to repeat what I said above, a supereminent prodigious power sent upon earth to encounter and master a giant evil.

And now, having thus described it, I profess my own absolute submission to its claim. I believe the whole revealed dogma as taught by the Apostles, as committed by the Apostles to the Church, and as declared by the Church to me. I receive it, as it is infallibly interpreted by the authority to whom it is thus committed, and (implicitly) as it shall be, in like manner, further interpreted by that same authority till the end of time. I submit, moreover, to the universally received traditions of the Church, in which lies the matter of those new dogmatic definitions which are from time to time made, and which in all times are the clothing and the illustration of the Catholic dogma as already defined. And I submit myself to those other decisions of the Holy See, theological or not, through the organs which it has itself appointed, which, waiving the question of their infallibility, on the lowest ground come to me with a claim to be accepted and obeyed. Also, I consider that, gradually and in the course of ages, Catholic inquiry has taken certain definite shapes, and has thrown itself into the form of a science, with a method and a phraseology of its own, under the intellectual handling of great minds, such as St. Athanasius, St. Augustine, and St. Thomas; and I feel no temptation at all to break in pieces the great legacy of thought thus committed to us for these latter days.

All this being considered as the profession which I make *ex animo*, as for myself, so also on the part of the Catholic body, as far as I know it, it will at first sight be said that the restless intellect of our common humanity is utterly weighed down, to the repression of all independent effort and action whatever, so that, if this is to be the mode of bringing it into order, it is brought into order only to be destroyed. But this is far from the result, far from what I conceive to be the intention of that high Providence who has provided a great remedy for a great evil,—far from borne out by the history of the conflict between Infallibility and Reason in the past, and the prospect of it in the future. The energy of the human intellect "does from opposition grow;" it thrives and is joyous, with a tough elastic strength, under the terrible blows of the divinely-fashioned weapon, and is never so much itself as when it has lately been overthrown. It is the custom with Protestant writers to consider that, whereas there are two great principles in action in the history of religion, Authority and Private Judgment, they have all the Private Judgment to themselves, and we have the full inheritance and the superincumbent oppression of Authority. But this is not so; it is the vast Catholic body itself, and it only, which affords an arena for both combatants in that awful, never-dying duel. It is necessary for the very life of religion, viewed in its large operations and its history, that the warfare should be incessantly carried on. Every exercise of Infallibility is brought out into act by an intense and varied operation of the Reason, both as its ally and as its opponent, and provokes again, when it has done its work, a reaction of Reason against it; and, as in a civil polity the State exists and endures by means of the rivalry and collision, the encroachments and defeats of its constituent parts, so in like manner Catholic Christendom is no simple exhibition of religious absolutism, but presents a continuous picture of Authority and Private Judgment alternately advancing and retreating as the ebb and flow of the tide;—it is a vast assemblage of human beings with wilful intellects and wild passions, brought together into one by the beauty and the Majesty of a Superhuman Power,—into what may be called a large reformatory or training-school, not as if into a hospital or into a prison, not in order to be sent to bed, not to be buried alive, but (if I may change my metaphor) brought together as if into some moral factory, for the melting, refining, and moulding, by an incessant, noisy process, of the raw material of human nature, so excellent, so dangerous, so capable of divine purposes.

St. Paul says in one place that his Apostolical power is given him to edification, and not to destruction. There can be no better account of the Infallibility of the Church. It is a supply for a need, and it does not go beyond that need. Its object is, and its effect also, not to enfeeble the freedom or vigour of human thought in religious speculation, but to resist and control its extravagance. What have been its great works? All of them in the distinct province of theology:—to put down Arianism, Eutychianism, Pelagianism, Manichæism, Lutheranism, Jansenism. Such is the broad result of its action in the past;—and now as to the securities which are given us that so it ever will act in time to come.

First, Infallibility cannot act outside of a definite circle of thought, and it must in all its decisions, or *definitions*, as they are called, profess to be keeping within it. The great truths of the moral law, of natural religion, and of Apostolical faith, are both its boundary and its foundation. It must not go beyond them, and it must ever appeal to them. Both its subject-matter, and its articles in that subject-matter, are fixed. And it must ever profess to be guided by Scripture and by tradition. It must refer to the particular Apostolic truth which it is enforcing, or (what is called) *defining*. Nothing, then, can be presented to me, in time to come, as part of the faith, but what I ought already to have received, and hitherto have been kept from receiving, (if so,) merely because it has not been brought home to me. Nothing can be imposed upon me different in kind from what I hold already,—much less contrary to it. The new truth which is promulgated, if it is to be called new, must be at least homogeneous, cognate, implicit, viewed relatively to the old truth. It must be what I may even have guessed, or wished, to be included in the Apostolic revelation; and at least it will be of such a character, that my thoughts readily concur in it or coalesce with it, as soon as I hear it. Perhaps I and others actually have always believed it, and the only question which is now decided in my behalf, is, that I have henceforth the satisfaction of having to believe, that I have only been holding all along what the Apostles held before me.

Let me take the doctrine which Protestants consider our greatest difficulty, that of the Immaculate Conception. Here I entreat the reader to recollect my main drift, which is this. I have no difficulty in receiving the doctrine; and that, because it so intimately harmonizes with that circle of recognized dogmatic truths, into which it has been recently received;—but if *I* have no difficulty, why may not another have no difficulty also? why may not a hundred? a thousand? Now I am sure that Catholics in general have not any intellectual difficulty at all on the subject of the Immaculate Conception; and that there is no reason why they should. Priests have no difficulty. You tell me that they *ought* to have a difficulty;—but they have not. Be large-minded enough to believe, that men may reason and feel very differently from yourselves; how is it that men, when left to themselves, fall into such various forms of religion, except that there are various types of mind among them, very distinct from each other? From my testimony then about myself, if you believe it, judge of others also who are Catholics: we do not find the difficulties which you do in the doctrines which we hold; we have no intellectual difficulty in that doctrine in particular, which you call a novelty of this day. We priests need not be hypocrites, though we be called upon to believe in the Immaculate Conception. To that large class of minds, who believe in Christianity after our manner,—in the particular temper, spirit, and light, (whatever word is used,) in which Catholics believe it,—there is no burden at all in holding that the Blessed Virgin was conceived without original sin; indeed, it is a simple fact to say, that Catholics have not come to believe it because it is defined, but that it was defined because they believed it.

So far from the definition in 1854 being a tyrannical infliction on the Catholic world, it was received every where on its promulgation with the greatest enthusiasm. It was in consequence of the unanimous petition, presented from all parts of the Church to the Holy See, in behalf of an *ex cathedrâ* declaration that the doctrine was Apostolic, that it was declared so to be. I never heard of one Catholic having difficulties in receiving that doctrine, whose faith on other grounds was not already suspicious. Of course there were grave and good men, who were made anxious by the doubt whether it could be formally proved to be Apostolical either by Scripture or tradition, and who accordingly, though believing it themselves, did not see how it could be defined by authority and imposed upon all Catholics as a matter of faith; but this is another matter. The point in question is, whether the doctrine is a burden. I believe it to be none. So far from it being so, I sincerely think that St.

Bernard and St. Thomas, who scrupled at it in their day, had they lived into this, would have rejoiced to accept it for its own sake. Their difficulty, as I view it, consisted in matters of words, ideas, and arguments. They thought the doctrine inconsistent with other doctrines; and those who defended it in that age had not that precision in their view of it, which has been attained by means of the long disputes of the centuries which followed. And in this want of precision lay the difference of opinion, and the controversy.

Now the instance which I have been taking suggests another remark; the number of those (so called) new doctrines will not oppress us, if it takes eight centuries to promulgate even one of them. Such is about the length of time through which the preparation has been carried on for the definition of the Immaculate Conception. This of course is an extraordinary case; but it is difficult to say what is ordinary, considering how few are the formal occasions on which the voice of Infallibility has been solemnly lifted up. It is to the Pope in Ecumenical Council that we look, as to the normal seat of Infallibility: now there have been only eighteen such Councils since Christianity was,—an average of one to a century,—and of these Councils some passed no doctrinal decree at all, others were employed on only one, and many of them were concerned with only elementary points of the Creed. The Council of Trent embraced a large field of doctrine certainly; but I should apply to its Canons a remark contained in that University Sermon of mine, which has been so ignorantly criticized in the Pamphlet which has been the occasion of this Volume;—I there have said that the various verses of the Athanasian Creed are only repetitions in various shapes of one and the same idea; and in like manner, the Tridentine Decrees are not isolated from each other, but are occupied in bringing out in detail, by a number of separate declarations, as if into bodily form, a few necessary truths. I should make the same remark on the various theological censures, promulgated by Popes, which the Church has received, and on their dogmatic decisions generally. I own that at first sight those decisions seem from their number to be a greater burden on the faith of individuals than are the Canons of Councils; still I do not believe that in matter of fact they are so at all, and I give this reason for it:—it is not that a Catholic, layman or priest, is indifferent to the subject, or, from a sort of recklessness, will accept any thing that is placed before him, or is willing, like a lawyer, to speak according to his brief, but that in such condemnations the Holy See is engaged, for the most part, in repudiating one or two great lines of error, such as Lutheranism or Jansenism, principally ethical not doctrinal, which are divergent from the Catholic mind, and that it is but expressing what any good Catholic, of fair abilities, though unlearned, would say himself, from common and sound sense, if the matter could be put before him.

Now I will go on in fairness to say what I think *is* the great trial to the Reason, when confronted with that august prerogative of the Catholic Church, of which I have been speaking. I enlarged just now upon the concrete shape and circumstances, under which pure infallible authority presents itself to the Catholic. That authority has the prerogative of an indirect jurisdiction on subject-matters which lie beyond its own proper limits, and it most reasonably has such a jurisdiction. It could not act in its own province, unless it had a right to act out of it. It could not properly defend religious truth, without claiming for that truth what may be called its *pomœria*; or, to take another illustration, without acting as we act, as a nation, in claiming as our own, not only the land on which we live, but what are called British waters. The Catholic Church claims, not only to judge infallibly on religious questions, but to animadvert on opinions in secular matters which bear upon religion, on matters of philosophy, of science, of literature, of history, and it demands our submission to her claim. It claims to censure books, to silence authors, and to forbid discussions. In this province, taken as a whole, it does not so much speak doctrinally, as enforce measures of discipline. It must of course be obeyed without a word, and perhaps in process of time it will tacitly recede from its own injunctions. In such cases the question of faith does not come in at all; for what is matter of faith is true for all times, and never can be unsaid. Nor does it at all follow, because there is a gift of infallibility in the Catholic Church, that therefore the parties who are in possession of it are in all their proceedings infallible. "O, it is excellent," says the poet, "to have a giant's strength, but tyrannous, to use it like a giant." I think history supplies us with instances in the Church, where legitimate power has been harshly used. To make such admission is no more than saying that the divine treasure, in the words of the Apostle, is "in earthen vessels;" nor does it follow that the substance of the acts of the ruling power is not right and expedient, because its manner may have been faulty. Such high authorities act by means of instruments; we know how such instruments claim for themselves the name of their principals, who thus get the credit of faults which really are not theirs. But granting all this to an extent greater than can with any show of reason be imputed to the ruling power in the Church, what difficulty is there in the fact of this want of prudence or moderation more than can be urged, with far greater justice, against Protestant communities and institutions? What is there in it to make us hypocrites, if it has not that effect upon Protestants? We are called upon, not to profess any thing, but to submit and be silent, as Protestant Churchmen have before now obeyed the royal command to abstain from certain theological questions. Such injunctions as I have been contemplating are laid merely upon our actions, not upon our thoughts. How, for instance, does it tend to make a man a hypocrite, to be forbidden to publish a libel? his thoughts are as free as before: authoritative prohibitions may tease and irritate, but they have no bearing whatever upon the exercise of reason.

So much at first sight; but I will go on to say further, that, in spite of all that the most hostile critic may urge about the encroachments or severities of high ecclesiastics, in times past, in the use of their power, I think that the event has shown after all, that they were mainly in the right, and that those whom they were hard upon were mainly in the wrong. I love, for instance, the name of Origen: I will not listen to the notion that so great a soul was lost; but I am quite sure that, in the contest between his doctrine and followers and the ecclesiastical power, his opponents were right, and he was wrong. Yet who can speak with patience of his enemy and the enemy of St. John Chrysostom, that Theophilus, bishop of Alexandria? who can admire or revere Pope Vigilius? And here another consideration presents itself to my thoughts. In reading ecclesiastical history, when I was an Anglican, it used to be forcibly brought home to me, how the initial error of what afterwards became heresy was the urging forward some truth against the prohibition of authority at an unseasonable time. There is a time for every thing, and many a man desires a reformation of an abuse, or the fuller development of a doctrine, or the adoption of a particular policy, but forgets to ask himself whether the right time for it is come: and, knowing that there is no one who will be doing any thing towards its accomplishment in his own lifetime unless he does it himself, he will not listen to the voice of authority, and he spoils a good work in his own century, in order that another man, as yet unborn, may not have the opportunity of bringing it happily to perfection in the next. He may seem to the world to be nothing else than a bold champion for the truth and a martyr to free opinion, when he is just one of those persons whom the competent authority ought to silence; and, though the case may not fall within that subject-matter in which that authority is infallible, or the formal conditions of the exercise of that gift may be wanting, it is clearly the duty of authority to act vigorously in the case. Yet its act will go down to posterity as an instance of a tyrannical interference with private judgment, and of the silencing of a reformer, and of a base love of corruption or error; and it will show still less to advantage, if the ruling power happens in its proceedings to evince any defect of prudence or consideration. And all those who take the part of that ruling authority will be considered as time-servers, or indifferent to the cause of uprightness and truth; while, on the other hand, the said authority may be accidentally supported by a violent ultra party, which exalts opinions into dogmas, and has it principally at heart to destroy every school of thought but its own.

Such a state of things may be provoking and discouraging at the time, in the case of two classes of persons; of moderate men who wish to make differences in religious opinion as little as they fairly can be made; and of such as keenly perceive, and are honestly eager to remedy, existing evils,—evils, of which divines in this or that foreign country know nothing at all, and which even at home, where they exist, it is not every one who has the means of estimating. This is a state of things both of past time and of the present. We live in a wonderful age; the enlargement of the circle of secular knowledge just now is simply a bewilderment, and the more so, because it has the promise of continuing, and that with greater rapidity, and more signal results. Now these discoveries, certain or probable, have in matter of fact an indirect bearing upon religious opinions, and the question arises how are the respective claims of revelation and of natural science to be adjusted. Few minds in earnest can remain at ease without some sort of rational grounds for their religious belief; to reconcile theory and fact is almost an instinct of the mind. When then a flood of facts, ascertained or suspected, comes pouring in upon us, with a multitude of others in prospect, all believers in Revelation, be they Catholic or not, are roused to consider their bearing upon themselves, both for the honour of God, and from tenderness for those many souls who, in consequence of the confident tone of the schools of secular knowledge, are in danger of being led away into a bottomless liberalism of thought.

I am not going to criticize here that vast body of men, in the mass, who at this time would profess to be liberals in religion; and who look towards the discoveries of the age, certain or in progress, as their informants, direct or indirect, as to what they shall think about the unseen and the future. The Liberalism which gives a colour to society now, is very different from that character of thought which bore the name thirty or forty years ago. Now it is scarcely a party; it is the educated lay world. When I was young, I knew the word first as giving name to a periodical, set up by Lord Byron and others. Now, as then, I have no sympathy with the philosophy of Byron. Afterwards, Liberalism was the badge of a theological school, of a dry and repulsive character, not very dangerous in itself, though dangerous as opening the door to evils which it did not itself either anticipate or comprehend. At present it is nothing else than that deep, plausible scepticism, of which I spoke above, as being the development of human reason, as practically exercised by the natural man.

The Liberal religionists of this day are a very mixed body, and therefore I am not intending to speak against them. There may be, and doubtless is, in the hearts of some or many of them a real antipathy or anger against revealed truth, which it is distressing to think of. Again; in many men of science or literature there may be an animosity arising from almost a personal feeling; it being a matter of party, a point of honour, the excitement of a game, or a satisfaction to the soreness or annoyance occasioned by the acrimony or narrowness of apologists for religion, to prove that Christianity or that Scripture is untrustworthy. Many scientific and literary men, on the other hand, go on, I am confident, in a straightforward impartial way, in their own province and on their own line of thought, without any disturbance from religious difficulties in themselves, or any wish at all to give pain to others by the result of their investigations. It would ill become me, as if I were afraid of truth of any kind, to blame those who pursue secular facts by means of the reason which God has given them, to their logical conclusions: or to be angry with science, because religion is bound in duty to take cognizance of its teaching. But putting these particular classes of men aside, as having no special call on the sympathy of the Catholic, of course he does most deeply enter into the feelings of a fourth and large class of men, in the educated portions of society, of religious and sincere minds, who are simply perplexed,—frightened or rendered desperate, as the case may be,—by the utter confusion into which late discoveries or speculations have thrown their most elementary ideas of religion. Who does not feel for such men? who can have one unkind thought of them? I take up in their behalf St. Augustine's beautiful words, "Illi in vos sæviant," &c. Let them be fierce with you who have no experience of the difficulty with which error is discriminated from truth, and the way of life is found amid the illusions of the world. How many a Catholic has in his thoughts followed such men, many of them so good, so true, so noble! how often has the wish risen in his heart that some one from among his own people should come forward as the champion of revealed truth against its opponents! Various persons, Catholic and Protestant, have asked me to do so myself; but I had several strong difficulties in the way. One of the greatest is this, that at the moment it is so difficult to say precisely what it is that is to be encountered and overthrown. I am far from denying that scientific knowledge is really growing, but it is by fits and starts; hypotheses rise and fall; it is difficult to anticipate which of them will keep their ground, and what the state of knowledge in relation to them will be from year to year. In this condition of things, it has seemed to me to be very undignified for a Catholic to commit himself to the work of chasing what might turn out to be phantoms, and, in behalf of some special objections, to be ingenious in devising a theory, which, before it was completed, might have to give place to some theory newer still, from the fact that those former objections had already come to nought under the uprising of others. It seemed to be specially a time, in which Christians had a call to be patient, in which they had no other way of helping those who were alarmed, than that of exhorting them to have a little faith and fortitude, and to "beware," as the poet says, "of dangerous steps." This seemed so clear to me, the more I thought of the matter, as to make me surmise, that, if I attempted what had so little promise in it, I should find that the highest Catholic Authority was against the attempt, and that I should have spent my time and my thought, in doing what either it would be imprudent to bring before the public at all, or what, did I do so, would only complicate matters further which were already complicated, without my interference, more than enough. And I interpret recent acts of that authority as fulfilling my expectation; I interpret them as tying the hands of a controversialist, such as I should be, and teaching us that true wisdom, which Moses inculcated on his people, when the Egyptians were pursuing them, "Fear ye not, stand still; the Lord shall fight for you, and ye shall hold your peace." And so far from finding a difficulty in obeying in this case, I have cause to be thankful and rejoice to have so clear a direction in a matter of difficulty.

But if we would ascertain with correctness the real course of a principle, we must look at it at a certain distance, and as history represents it to us. Nothing carried on by human instruments, but has its irregularities, and affords ground for criticism, when minutely scrutinized in matters of detail. I have been speaking of that aspect of the action of an infallible authority, which is most open to invidious criticism from those who view it from without; I have tried to be fair, in estimating what can be said to its disadvantage, as witnessed at a particular time in the Catholic Church, and now I wish its adversaries to be equally fair in their judgment upon its historical character. Can, then, the infallible authority, with any show of reason, be said in fact to have destroyed the energy of the Catholic intellect? Let it be observed, I have not here to speak of any conflict which ecclesiastical authority has had with science, for this simple reason, that conflict there has been none; and that, because the secular sciences, as they now exist, are a novelty in the world, and there has been no time yet for a history of relations between theology and these new methods of knowledge, and indeed the Church may be said to have kept clear of them, as is proved by the constantly cited case of Galileo. Here "exceptio probat regulam:" for it is the one stock argument. Again, I have not to speak of any relations of the Church to the new sciences, because my simple question all along has been whether the assumption of infallibility by the proper authority is adapted to make me a hypocrite, and till that authority passes decrees on pure physical subjects and calls on me to subscribe them, (which it never will do, because it has not the power,) it has no tendency to interfere by any of its acts with my private judgment on those points. The simple question is, whether authority has so acted upon the reason of individuals, that they can have no opinion of their own, and have but an alternative of slavish superstition or secret rebellion of heart; and I think the whole history of theology puts an absolute negative upon such a supposition.

It is hardly necessary to argue out so plain a point. It is individuals, and not the Holy See, that have taken the initiative, and given the lead to the Catholic mind, in theological inquiry. Indeed, it is one of the reproaches urged against the Roman Church, that it has originated nothing, and has only served as a sort of *remora* or break in the development of doctrine. And it is an objection which I really embrace as a truth; for such I conceive to be the main purpose of its extraordinary gift. It is said, and truly, that the Church of Rome possessed no great mind in the whole period of persecution. Afterwards for a long while, it has not a single doctor to show; St Leo, its first, is the teacher of one point of doctrine; St. Gregory, who stands at the very extremity of the first age of the Church, has no place in dogma or philosophy. The great luminary of the western world is, as we know, St. Augustine; he, no infallible teacher, has formed the intellect of Christian Europe; indeed to the African Church generally we must look for the best early exposition of Latin ideas. Moreover, of the African divines, the first in order of time, and not the least influential, is the strong-minded and heterodox Tertullian. Nor is the Eastern intellect, as such, without its share in the formation of the Latin teaching. The free thought of Origen is visible in the writings of the Western Doctors, Hilary and Ambrose; and the independent mind of Jerome has enriched his own vigorous commentaries on Scripture, from the stores of the scarcely orthodox Eusebius. Heretical questions have been transmuted by the living power of the Church into salutary truths. The case is the same as regards the Ecumenical Councils. Authority in its most imposing exhibition, grave bishops, laden with the traditions and rivalries of particular nations or places, have been guided in their decisions by the commanding genius of individuals, sometimes young and of inferior rank. Not that uninspired intellect overruled the super-human gift which was committed to the Council, which would be a self-contradictory assertion, but that in that process of inquiry and deliberation, which ended in an infallible enunciation, individual reason was paramount. Thus Malchion, a mere presbyter, was the instrument of the great Council of Antioch in the third century in meeting and refuting, for the assembled Fathers, the heretical Patriarch of that see. Parallel to this instance is the influence, so well known, of a young deacon, St. Athanasius, with the 318 Fathers at Nicæa. In mediæval times we read of St. Anselm at Bari, as the champion of the Council there held, against the Greeks. At Trent, the writings of St. Bonaventura, and, what is more to the point, the address of a Priest and theologian, Salmeron, had a critical effect on some of the definitions of dogma. In some of these cases the influence might be partly moral, but in others it was that of a discursive knowledge of ecclesiastical writers, a scientific acquaintance with theology, and a force of thought in the treatment of doctrine.

There are of course intellectual habits which theology does not tend to form, as for instance the experimental, and again the philosophical; but that is because it *is* theology, not because of the gift of infallibility. But, as far as this goes, I think it could be shown that physical science on the other hand, or again mathematical, affords but an imperfect training for the intellect. I do not see then how any objection about the narrowness of theology comes into our question, which simply is, whether the belief in an infallible authority destroys the independence of the mind; and I consider that the whole history of the Church, and especially the history of the theological schools, gives a negative to the accusation. There never was a time when the intellect of the educated class was more active, or rather more restless, than in the middle ages. And then again all through Church history from the first, how slow is authority in interfering! Perhaps a local teacher, or a doctor in some local school, hazards a proposition, and a controversy ensues. It smoulders or burns in one place, no one interposing; Rome simply lets it alone. Then it comes before a Bishop; or some priest, or some professor in some other seat of learning takes it up; and then there is a second stage of it. Then it comes before a University, and it may be condemned by the theological faculty. So the controversy proceeds year after year, and Rome is still silent. An appeal perhaps is next made to a seat of authority inferior to Rome; and then at last after a long while it comes before the supreme power. Meanwhile, the question has been ventilated and turned over and over again, and viewed on every side of it, and authority is called upon to pronounce a decision, which has already been arrived at by reason. But even then, perhaps the supreme authority hesitates to do so, and nothing is determined on the point for years: or so generally and vaguely, that the whole controversy has to be gone through again, before it is ultimately determined. It is manifest how a mode of proceedings, such as this, tends not only to the liberty, but to the courage, of the individual theologian or controversialist. Many a man has ideas, which he hopes are true, and useful for his day, but he is not confident about them, and wishes to have them discussed. He is willing, or rather would be thankful, to give them up, if they can be proved to be erroneous or dangerous, and by means of controversy he obtains his end. He is answered, and he yields; or on the contrary he finds that he is considered safe. He would not dare to do this, if he knew an authority, which was supreme and final, was watching every word he said, and made signs of assent or dissent to each sentence, as he uttered it. Then indeed he would be fighting, as the Persian soldiers, under the lash, and the freedom of his intellect might truly be said to be beaten out of him. But this has not been so:—I do not mean to say that, when controversies run high, in schools or even in small portions of the Church, an interposition may not advisably take place; and again, questions may be of that urgent nature, that an appeal must, as a matter of duty, be made at once to the highest authority in the Church; but if we look into the history of controversy, we shall find, I think, the general run of things to be such as I have represented it. Zosimus treated Pelagius and Cœlestius with extreme forbearance; St. Gregory VII. was equally indulgent with Berengarius:—by reason of the very power of the Popes they have commonly been slow and moderate in their use of it.

And here again is a further shelter for the legitimate exercise of the reason:—the multitude of nations which are within the fold of the Church will be found to have acted for its protection, against any narrowness, on the supposition of narrowness, in the various authorities at Rome, with whom lies the practical decision of controverted questions. How have the Greek traditions been respected and provided for in the later Ecumenical Councils, in spite of the countries that held them being in a state of schism! There are important points of doctrine which have been (humanly speaking) exempted from the infallible sentence, by the tenderness with which its instruments, in framing it, have treated the opinions of particular places. Then, again, such national influences have a providential effect in moderating the bias which the local influences of Italy may exert upon the See of St. Peter. It stands to reason that, as the Gallican Church has in it a French element, so Rome must have in it an element of Italy; and it is no prejudice to the zeal and devotion with which we submit ourselves to the Holy See to admit this plainly. It seems to me, as I have been saying, that Catholicity is not only one of the notes of the Church, but, according to the divine purposes, one of its securities. I think it would be a very serious evil, which Divine Mercy avert! that the Church should be contracted in Europe within the range of particular nationalities. It is a great idea to introduce Latin civilization into America, and to improve the Catholics there by the energy of French devotedness; but I trust that all European races will ever have a place in the Church, and assuredly I think that the loss of the English, not to say the German element, in its composition has been a most serious misfortune. And certainly, if there is one consideration more than another which should make us English grateful to Pius the Ninth, it is that, by giving us a Church of our own, he has prepared the way for our own habits of mind, our own manner of reasoning, our own tastes, and our own virtues, finding a place and thereby a sanctification, in the Catholic Church.

[ch. 5]

A GRAMMAR OF ASSENT

Religion and Theology

A dogma is a proposition; it stands for a notion or for a thing; and to believe it is to give the assent of the mind to it, as it stands for the one or for the other. To give a real assent to it is an act of religion; to give a notional, is a theological act. It is discerned, rested in, and appropriated as a reality, by the religious imagination; it is held as a truth, by the theological intellect.

Not as if there were in fact, or could be, any line of demarcation or party-wall between these two modes of assent, the religious and the theological. As intellect is common to all men as well as imagination, every religious man is to a certain extent a theologian, and no theology can start or thrive without the initiative and abiding presence of religion. As in matters of this world, sense, sensation, instinct, intuition, supply us with facts, and the intellect uses them; so, as regards our relations with the Supreme Being, we get our facts from the witness, first of nature, then of revelation, and our doctrines, in which they issue, through the exercise of abstraction and inference. This is obvious; but it does not interfere with holding that there is a theological habit of mind, and a religious, each distinct from each, religion using theology, and theology using religion. . . .

Here we have the solution of the common mistake of supposing that there is a contrariety and antagonism between a dogmatic creed and vital religion. People urge that salvation consists, not in believing the propositions that there is a God, that there is a Saviour, that our Lord is God, that there is a Trinity, but in believing in God, in a Saviour, in a Sanctifier; and they object that such propositions are but a formal and human medium destroying all true reception of the Gospel, and making religion a matter of words or of logic, instead of its having its seat in the heart. They are right so far as this, that men can and sometimes do rest in the propositions themselves as expressing intellectual notions; they are wrong, when they maintain that men need do so or always do so. The propositions may and must be used, and can easily be used, as the expression of facts, not notions, and they are necessary to the mind in the same way that language is ever necessary for denoting facts, both for ourselves as individuals, and for our intercourse with others. Again, they are useful in their dogmatic aspect as ascertaining and making clear for us the truths on which the religious imagination has to rest. Knowledge must ever precede the exercise of the affections. We feel gratitude and love, we feel indignation and dislike, when we have the informations actually put before us which are to kindle those several emotions. We love our parents, as our parents, when we know them to be our parents; we must know concerning God, before we can feel love, fear, hope, or trust towards Him. Devotion must have its objects; those objects, as being supernatural, when not represented to our senses by material symbols, must be set before the mind in propositions. The formula, which embodies a dogma for the theologian, readily suggests an object for the worshipper. It seems a truism to say, yet it is all that I have been saying, that in religion the imagination and affections should always be under the control of reason. Theology may stand as a substantive science, though it be without the life of religion; but religion cannot maintain its ground at all without theology. Sentiment, whether imaginative or emotional, falls back upon the intellect for its stay, when sense cannot be called into exercise; and it is in this way that devotion falls back upon dogma.

[ch. 5]

A LETTER TO THE DUKE OF NORFOLK

Conscience

It seems, then, that there are extreme cases in which Conscience may come into collision with the word of a Pope, and is to be followed in spite of that word. Now I wish to place this proposition on a broader basis, acknowledged by all Catholics, and, in order to do this satisfactorily, as I began with the prophecies of Scripture and the primitive Church, when I spoke of the Pope's prerogatives, so now I must begin with the Creator and His creature, when I would draw out the prerogatives and the supreme authority of Conscience.

I say, then, that the Supreme Being is of a certain character, which, expressed in human language, we call ethical. He has the attributes of justice, truth, wisdom, sanctity, benevolence and mercy, as eternal characteristics in His nature, the very Law of His being, identical with Himself; and next, when He became Creator, He implanted this Law, which is Himself, in the intelligence of all His rational creatures. The Divine Law, then, is the rule of ethical truth, the standard of right and wrong, a sovereign, irreversible, absolute authority in the presence of men and angels. . . . This law, as apprehended in the minds of individual men, is called "conscience;" and though it may suffer refraction in passing into the intellectual medium of each, it is not therefore so affected as to lose its character of being the Divine Law, but still has, as such, the prerogative of commanding obedience. . . .

This view of conscience, I know, is very different from that ordinarily taken of it, both by the science and literature, and by the public opinion, of this day. It is founded on the doctrine that conscience is the voice of God, whereas it is fashionable on all hands now to consider it in one way or another a creation of man. Of course, there are great and broad exceptions to this statement. It is not true of many or most religious bodies of men; especially not of their teachers and ministers. When Anglicans, Wesleyans, the various Presbyterian sects in Scotland, and other denominations among us, speak of conscience, they mean what we mean, the voice of God in the nature and heart of man, as distinct from the voice of Revelation. They speak of a principle planted within us, before we have had any training, although training and experience are necessary for its strength, growth, and due formation. They consider it a constituent element of the mind, as our perception of other ideas may be, as our powers of reasoning, as our sense of order and the beautiful, and our other intellectual endowments. They consider it, as Catholics consider it, to be the internal witness of both the existence and the law of God. They think it holds of God, and not of man, as an Angel walking on the earth would be no citizen or dependent of the Civil Power. They would not allow, any more than we do, that it could be resolved into any combination of principles in our nature, more elementary than itself; nay, though it may be called, and is, a law of the mind, they would not grant that it was nothing more; I mean, that it was not a dictate, nor conveyed the notion of responsibility, of duty, of a threat and a promise, with a vividness which discriminated it from all other constituents of our nature.

This, at least, is how I read the doctrine of Protestants as well as of Catholics. The rule and measure of duty is not utility, nor expedience, nor the happiness of the greatest number, nor State convenience, nor fitness, order, and the *pulchrum*. Conscience is not a long-sighted selfishness, nor a desire to be consistent with oneself; but it is a messenger from Him, who, both in nature and in grace, speaks to us behind a veil, and teaches and rules us by His representatives. Conscience is the aboriginal Vicar of Christ, a prophet in its informations, a monarch in its peremptoriness, a priest in its blessings and anathemas, and, even though the eternal priesthood throughout the Church could cease to be, in it the sacerdotal principle would remain and would have a sway.

Words such as these are idle empty verbiage to the great world of philosophy now. All through my day there has been a resolute warfare, I had almost said conspiracy against the rights of conscience, as I have described it. Literature and science have been embodied in great institutions in order to put it down. Noble buildings have been reared as fortresses against that spiritual, invisible influence which is too subtle for science and too profound for literature. Chairs in Universities have been made the seats of an antagonist tradition. Public writers, day after day, have indoctrinated the minds of innumerable readers with theories subversive of its claims. As in Roman times, and in the middle age, its supremacy was assailed by the arm of physical force, so now the intellect is put in operation to sap the foundations of a power which the sword could not destroy. We are told that conscience is but a twist in primitive and untutored man; that its dictate is an imagination; that the very notion of guiltiness, which that dictate enforces, is simply irrational, for how can there possibly be freedom of will, how can there be consequent responsibility, in that infinite eternal network of cause and effect, in which we helplessly lie? and what retribution have we to fear, when we have had no real choice to do good or evil?

So much for philosophers; now let us see what is the notion of conscience in this day in the popular mind. There, no more than in the intellectual world, does "conscience" retain the old, true, Catholic meaning of the word. There too the idea, the presence of a Moral Governor is far away from the use of it, frequent and emphatic as that use of it is. When men advocate the rights of conscience, they in no sense mean the rights of the Creator, nor the duty to Him, in thought and deed, of the creature; but the right of thinking, speaking, writing, and acting, according to their judgment or their humour, without any thought of God at all. They do not even pretend to go by any moral rule, but they demand, what they think is an Englishman's prerogative, for each to be his own master in all things, and to profess what he pleases, asking no one's leave, and accounting priest or preacher, speaker or writer, unutterably impertinent, who dares to say a word against his going to perdition, if he like it, in his own way. Conscience has rights because it has duties; but in this age, with a large portion of the public, it is the very right and freedom of conscience to dispense with conscience, to ignore a Lawgiver and Judge, to be independent of unseen obligations. It becomes a licence to take up any or no religion, to take up this or that and let it go again, to go to church, to go to chapel, to boast of being above all religions and to be an impartial critic of each of them. Conscience is a stern monitor, but in this century it has been superseded by a counterfeit, which the eighteen centuries prior to it never heard of, and could not have mistaken for it, if they had. It is the right of self-will.

And now I shall turn aside for a moment to show how it is that the Popes of our century have been misunderstood by the English people, as if they really were speaking against conscience in the true sense of the word, when in fact they were speaking against it in the various false senses, philosophical or popular, which in this day are put upon the word. . . .

So indeed it is; did the Pope speak against Conscience in the true sense of the word, he would commit a suicidal act. He would be cutting the ground from under his feet. His very mission is to proclaim the moral law, and to protect and strengthen that "Light which enlighteneth every man that cometh into the world." On the law of conscience and its sacredness are founded both his authority in theory and his power in fact. Whether this or that particular Pope in this bad world always kept this great truth in view in all he did, it is for history to tell. I am considering here the Papacy in its office and duties, and in reference to those who acknowledge its claims. They are not bound by the Pope's personal character or private acts, but by his formal teaching. Thus viewing his position, we shall find that it is by the universal sense of right and wrong, the consciousness of transgression, the pangs of guilt, and the dread of retribution, as first principles deeply lodged in the hearts of men, it is thus and only thus, that he has gained his footing in the world and achieved his success. It is his claim to come from the Divine Lawgiver, in order to elicit, protect, and enforce those truths which the Lawgiver has sown in our very nature, it is this and this only that is the explanation of his length of life more than antediluvian. The championship of the Moral Law and of conscience is his *raison d'être*. The fact of his mission is the answer to the complaints of those who feel the insufficiency of the natural light; and the insuficiency of that light is the justification of his mission.

All sciences, except the science of Religion, have their certainty in themselves; as far as they are sciences, they consist of necessary conclusions from undeniable premises, or of phenomena manipulated into general truths by an irresistible induction. But the sense of right and wrong, which is the first element in religion, is so delicate, so fitful, so easily puzzled, obscured, perverted, so subtle in its argumentative methods, so impressible by education, so biassed by pride and passion, so unsteady in its course, that, in the struggle for existence amid the various exercises and triumphs of the human intellect, this sense is at once the highest of all teachers, yet the least luminous; and the Church, the Pope, the Hierarchy are, in the Divine purpose, the supply of an urgent demand. National Religion, certain as are its grounds and its doctrines as addressed to thoughtful, serious minds, needs, in order that it may speak to mankind with effect and subdue the world, to be sustained and completed by Revelation.

In saying all this, of course I must not be supposed to be limiting the Revelation of which the Church is the keeper to a mere republication of the Natural Law; but still it is true, that, though Revelation is so distinct from the teaching of nature and beyond it, yet it is not independent of it, nor without relations towards it, but is its complement, reassertion, issue, embodiment, and interpretation. The Pope, who comes of Revelation, has no jurisdication over Nature. If, under the plea of his revealed prerogatives, he neglected his mission of preaching truth, justice, mercy, and peace, much more if he trampled on the consciences of his subjects,—if he had done so all along, as Protestants say, then he could not have lasted all these many centuries till now, so as to supply a mark for their reprobation. . . .

. . . conscience is not a judgment upon any speculative truth, any abstract doctrine, but bears immediately on conduct, on something to be done or not done. . . . Hence conscience cannot come into direct collision with the Church's or the Pope's infallibility; which is engaged on general propositions, and in the condemnation of particular and given errors.

Next, I observe that, conscience being a practical dictate, a collision is possible between it and the Pope's authority only when the Pope legislates, or gives particular orders, and the like. But a Pope is not infallible in his laws, nor in his commands, nor in his acts of state, nor in his administration, nor in his public policy. . . .

. . . When it has the right of opposing the supreme, though not infallible Authority of the Pope, it must be something more than that miserable counterfeit which, as I have said above, now goes by the name. If in a particular case it is to be taken as a sacred and sovereign monitor, its dictate, in order to prevail against the voice of the Pope, must follow upon serious thought, prayer, and all available means of arriving at a right judgment on the matter in question. And further, obedience to the Pope is what is called "in possession;" that is, the *onus probandi* of establishing a case against him lies, as in all cases of exception, on the side of conscience. Unless a man is able to say to himself, as in the Presence of God, that he must not, and dare not, act upon the Papal injunction, he is bound to obey it, and would commit a great sin in disobeying it. *Primâ-facie* it is his bounden duty, even from a sentiment of loyalty, to believe the Pope right and to act accordingly. He must vanquish that mean, ungenerous, selfish, vulgar spirit of his nature, which, at the very first rumour of a command, places itself in opposition to the Superior who gives it, asks itself whether he is not exceeding his right, and rejoices, in a moral and practical matter to commence with scepticism. He must have no wilful determination to exercise a right of thinking, saying, doing just what he pleases, the question of truth and falsehood, right and wrong, the duty if possible of obedience, the love of speaking as his Head speaks, and of standing in all cases on his Head's side, being simply discarded. If this necessary rule were observed, collisions between the Pope's authority and the authority of conscience would be very rare. On the other hand, in the fact that, after all, in extraordinary cases, the conscience of each individual is free, we have a safeguard and security, were security necessary (which is a most gratuitous supposition), that no Pope ever will be able, as the objection supposes, to create a false conscience for his own ends. . . .

Thus, if the Pope told the English Bishops to order their priests to stir themselves energetically in favour of teetotalism, and a particular priest was fully persuaded that abstinence from wine, &c., was practically a Gnostic error, and therefore felt he could not so exert himself without sin; or suppose there was a Papal order to hold lotteries in each mission for some religious object, and a priest could say in God's sight that he believed lotteries to be morally wrong, that priest in either of these cases could commit a sin *hic et nunc* if he obeyed the Pope, whether he was right or wrong in his opinion, and, if wrong, although he had not taken proper pains to get at the truth of the matter. . . .

I add one remark. Certainly, if I am obliged to bring religion into after-dinner toasts, (which indeed does not seem quite the thing) I shall drink—to the Pope, if you please,—still, to Conscience first, and to the Pope afterwards.

[sect. 5]

Papal Infallibility

The infallibility, whether of the Church or of the Pope, acts principally or solely in two channels, in direct statements of truth, and in the condemnation of error. The former takes the shape of doctrinal definitions, the latter stigmatizes propositions as heretical, next to heresy, erroneous, and the like. In each case the Church, as guided by her Divine Master, has made provision for weighing as lightly as possible on the faith and conscience of her children.

As to the condemnation of propositions all she tells us is, that the thesis condemned when taken as a whole, or, again, when viewed in its context, is heretical, or blasphemous, or impious, or whatever like epithet she affixes to it. We have only to trust her so far as to allow ourselves to be warned against the thesis, or the work containing it.

Theologians employ themselves in determining what precisely it is that is condemned in that thesis or treatise; and doubtless in most cases they do so with success; but that determination is not *de fide*; all that is of faith is that there is in that thesis itself, which is noted, heresy or error, or other like peccant matter, as the case may be, such, that the censure is a peremptory command to theologians, preachers, students, and all other whom it concerns, to keep clear of it. But so light is this obligation, that instances frequently occur, when it is successfully maintained by some new writer, that the Pope's act does not imply what it has seemed to imply, and questions which seemed to be closed, are after a course of years re-opened. In discussions such as these, there is a real exercise of private judgment and an allowable one; the act of faith, which cannot be superseded or trifled with, being, I repeat, the unreserved acceptance that the thesis in question is heretical, or the like, as the Pope or the Church has spoken of it.

In these cases which in a true sense may be called the Pope's *negative* enunciations, the opportunity of a legitimate minimizing lies in the intensely concrete character of the matters condemned; in his affirmative enunciations a like opportunity is afforded by their being more or less abstract. Indeed, excepting such as relate to persons, that is, to the Trinity in Unity, the Blessed Virgin, the Saints, and the like, all the dogmas of Pope or of Council are but general, and so far, in consequence, admit of exceptions in their actual application,—these exceptions being determined either by other authoritative utterances, or by the scrutinizing vigilance, acuteness, and subtlety of the *Schola Theologorum*.

One of the most remarkable instances of what I am insisting on is found in a dogma, which no Catholic can ever think of disputing, viz., that "Out of the Church, and out of the faith, is no salvation." Not to go to Scripture, it is the doctrine of St. Ignatius, St. Irenæus, St. Cyprian in the first three centuries, as of St. Augustine and his contemporaries in the fourth and fifth. It can never be other than an elementary truth of Christianity; and the present Pope has proclaimed it as all Popes, doctors, and bishops before him. But that truth has two aspects, according as the force of the negative falls upon the "Church" or upon the "salvation." The main sense is, that there is no other communion or so-called Church, but the Catholic, in which are stored the promises, the sacraments, and other means of salvation; the other and derived sense is, that no one can be saved who is not *in* that one and only Church. But it does not follow, because there is no Church but one, which has the Evangelical gifts and privileges to bestow, that therefore no one can be saved without the intervention of that one Church.

[sect. 9]

PREFACE TO THE VIA MEDIA

Triple Office of the Church

. . . Two broad charges are brought against the Catholic Religion . . . One is the contrast which modern Catholicism is said to present with the religion of the Primitive Church, in teaching, conduct, worship, and polity, and this difficulty I have employed myself in discussing and explaining at great length in my Essay on Development and Doctrine, published in 1845.

The other, which is equally obvious and equally serious, is the difference which at first sight presents itself between its formal teaching and its popular and political manifestations; for instance, between the teaching of the Breviary and of the Roman Catechism on the one hand, and the spirit and tone of various manuals of Prayer and Meditation and of the Sermons or Addresses of ecclesiastics in high position on the other. This alleged discordance I have nowhere treated from a Catholic point of view; yet it certainly has a claim to be explained; and, as I have said, at least I can show how I explain it to myself, even though others refuse to take my explanations.

My answer shall be this:—that from the nature of the case, such an apparent contrariety between word and deed, the abstract and the concrete, could not but take place, supposing the Church to be gifted with those various prerogatives, and charged with those independent and conflicting duties, which Anglicans, as well as ourselves, recognize as belonging to her. Her organization cannot be otherwise than complex, considering the many functions which she has to fulfil, the many aims to keep in view, the many interests to secure,—functions, aims, and interests, which in their union and divergence remind us of the prophet's vision of the Cherubim, in whom "the wings of one were joined to the wings of another," yet "they turned not, when they went, but every one went straight forward." Or, to speak without figure, we know in matters of this world, how difficult it is for one and the same man to satisfy independent duties and incommensurable relations; to act at once as a parent and a judge, as a soldier and a minister of religion, as a philosopher and a statesman, as a courtier or a politician and a Catholic; the rules of conduct in these various positions being so distinct, and the obligations so contrary. Prudent men keep clear, if they can, of such perplexities; but as to the Church, gifted as she is with grace up to the measure of her responsibilities, if she has on her an arduous work, it is sufficient to refer to our Lord's words, "What is impossible with men, is possible with God," in order to be certain (in spite of appearances) of her historical uprightness and consistency. At the same time it may undeniably have happened before now that her rulers and authorities, as men, on certain occasions have come short of what was required of them, and have given occasion to criticism, just or unjust, on account of the special antagonisms or compromises by means of which her many-sided mission under their guidance has been carried out.

With this introduction I remark as follows:—When our Lord went up on high, He left His representative behind Him. This was Holy Church, His mystical Body and Bride, a Divine Institution, and the shrine and organ of the Paraclete, who speaks through her till the end comes. She, to use an Anglican poet's words, is "His very self below," as far as men on earth are equal to the discharge and fulfilment of high offices, which primarily and supremely are His.

These offices, which specially belong to Him as Mediator, are commonly considered to be three; He is Prophet, Priest, and King; and after His pattern, and in human measure, Holy Church has a triple office too; not the Prophetical alone and in isolation, as these Lectures virtually teach, but three offices, which are indivisible, though diverse, viz. teaching, rule, and sacred ministry. This then is the point on which I shall now insist, the very title of the Lectures I am to criticize suggesting to me how best to criticize them.

I will but say in passing, that I must not in this argument be supposed to forget that the Pope, as the Vicar of Christ, inherits these offices and acts for the Church in them. This is another matter; I am speaking here of the Body of Christ, and the sovereign Pontiff would not be the visible head of that Body, did he not first belong to it. He is not himself the Body of Christ, but the chief part of the Body; I shall have quite opportunities enough in what is to come to show that I duly bear him in mind.

Christianity, then, is at once a philosophy, a political power, and a religious rite: as a religion, it is Holy; as a philosophy, it is Apostolic; as a political power, it is imperial, that is, One and Catholic. As a religion, its special centre of action is pastor and flock; as a philosophy, the Schools; as a rule, the Papacy and its Curia.

Though it has exercised these three functions in substance from the first, they were developed in their full proportions one after another, in a succession of centuries; first, in the primitive time it was recognized as a worship, springing up and spreading in the lower ranks of society, and among the ignorant and dependent, and making its power felt by the heroism of its Martyrs and confessors. Then it seized upon the intellectual and cultivated class, and created a theology and schools of learning. Lastly it seated itself, as an ecclesiastical polity, among princes, and chose Rome for its centre.

Truth is the guiding principle of theology and theological inquiries; devotion and edification, of worship; and of government, expedience.

The instrument of theology is reasoning; of worship, our emotional nature; of rule, command and coercion. Further, in man as he is, reasoning tends to rationalism; devotion to superstition and enthusiasm; and power to ambition and tyranny.

Arduous as are the duties involved in these three offices, to discharge one by one, much more arduous are they to administer, when taken in combination. Each of the three has its separate scope and direction; each has its own interests to promote and further; each has to find room for the claims of the other two; and each will find its own line of action influenced and modified by the others, nay, sometimes in a particular case the necessity of the others converted into a rule of duty for itself.

"Who," in St. Paul's words, "is sufficient for these things?" Who, even with divine aid, shall successfully administer offices so independent of each other, so divergent, and so conflicting? What line of conduct, except on the long, the very long run, is at once edifying, expedient, and true? Is it not plain, that, if one determinate course is to be taken by the Church, acting at once in all three capacities, so opposed to each other in their ideas, that course must, as I have said, be deflected from the line which would be traced out by any one of them, if viewed by itself, or else the requirements of one or two sacrificed to the interests of the third? What, for instance, is to be done in a case when to enforce a theological point, as the Schools determine it, would make a particular population less religious, not more so, or cause riots or risings? Or when to defend a champion of ecclesiastical liberty in one country would encourage an Anti-Pope, or hazard a general persecution, in another? or when either a schism is to be encountered or an opportune truth left undefined?

All this was foreseen certainly by the Divine Mind, when He committed to His Church so complex a mission; and, by promising her infallibility in her formal teaching, He indirectly protected her from serious error in worship and political action also. This aid, however, great as it is, does not secure her from all dangers as regards the problem which she has to solve; nothing but the gift of impeccability granted to her authorities would secure them from all liability to mistake in their conduct, policy, words and decisions, in her legislative and her executive, in ecclesiastical and disciplinarian details; and such a gift they have not received. In consequence, however well she may perform her duties on the whole, it will always be easy for her enemies to make a case against her, well founded or not, from the action or interaction, or the chronic collisions or contrasts, or the temporary suspense or delay, of her administration, in her three several departments of duty,—her government, her devotions, and her schools,—from the conduct of her rulers, her divines, her pastors, or her people.

It is this difficulty lying in the nature of the case, which supplies the staple of those energetic charges and vivid pictures of the inconsistency, double-dealing, and deceit of the Church of Rome, as found in Protestant writings . . .

I am to apply then the doctrine of the triple office of the Church in explanation of this phenomenon, which gives so much offence to Protestants; and I begin by admitting the general truth of the facts alleged against us;—at the same time . . . there is one misconception of fact which needs to be corrected before I proceed . . . ambition, craft, cruelty, and superstition are not commonly the characteristic of theologians. . . . Nor, again, is it even accurate to say . . . that those so-called corruptions are at least the result and development of those abstract decrees: on the contrary, they bear on their face the marks of having a popular or a political origin, and in fact theology, so far from encouraging them, has restrained and corrected such extravagances as have been committed, through human infirmity, in the exercise of the regal and sacerdotal powers; nor is religion ever in greater danger than when, in consequence of national or international troubles, the Schools of theology have been broken up and ceased to be.

And this will serve as a proposition with which to begin. I say, then, Theology is the fundamental and regulating principle of the whole Church system. It is commensurate with Revelation, and Revelation is the initial and essential idea of Christianity. It is the subject-matter, the formal cause, the expression, of the Prophetical Office, and, as being such, has created both the Regal Office and the Sacerdotal. And it has in a certain sense a power of jurisdiction over those offices, as being its own creations, theologians being ever in request and in employment in keeping within bounds both the political and popular elements in the Church's constitution,—elements which are far more congenial than itself to the human mind, are far more liable to excess and corruption, and are ever struggling to liberate themselves from those restraints which are in truth necessary for their well-being. . . .

Yet theology cannot always have its own way; it is too hard, too intellectual, too exact, to be always equitable, or to be always compassionate; and it sometimes has a conflict or overthrow, or has to consent to a truce or a compromise, in consequence of the rival force of religious sentiment or ecclesiastical interests; and that, sometimes in great matters, sometimes in unimportant. . . .

. . . Theology lays down the undeniable truth . . . that our good works have merit and are a ground of confidence for us in God's judgment of us. This dogma shocks good Protestants, who think that, in the case of an individual Catholic, it is the mark of a self-righteous spirit, and incompatible with his renunciation of his own desert and with a recourse to God's mercy. But they confuse an intellectual view with a personal sentiment. . . .

Again, I have already referred to the dilemma which has occurred before now in the history of the Church, when a choice had to be made between leaving a point of faith at a certain moment undefined, and indirectly opening the way to some extended and permanent schism. Here her Prophetical function is impeded for a while in its action, perhaps seriously, by the remonstrances of charity and of the spirit of peace.

In another familiar instance which may be given, the popular and scholastic elements in the Church seem to change parts, and theology to be kind and sympathetic and religion severe. I mean, whereas the whole School with one voice speaks of freedom of conscience as a personal prerogative of each individual, on the other hand the vow of obedience may sometimes in particular cases be enforced by Religious Superiors in some lesser matter to the conceivable injury of such sacred freedom of thought.

Another instance of collision in a small matter is before us just at this time, the theological and religious element of the Church being in antagonism with the political. Humanity, a sense of morality, hatred of a special misbelief, views of Scripture prophecy, a feeling of brotherhood with Russians, Greeks, and Bulgarians, though schismatics, have determined some of us against the Turkish cause; and a dread lest Russia, if successful, should prove a worse enemy to the Church than Turks can be, determines others of us in favour of it.

But I will come to illustrations which involve more difficult questions. Truth is the principle on which all intellectual, and therefore all theological inquiries proceed, and is the motive power which gives them effect; but the principle of popular edification, quickened by a keen sensitiveness of the chance of scandals, is as powerful as Truth, when the province is Religion. To the devotional mind what is new and strange is as repulsive, often as dangerous, as falsehood is to the scientific. Novelty is often error to those who are unprepared for it, from the refraction with which it enters into their conceptions. Hence popular ideas on religion are practically a match for the clearest *dicta*, deductions, and provisos of the Schools, and will have their way in cases when the particular truth, which is the subject of them, is not of vital or primary importance. Thus, in a religion, which embraces large and separate classes of adherents, there always is of necessity to a certain extent an exoteric and an esoteric doctrine.

The history of the Latin versions of the Scriptures furnishes a familiar illustration of this conflict between popular and educated faith. The Gallican version of the Psalter, St. Jerome's earlier work, got such possession of the West, that to this day we use it instead of his later and more correct version from the Hebrew. Devotional use prevailed over scholastic accuracy in a matter of secondary concern. . . . A parallel anxiety for the same reason is felt at this time within the Anglican communion, upon the proposal to amend King James's Translation of the Scriptures.

Here we see the necessary contrast between religious inquiry or teaching, and investigation in purely secular matters. Much is said in this day by men of science about the duty of honesty in what is called the pursuit of truth,—by "pursuing truth" being meant the pursuit of facts. It is just now reckoned a great moral virtue to be fearless and thorough in inquiry into facts; and, when science crosses and breaks the received path of Revelation, it is reckoned a serious imputation upon the ethical character of religious men, whenever they show hesitation to shift at a minute's warning their position, and to accept as truths shadowy views at variance with what they have ever been taught and have held. But the contrast between the cases is plain. The love and pursuit of truth in the subject-matter of religion, if it be genuine, must always be accompanied by the fear of error, of error which may be sin. An inquirer in the province of religion is under a responsibility for his reasons and for their issue. But, whatever be the real merits, nay, virtues, of inquirers into physical or historical facts, whatever their skill, their acquired caution, their experience, their dispassionateness and fairness of mind, they do not avail themselves of these excellent instruments of inquiry as a matter of conscience, but because it is expedient, or honest, or beseeming, or praiseworthy, to use them; nor, if in the event they were found to be wrong as to their supposed discoveries, would they, or need they, feel aught of the remorse and self-reproach of a Catholic, on whom it breaks that he has been violently handling the text of Scripture, misinterpreting it, or superseding it, on an hypothesis which he took to be true, but which turns out to be untenable.

Let us suppose in his defence that he was challenged either to admit or to refute what was asserted, and to do so without delay; still it would have been far better could he have waited awhile, as the event has shown,—nay, far better, even though the assertion has proved true. Galileo might be right in his conclusion that the earth moves; to consider him a heretic might have been wrong; but there was nothing wrong in censuring abrupt, startling, unsettling, unverified disclosures, if such they were, disclosures at once uncalled for and inopportune, at a time when the limits of revealed truth had not as yet been ascertained. A man ought to be very sure of what he is saying, before he risks the chance of contradicting the word of God. It was safe, not dishonest, to be slow in accepting what nevertheless turned out to be true. Here is an instance in which the Church obliges Scripture expositors, at a given time or place, to be tender of the popular religious sense.

I have been led on to take a second view of this matter. That jealousy of originality in the matter of religion, which is the instinct of piety, is, in the case of questions which excite the popular mind, the dictate of charity also. Galileo's truth is said to have shocked and scared the Italy of his day. It revolutionized the received system of belief as regards heaven, purgatory, and hell, to say that the earth went round the sun, and it forcibly imposed upon categorical statements of Scripture, a figurative interpretation. Heaven was no longer above, and earth below; the heavens no longer literally opened and shut; purgatory and hell were not for certain under the earth. The catalogue of theological truths was seriously curtailed. Whither did our Lord go on His ascension? If there is to be a plurality of worlds, what is the special importance of this one? and is the whole visible universe with its infinite spaces, one day to pass away? We are used to these questions now, and reconciled to them; and on that account are not fit judges of the disorder and dismay, which the Galilean hypothesis would cause to good Catholics, as far as they became cognizant of it, or how necessary it was in charity, especially then, to delay the formal reception of a new interpretation of Scripture, till their imaginations should gradually get accustomed to it.

As to the particular measures taken at the time with this end, I neither know them accurately, nor have I any anxiety to know them. They do not fall within the scope of my argument; I am only concerned with the principle on which they were conducted. All I say is, that not all knowledge is suited to all minds; a proposition may be ever so true, yet at a particular time and place may be "temerarious, offensive to pious ears, and scandalous," though not "heretical" nor "erroneous." It must be recollected what very strong warnings we have from our Lord and St. Paul against scandalizing the weak and unintellectual. The latter goes into detail upon the point. He says, that, true as it may be that certain meats are allowable, this allowance cannot in charity be used in a case in which it would be of spiritual injury to others. . . .

Now, while saying this, I know well that "all things have their season," and that there is not only "a time to keep silence," but "a time to speak," and that, in some states of society, such as our own, it is the worst charity, and the most provoking, irritating rule of action, and the most unhappy policy, not to speak out, not to suffer to be spoken out, all that there is to say. Such speaking out is under such circumstances the triumph of religion, whereas concealment, accommodation, and evasion is to co-operate with the spirit of error;—but it is not always so. . . .

In truth we recognize the duty of concealment, or what may be called evasion, not in religious matters only, but universally. It is very well for sublime sciences, which work out their problems apart from the crowding and jostling, the elbowing and the toe-treading of actual life, to care for nobody and nothing but themselves, and to preach and practise the cheap virtue of devotion to what they call truth, meaning of course facts; but a liberty to blurt out all things whatever without self-restraint is not only forbidden by the Church, but by Society at large; of which such liberty, if fully carried out, would certainly be the dissolution. Veracity, like other virtues, lies in a mean. Truth indeed, but not necessarily the whole truth, is the rule of Society. Every class and profession has its secrets; the family lawyer, the medical adviser, the politician, as well as the priest. The physician often dares not tell the whole truth to his patient about his case, knowing that to do so would destroy his chance of recovery. Statesmen in Parliament, I suppose, fight each other with second-best arguments, the real reasons for the policy which they are respectively advocating being, as each is conscious to each, not these, but reasons of state, secrets whether of her Majesty's Privy Council or of diplomacy. As to the polite world, which, to be sure, is in itself not much of an authority, I think an authoress of the last century illustrates in a tale how it would not hold together, if every one told the whole truth to every one, as to what he thought of him. From the time that the Creator clothed Adam, concealment is in some sense the necessity of our fall.

This, then, is one cause of that twofold or threefold aspect of the Catholic Church, which I have set myself to explain. Many popular beliefs and practices have, in spite of theology, been suffered by Catholic prelates, lest, "in gathering up the weeds," they should "root up the wheat with them." We see the operation of this necessary economy in the instance of the Old Covenant, in the gradual disclosures made, age after age, to the chosen people. The most striking of these accommodations is the long sufferance of polygamy, concubinage, and divorce.

This indeed is the great principle of Economy, as advocated in the Alexandrian school, which is in various ways sanctioned in Scripture. In some fundamental points indeed, in the Unity and Omnipotence of God, the Mosaic Law, so tolerant of barbaric cruelty, allowed of no condescension to the ethical state of the times; indeed the very end of the Dispensation was to denounce idolatry, and the sword was its instrument of denunciation; but where the mission of the chosen people was not directly concerned, and amid the heathen populations, even idolatry itself was suffered with something of a Divine sanction, as if a deeper sentiment might lie hid under it. . . .

From the time that the Apostles preached, such toleration in primary matters of faith and morals is at an end as regards Christendom. Idolatry is a sin against light; and, while it would involve heinous guilt, or rather is impossible, in a Catholic, it is equally inconceivable in even the most ignorant sectary who claims the Christian name; nevertheless, the principle and the use of the Economy has a place, and is a duty still among Catholics, though not as regards the first elements of Revelation. We have still, as Catholics, to be forbearing and to be silent in many cases, amid the mistakes, excesses, and superstitions of individuals and of classes of our brethren, which we come across. Also in the case of those who are not Catholic, we feel it a duty sometimes to observe the rule of silence, even when so serious a truth as the "Extra Ecclesiam nulla salus" comes into consideration. This truth, indeed, must ever be upheld, but who will venture to blame us, or reproach us with double-dealing, for holding it to be our duty, though we thus believe, still, in a case when a Protestant, near death and to all appearance in good faith, is sure, humanly speaking, not to accept Catholic truth, if urged upon him, to leave such a one to his imperfect Christianity, and to the mercy of God, and to assist his devotions as far as he will let us carry him, rather than to precipitate him at such a moment into controversy which may ruffle his mind, dissipate his thoughts, unsettle such measure of faith as he has, and rouse his slumbering prejudices and antipathies against the Church? Yet this might be represented as countenancing a double aspect of Catholic doctrine and as evasive and shuffling, theory saying one thing, and practice sanctioning another.

I shelter what I go on to say of the Church's conduct occasionally towards her own children, under this rule of her dealing with strangers:—The rule is the same in its principle as that of Moses or St. Paul, or the Alexandrians, or St. Augustine, though it is applied to other subject-matters. Doubtless, her abstract standard of religion and morals in the Schools is higher than that which we witness in her children in particular countries or at particular times; but doubtless also, she, like the old prophets before her, from no fault of hers, is not able to enforce it. Human nature is in all ages one and the same: as it showed itself in the Israelites, so it shows itself in the world at large now, though one country may be better than another. At least, in some countries, truth and error in religion may be so intimately connected as not to admit of separation. I have already referred to our Lord's parable of the wheat and the cockle. For instance, take the instance of relics; modern divines and historians may have proved that certain recognized relics, though the remains of some holy man, still do not certainly belong to the Saint to whom they are popularly appropriated; and in site of this, a bishop may have sanctioned a public veneration of them, which has arisen out of this unfounded belief. . . .

The difficulty of course is to determine the point at which such religious manifestations become immoderate, and an allowance of them wrong; it would be well, if all suspicious facts could be got rid of altogether. Their tolerance may sometimes lead to pious frauds, which are simply wicked. An ecclesiastical superior certainly cannot sanction alleged miracles or prophecies which he knows to be false, or by his silence connive at a tradition of them being started among his people. Nor can he be dispensed of the duty, when he comes into an inheritance of error or superstition, which is immemorial, of doing what he can to alleviate and dissipate it, though to do this without injury to what is true and good, can after all be only a gradual work. Errors of fact may do no harm, and their removal may do much.

As neither the local rulers nor the pastors of the Church are impeccable in act nor infallible in judgment, I am not obliged to maintain that all ecclesiastical measures and permissions have ever been praiseworthy and safe precedents. But as to the mere countenancing of superstitions, it must not be forgotten, that our Lord Himself, on one occasion, passed over the superstitious act of a woman who was in great trouble, for the merit of the faith which was the real element in it. She was under the influence of what would be called, were she alive now, a "corrupt" religion, yet she was rewarded by a miracle. She came behind our Lord and touched Him, hoping "virtue would go out of Him," without His knowing it. She paid a sort of fetish reverence to the hem of His garment; she stole, as she considered, something from Him, and was much disconcerted at being found out. When our Lord asked who had touched Him, "fearing and trembling," says St. Mark, "knowing what was done in her, she came and fell down before Him, and told Him all the truth," as if there were anything to tell to the All-knowing. What was our Lord's judgment on her? "Daughter, thy faith hath made thee whole; go in peace." Men talk of our double aspect now; has not the first age a double aspect? Do not such incidents in the Gospel as this, and the miracle on the swine, the pool of Bethesda, the restoration of the servant's ear, the changing water into wine, the coin in the fish's mouth, and the like, form an aspect of Apostolic Christianity very different from that presented by St. Paul's Pastoral Epistles and the Epistle General of St. John? Need men wait for the Medieval Church in order to make their complaint that the theology of Christianity does not accord with its religious manifestations?

This woman, who is so prominently brought before us by three evangelists, doubtless understood that, if the garment had virtue, this arose from its being Christ's; and so a poor Neapolitan crone, who chatters to the crucifix, refers that crucifix in her deep mental consciousness to an original who once hung upon a cross in flesh and blood; but if, nevertheless she is puzzle-headed enough to assign virtue to it in itself, she does no more than the woman in the Gospel, who preferred to rely for a cure on a bit of cloth, which was our Lord's, to directly and honestly addressing Him. Yet He praised her before the multitude, praised her for what might, not without reason, be called an idolatrous act; for in His new law He was opening the meaning of the word "idolatry," and applying it to various sins, the adoration paid to rich men, to the thirst after gain, to ambition, and the pride of life, idolatries worse in His judgment than the idolatry of ignorance, but not commonly startling or shocking to educated minds.

And may I not add that this aspect of our Lord's teaching is quite in keeping with the general drift of His discourses? Again and again He insists on the necessity of faith; but where does He insist on the danger of superstition, an infirmity, which, taking human nature as it is, is the sure companion of faith, when vivid and earnest? Taking human nature as it is, we may surely concede a little superstition, as not the worst of evils, if it be the price of making sure of faith. Of course it need not be the price; and the Church, in her teaching function, will ever be vigilant against the inroad of what is a degradation both of faith and of reason: but considering, as Anglicans will allow, how intimately the sacramental system is connected with Christianity, and how feeble and confused is at present the ethical intelligence of the world at large, it is a distant day, at which the Church will find it easy, in her oversight of her populations, to make her Sacerdotal office keep step with her Prophetical. Just now I should be disposed to doubt whether that nation really had the faith, which is free in all its ranks and classes from all kinds and degrees of what is commonly considered superstition. . . .

It must be recollected, that, while the Catholic Church is ever most precise in her enunciation of doctrine, and allows no liberty of dissent from her decisions, (for on such objective matters she speaks with the authority of infallibility,) her tone is different, in the sanction she gives to devotions, as they are of a subjective and personal nature. Here she neither prescribes measure, nor forbids choice, nor, except so far as they imply doctrine, is she infallible in her adoption or use of them. This is an additional reason why the formal decrees of Councils and statements of theologians differ in their first aspect from the religion of the uneducated classes; the latter represents the wayward popular taste, and the former the critical judgments of clear heads and holy hearts.

This contrast will be the greater, when, as sometimes happens, ecclesiastical authority takes part with the popular sentiment against a theological decision. Such, we know, was the case, when St. Peter himself committed an error in conduct, in the countenance he gave to the Mosaic rites in consequence of the pressure exerted on him by the Judaic Christians. On that occasion St. Paul withstood him, "because he was to be blamed." A fault, which even the first Pope incurred, may in some other matter of rite or devotion find a place now and then in the

history of holy and learned ecclesiastics who were not Popes. Such an
instance seems presented to us in the error of judgment which was
committed by the Fathers of the Society of Jesus in China, in their
adoption of certain customs which they found among the heathen there;
and Protestant writers in consequence have noted it as a signal instance
of the double-faced conduct of Catholics, as if they were used to present
their religion under various aspects according to the expedience of the
place or time. But that there is a religious way of thus accommodating
ourselves to those among whom we live, and whom it is our duty, if
possible, to convert, is plain from St. Paul's own rule of life, considering
he "became to the Jews as a Jew, that he might gain the Jews, and to
them that were without the law, as if he were without the law, and
became all things to all men that he might save all." Or what shall we
say to the commencement of St. John's Gospel, in which the Evangelist
may be as plausibly represented to have used the language of heathen
classics with the purpose of interesting and gaining the Platonizing Jews,
as the Jesuits be charged with duplicity and deceit in aiming at the
conversion of the heathen in the East by an imitation of their customs.
St. Paul on various occasions acts in the same spirit of economy, as did
the great Missionary Church of Alexandria in the centuries which
followed; its masters did but carry out, professedly, a principle of action,
of which they considered they found examples in Scripture. Anglicans
who appeal to the Ante-nicene period as especially their own, should be
tender of the memories of Theonas, Clement, Origen, and Gregory
Thaumaturgus. . . .

I observe then that Apostolicity of doctrine and Sanctity of worship,
as attributes of the Church, are differently circumstanced from her regal
autocracy. Tradition in good measure is sufficient for doctrine, and
popular custom and conscience for worship, but tradition and custom
cannot of themselves secure independence and self-government. The
Greek Church shows this, which has lost its political life, while its
doctrine, and its ritual and devotional system, have little that can be
excepted against. If the Church is to be regal, a witness for Heaven,
unchangeable amid secular changes, if in every age she is to hold her
own, and proclaim as well as profess the truth, if she is to thrive without
or against the civil power, if she is to be resourceful and self-recuperative
under all fortunes, she must be more than Holy and Apostolic; she must
be Catholic. Hence it is that, first, she has ever from her beginning
onwards had a hierarchy and a head, with a strict unity of polity, the
claim of an exclusive divine authority and blessing, the trusteeship of the
gospel gifts, and the exercise over her members of an absolute and almost
despotic rule. And next, as to her work, it is her special duty, as a
sovereign State, to consolidate her several portions, to enlarge her

territory, to keep up and to increase her various populations in this ever-dying, ever-nascent world, in which to be stationary is to lose ground, and to repose is to fail. It is her duty to strengthen and facilitate the intercourse of city with city, and race with race, so that an injury done to one is felt to be an injury to all, and the act of individuals has the energy and momentum of the whole body. It is her duty to have her eyes upon the movements of all classes in her wide dominion, on eclesiastics and laymen, on the regular clergy and secular, on civil society, and political movements. She must be on the watch-tower, discerning in the distance and providing against all dangers; she has to protect the ignorant and weak, to remove scandals, to see to the education of the young, to administer temporalities, to initiate, or at least to direct all Christian work, and all with a view to the life, health, and strength of Christianity, and the salvation of souls.

It is easy to understand how from time to time such serious interests and duties involve, as regards the parties who have the responsibility of them, the risk, perhaps the certainty, at least the imputation, of ambition or other selfish motive, and still more frequently of error in judgment, or violent action, or injustice. . . .

To conclude:—whatever is great refuses to be reduced to human rule, and to be made consistent in its many aspects with itself. Who shall reconcile with each other the various attributes of the Infinite God? and, as He is, such in there several degrees are His works. This living world to which we belong, how self-contradictory it is, when we attempt to measure and master its meaning and scope! And how full of incongruities, that is, of mysteries, in its higher and finer specimens, is the soul of man, viewed in its assemblage of opinions, tastes, habits, powers, aims, and doings! We need not feel surprise then, if Holy Church too, the supernatural creation of God, is an instance of the same law, presenting to us an admirable consistency and unity in word and deed, as her general characteristic, but crossed and discredited now and then by apparent anomalies which need, and which claim, at our hands an exercise of faith.

[*Via Media*, vol. i]

5
THE WRITER

It is the literary aspect of Newman's genius which more than any other calls out for recognition and revaluation. It is true that his reputation as a rhetorician and stylist is well established, but there is more to Newman as a writer than the luminous eloquence of the sermons and *The Idea of a University*, and the autobiographical charm and sincerity of the *Apologia*. The standard literary guides also refer to the verse and the two novels, but they rightly imply that these belong to the category of 'minor' works. It is no exaggeration to say that Newman's most distinctive strength as a writer continues to be virtually ignored, as is indicated by the fact that the work which he himself considered his 'best written book', *Lectures on the Present Position of Catholics* (1851), remains almost unknown.

The 'occasional' nature of nearly all of Newman's writings should be seen as inseparable from the fact that as a writer he was above all a controversialist. Not only did he derive so much of his intellectual stimulus from controversy, but it was controversy that stimulated so much of his most creative writing. For the great controversialist was also a great satirist. Not only does a satirical element run through most of his writings, but there are three specific works that may reasonably be called 'satires'. The first, *The Tamworth Reading Room* (1841), which belongs to his Anglican period, is an attack on the secularization of education and morality. *Lectures on certain Difficulties felt by Anglicans in submitting to the Catholic Church* (1850), which is an attempt to persuade Anglo-Catholics to leave the Church of England after the famous 'Gorham judgement' against baptismal regeneration, and *Lectures on the Present Position of Catholics*, written in the face of a violent outbreak of anti-Catholicism following the restoration of the Catholic hierarchy, are the exuberant fruits of his honeymoon period as a Catholic. The latter is not only his most brilliant achievement as a satirist, but it is surely one of the masterpieces of prose satire in the English language.

The book on which Newman's literary reputation is usually presumed particularly to depend is the *Apologia*, which is indeed a classic comparable to St Augustine's *Confessions*. The work was the result of the controversy with Charles Kingsley, who had questioned Newman's commitment to the truth, but, in order to vindicate finally and fully his integrity, Newman abandons the art of the controversialist in favour of the art of the autobiographer. And so the anomaly is that the *Apologia*, although it is the record of much past controversy, is itself for the most part so unlike the rest of the works: with the signal exception of the last chapter, calmly dispassionate documentary history replaces the usual rhetoric of argument and polemic, while a limpid prose style, as compelling as it is uncharacteristic, conveys the almost confidential voice of an author anxious to convince the reader of his integrity and sincerity.

Like Cicero, whom Newman greatly admired both as a controversialist and as a master of style, it is hardly possible to imagine Newman without his letters, so integral do they seem to his artistic and intellectual achievement. Not only does the vast corpus of correspondence provide a detailed and extended commentary on the published works, but it is in itself a marvellous manifestation of Newman's powers as an 'occasional' writer. There has been found room here for only a very few letters and excerpts from letters, but between them they are intended to suggest something of the range of a body of correspondence which is surely unique. Here is to be found the descriptive writer with an extraordinary responsiveness to the concrete and the sensuous; the conversationalist whose strikingly colloquial tone of voice is wonderfully echoed in the written word; the humorist whose exuberance belies the stereotyped picture of the sad, hypersensitive recluse; and finally the controversialist, whose superbly sarcastic snubs have a classic quality of their own.

APOLOGIA PRO VITA SUA

Conversion to Rome

The Long Vacation of 1839 began early. There had been a great many visitors to Oxford from Easter to Commemoration; and Dr. Pusey's party had attracted attention, more, I think, than in any former year. I had put away from me the controversy with Rome for more than two years. In my Parochial Sermons the subject had at no time been introduced: there had been nothing for two years, either in my Tracts or in the British Critic, of a polemical character. I was returning, for the Vacation, to the course of reading which I had many years before chosen as especially my own. I have no reason to suppose that the thoughts of Rome came across my mind at all. About the middle of June I began to study and master the history of the Monophysites. I was absorbed in the doctrinal question. This was from about June 13th to August 30th. It was during this course of reading that for the first time a doubt came upon me of the tenableness of Anglicanism. I recollect on the 30th of July mentioning to a friend, whom I had accidentally met, how remarkable the history was; but by the end of August I was seriously alarmed.

I have described in a former work, how the history affected me. My stronghold was Antiquity; now here, in the middle of the fifth century, I found, as it seemed to me, Christendom of the sixteenth and the nineteenth centuries reflected. I saw my face in that mirror, and I was a Monophysite. The Church of the *Via Media* was in the position of the Oriental communion, Rome was, where she now is; and the Protestants were the Eutychians. Of all passages of history, since history has been, who would have thought of going to the sayings and doings of old Eutyches, that *delirus senex*, as (I think) Petavius calls him, and to the enormities of the unprincipled Dioscorus, in order to be converted to Rome! . . .

Hardly had I brought my course of reading to a close, when the Dublin Review of that same August was put into my hands, by friends who were more favourable to the cause of Rome than I was myself. There was an article in it on the "Anglican Claim" by Dr. Wiseman. This was about the middle of September. It was on the Donatists, with an application to Anglicanism. I read it, and did not see much in it. The Donatist controversy was known to me for some years, as has appeared already. The case was not parallel to that of the Anglican Church. St. Augustine in Africa wrote against the Donatists in Africa. They were a furious party who made a schism within the African Church, and not beyond its limits. It was a case of Altar against Altar, of two occupants of the same See, as that between the Non-jurors in England and the

Established Church; not the case of one Church against another, as of Rome against the Oriental Monophysites. But my friend, an anxiously religious man, now, as then, very dear to me, a Protestant still, pointed out the palmary words of St. Augustine, which were contained in one of the extracts made in the Review, and which had escaped my observation. "Securus judicat orbis terrarum."* He repeated these words again and again, and, when he was gone, they kept ringing in my ears. "Securus judicat orbis terrarum;" they were words which went beyond the occasion of the Donatists: they applied to that of the Monophysites. They gave a cogency to the Article, which had escaped me at first. They decided ecclesiastical questions on a simpler rule than that of Antiquity; nay, St. Augustine was one of the prime oracles of Antiquity; here then Antiquity was deciding against itself. What a light was hereby thrown upon every controversy in the Church! not that, for the moment, the multitude may not falter in their judgment,—not that, in the Arian hurricane, Sees more than can be numbered did not bend before its fury, and fall off from St. Athanasius,—not that the crowd of Oriental Bishops did not need to be sustained during the contest by the voice and the eye of St. Leo; but that the deliberate judgment, in which the whole Church at length rests and acquiesces, is an infallible prescription and a final sentence against such portions of it as protest and secede. Who can account for the impressions which are made on him? For a mere sentence, the words of St. Augustine, struck me with a power which I never had felt from any words before. To take a familiar instance, they were like the "Turn again Whittington" of the chime; or, to take a more serious one, they were like the "Tolle, lege,—Tolle, lege," of the child, which converted St. Augustine himself. "Securus judicat orbis terrarum!" By those great words of the ancient Father, interpreting and summing up the long and varied course of ecclesiastical history, the theory of the *Via Media* was absolutely pulverized.

I became excited at the view thus opened upon me. I was just starting on a round of visits; and I mentioned my state of mind to two most intimate friends: I think to no others. After a while, I got calm, and at length the vivid impression upon my imagination faded away. What I thought about it on reflection, I will attempt to describe presently. I had to determine its logical value, and its bearing upon my duty. Meanwhile, so far as this was certain,—I had seen the shadow of a hand upon the wall. It was clear that I had a good deal to learn on the question of the Churches, and that perhaps some new light was coming upon me. He who has seen a ghost, cannot be as if he had never seen it.

---

* Newman's own (free) translation is: 'The universal Church is in its judgments secure of truth.' [Editor's note]

The heavens had opened and closed again. The thought for the moment had been, "The Church of Rome will be found right after all;" and then it had vanished. My old convictions remained as before. . . .

In the summer of 1841, I found myself at Littlemore without any harass or anxiety on my mind. I had determined to put aside all controversy, and I set myself down to my translation of St. Athanasius; but, between July and November, I received three blows which broke me.

1. I had got but a little way in my work, when my trouble returned on me. The ghost had come a second time. In the Arian History I found the very same phenomenon, in a far bolder shape, which I had found in the Monophysite. I had not observed it in 1832. Wonderful that this should come upon me! I had not sought it out; I was reading and writing in my own line of study, far from the controversies of the day, on what is called a "metaphysical" subject; but I saw clearly, that in the history of Arianism, the pure Arians were the Protestants, the semi-Arians were the Anglicans, and that Rome now was what it was then. The truth lay, not with the *Via Media*, but with what was called "the extreme party." As I am not writing a work of controversy, I need not enlarge upon the argument; I have said something on the subject in a Volume, from which I have already quoted.

2. I was in the misery of this new unsettlement, when a second blow came upon me. The Bishops one after another began to charge against me. It was a formal, determinate movement. This was the real "understanding;" that, on which I had acted on the first appearance of Tract 90, had come to nought. I think the words, which had then been used to me, were, that "perhaps two or three of them might think it necessary to say something in their charges;" but by this time they had tided over the difficulty of the Tract, and there was no one to enforce the "understanding." They went on in this way, directing charges at me, for three whole years. I recognized it as a condemnation; it was the only one that was in their power. At first I intended to protest; but I gave up the thought in despair. . . .

3. As if all this were not enough, there came the affair of the Jerusalem Bishopric; and, with a brief mention of it, I shall conclude.

I think I am right in saying that it had been long a desire with the Prussian Court to introduce Episcopacy into the new Evangelical Religion, which was intended in that country to embrace both the Lutheran and Calvinistic bodies. I almost think I heard of the project, when I was at Rome in 1833, at the Hotel of the Prussian Minister, M. Bunsen, who was most hospitable and kind, as to other English visitors, so also to my friends and myself. The idea of Episcopacy, as the Prussian king understood it, was, I suppose, very different from that taught in the

Tractarian School: but still, I suppose also, that the chief authors of that school would have gladly seen such a measure carried out in Prussia, had it been done without compromising those principles which were necessary to the being of a Church. About the time of the publication of Tract 90, M. Bunsen and the then Archbishop of Canterbury were taking steps for its execution, by appointing and consecrating a Bishop for Jerusalem. Jerusalem, it would seem, was considered a safe place for the experiment; it was too far from Prussia to awaken the susceptibilities of any party at home; if the project failed, it failed without harm to any one; and, if it succeeded, it gave Protestantism a *status* in the East, which, in association with the Monophysite or Jacobite and the Nestorian bodies, formed a political instrument for England, parallel to that which Russia had in the Greek Church, and France in the Latin. . . .

Now here, at the very time that the Anglican Bishops were directing their censure upon me for avowing an approach to the Catholic Church not closer than I believed the Anglican formularies would allow, they were on the other hand, fraternizing, by their act or by their sufferance, with Protestant bodies, and allowing them to put themselves under an Anglican Bishop, without any renunciation of their errors or regard to their due reception of baptism and confirmation; while there was great reason to suppose that the said Bishop was intended to make converts from the orthodox Greeks, and the schismatical Oriental bodies, by means of the influence of England. This was the third blow, which finally shattered my faith in the Anglican Church. That Church was not only forbidding any sympathy or concurrence with the Church of Rome, but it actually was courting an intercommunion with Protestant Prussia and the heresy of the Orientals. The Anglican Church might have the Apostolical succession, as had the Monophysites; but such acts as were in progress led me to the gravest suspicion, not that it would soon cease to be a Church, but that, since the 16th century, it had never been a Church all along.

[ch. 3]

While my old and true friends were thus in trouble about me, I suppose they felt not only anxiety but pain, to see that I was gradually surrendering myself to the influence of others, who had not their own claims upon me, younger men, and of a cast of mind in no small degree uncongenial to my own. A new school of thought was rising, as is usual in doctrinal inquiries, and was sweeping the original party of the Movement aside, and was taking its place. The most prominent person in it, was a man of elegant genius, of classical mind, of rare talent in literary composition:—Mr. Oakeley. He was not far from my own age; I had long known him, though of late years he had not been in residence at

Oxford; and quite lately, he had been taking several signal occasions of renewing that kindness, which he ever showed towards me when we were both in the Anglican Church. His tone of mind was not unlike that which gave a character to the early Movement; he was almost a typical Oxford man, and, as far as I recollect, both in political and ecclesiastical views, would have been of one spirit with the Oriel party of 1826–1833. But he had entered late into the Movement; he did not know its first years; and, beginning with a new start, he was naturally thrown together with that body of eager, acute, resolute minds who had begun their Catholic life about the same time as he, who knew nothing about the *Via Media*, but had heard much about Rome. This new party rapidly formed and increased, in and out of Oxford, and, as it so happened, contemporaneously with that very summer, when I received so serious a blow to my ecclesiastical views from the study of the Monophysite controversy. These men cut into the original Movement at an angle, fell across its line of thought, and then set about turning that line in its own direction. They were most of them keenly religious men, with a true concern for their souls as the first matter of all, with a great zeal for me, but giving little certainty at the time as to which way they would ultimately turn. Some in the event have remained firm to Anglicanism, some have become Catholics, and some have found a refuge in Liberalism. Nothing was clearer concerning them, than that they needed to be kept in order; and on me who had had so much to do with the making of them, that duty was as clearly incumbent; and it is equally clear, from what I have already said, that I was just the person, above all others, who could not undertake it. There are no friends like old friends; but of those old friends, few could help me, few could understand me, many were annoyed with me, some were angry, because I was breaking up a compact party, and some, as a matter of conscience, could not listen to me. When I looked round for those whom I might consult in my difficulties, I found the very hypothesis of those difficulties acting as a bar to their giving me their advice. Then I said, bitterly, "You are throwing me on others, whether I will or no." Yet still I had good and true friends around me of the old sort, in and out of Oxford too, who were a great help to me. But on the other hand, though I neither was so fond (with a few exceptions) of the persons, nor of the methods of thought, which belonged to this new school, as of the old set, though I could not trust in their firmness of purpose, for, like a swarm of flies, they might come and go, and at length be divided and dissipated, yet I had an intense sympathy in their object and in the direction in which their path lay, in spite of my old friends, in spite of my old life-long prejudices. In spite of my ingrained fears of Rome, and the decision of my reason and conscience against her usages, in spite of my affection for Oxford and Oriel, yet I had a secret longing love of Rome the Mother of English Christianity, and I had a true devotion to the Blessed Virgin, in whose College I lived, whose Altar I served, and whose Immaculate Purity I had in one of my earliest printed Sermons made much of. And it was the consciousness of this bias in myself, if it is so to be called, which made me preach so earnestly against the danger of being swayed in religious inquiry by our sympathy rather than by our reason. And moreover, the members of this new school looked up to me, as I have said, and did me true kindnesses, and really loved me, and stood by me in trouble, when others went away, and for all this I was grateful; nay, many of them were in trouble themselves, and in the same boat with me, and that was a further cause of sympathy between us; and hence it was, when the new school came on in force, and into collision with the old, I had not the heart, any more than the power, to repel them; I was in great perplexity, and hardly knew where I stood; I took their part; and, when I wanted to be in peace and silence, I had to speak out, and I incurred the charge of weakness from some men, and of mysteriousness, shuffling, and underhand dealing from the majority.

Now I will say here frankly, that this sort of charge is a matter which I cannot properly meet, because I cannot duly realize it. I have never had any suspicion of my own honesty; and, when men say that I was dishonest, I cannot grasp the accusation as a distinct conception, such as it is possible to encounter. If a man said to me, "On such a day and before such persons you said a thing was white, when it was black," I understand what is meant well enough, and I can set myself to prove an *alibi* or to explain the mistake; or if a man said to me, "You tried to gain me over to your party, intending to take me with you to Rome, but you did not succeed," I can give him the lie, and lay down an assertion of my own as firm and as exact as his, that not from the time that I was first unsettled, did I ever attempt to gain any one over to myself or to my Romanizing opinions, and that it is only his own coxcombical fancy which has bred such a thought in him: but my imagination is at a loss in presence of those vague charges, which have commonly been brought against me, charges which are made up of impressions, and understandings, and inferences, and hearsay, and surmises. Accordingly, I shall not make the attempt, for, in doing so, I should be dealing blows in the air; what I shall attempt is to state what I know of myself and what I recollect, and leave to others its application.

While I had confidence in the *Via Media*, and thought that nothing could overset it, I did not mind laying down large principles, which I saw would go further than was commonly perceived. I considered that to make the *Via Media* concrete and substantive, it must be much more than it was in outline; that the Anglican Church must have a ceremonial, a ritual, and a fulness of doctrine and devotion, which it had not at present, if it were to compete with the Roman Church with any prospect of success. Such additions would not remove it from its proper basis, but would merely strengthen and beautify it: such, for instance, would be confraternities, particular devotions, reverence for the Blessed Virgin, prayers for the dead, beautiful churches, munificent offerings to them and in them, monastic houses, and many other observances and institutions, which I used to say belonged to us as much as to Rome, though Rome had appropriated them and boasted of them, by reason of our having let them slip from us. . . . With these feelings I frankly admit, that, while I was working simply for the sake of the Anglican Church, I did not at all mind, though I found myself laying down principles in its defence, which went beyond that particular kind of defence which high-and-dry men thought perfection, and even though I ended in framing a kind of defence, which they might call a revolution, while I thought it a restoration. . . .

To so much I confess; but I do not confess, I simply deny that I ever said any thing which secretly bore against the Church of England, knowing it myself, in order that others might unwarily accept it. It was indeed one of my great difficulties and causes of reserve, as time went on, that I at length recognized in principles which I had honestly preached as if Anglican, conclusions favourable to the cause of Rome. Of course I did not like to confess this; and, when interrogated, was in consequence in perplexity. The prime instance of this was the appeal to Antiquity; St. Leo had overset, in my own judgment, its force as the special argument for Anglicanism; yet I was committed to Antiquity, together with the whole Anglican school; what then was I to say, when acute minds urged this or that application of it against the *Via Media*? it was impossible that, in such circumstances, any answer could be given which was not unsatisfactory, or any behaviour adopted which was not mysterious. Again, sometimes in what I wrote I went just as far as I saw, and could as little say more, as I could see what is below the horizon; and therefore, when asked as to the consequences of what I had said, I had no answer to give. Again, sometimes when I was asked, whether certain conclusions did not follow from a certain principle, I might not be able to tell at the moment, especially if the matter were complicated; and for this reason, if for no other, because there is great difference between a conclusion in the abstract and a conclusion in the concrete, and because a conclusion may be modified in fact by a conclusion from some opposite principle. Or it might so happen that my head got simply confused, by the very strength of the logic which was administered to me, and thus I gave my sanction to conclusions which really were not mine; and when the report of those conclusions came round to me through others, I had to unsay them. And then again, perhaps I did not like to see men scared or scandalized by unfeeling logical inferences, which would not have troubled them to the day of their death, had they not been forced to recognize them. And then I felt altogether the force of the maxim of St. Ambrose, "Non in dialecticâ complacuit Deo salvum facere populum suum;"* I had a great dislike of paper logic. For myself, it was not logic that carried me on; as well might one say that the quicksilver in the barometer changes the weather. It is the concrete being that reasons; pass a number of years, and I find my mind in a new place; how? the whole man moves; paper logic is but the record of it. All the logic in the world would not have made me move faster towards Rome than I did; as well might you say that I have arrived at the end of my journey, because I see the village church before me, as venture to assert that the miles, over which my soul had to pass before it got to Rome, could be annihilated, even though I had been in possession of some far clearer view than I then had, that Rome was my ultimate destination. Great acts take time. At least this is what I felt in my own case; and therefore to come to me with methods of logic had in it the nature of a provocation, and, though I do not think I ever showed it, made me somewhat indifferent how I met them, and perhaps led me, as a means of relieving my impatience, to be mysterious or irrelevant, or to give in because I could not meet them to my satisfaction. And a greater trouble still than these logical mazes, was the introduction of logic into every subject whatever, so far, that is, as this was done. Before I was at Oriel, I recollect an acquaintance saying to me that "the Oriel Common Room stank of Logic." One is not at all pleased when poetry, or eloquence, or devotion, is considered as if chiefly intended to feed syllogisms. Now, in saying all this, I am saying nothing against the deep piety and earnestness which were characteristics of this second phase of the Movement, in which I had taken so prominent a part. What I have been observing is, that this phase had a tendency to bewilder and to upset me; and, that, instead of saying so, as I ought to have done, perhaps from a sort of laziness I gave answers at random, which have led to my appearing close or inconsistent. . . .

There was another source of the perplexity with which at this time I was encompassed, and of the reserve and mysteriousness, of which that perplexity gained for me the credit. After Tract 90 the Protestant world would not let me alone; they pursued me in the public journals to Littlemore. Reports of all kinds were circulated about me. "Imprimis, why did I go up to Littlemore at all? For no good purpose certainly; I

---

* 'It is not by Logic that God has decided to save his people.' *De Fide ad Gratianum Augustum*, 1. 5. 42. [Editor's note]

dared not tell why." Why, to be sure, it was hard that I should be obliged to say to the Editors of newspapers that I went up there to say my prayers; it was hard to have to tell the world in confidence, that I had a certain doubt about the Anglican system, and could not at that moment resolve it, or say what would come of it; it was hard to have to confess that I had thought of giving up my Living a year or two before, and that this was a first step to it. It was hard to have to plead, that, for what I knew, my doubts would vanish, if the newspapers would be so good as to give me time and let me alone. Who would ever dream of making the world his confidant? yet I was considered insidious, sly, dishonest, if I would not open my heart to the tender mercies of the world. But they persisted: "What was I doing at Littlemore?" Doing there! have I not retreated from you? have I not given up my position and my place? am I alone, of Englishmen, not to have the privilege to go where I will, no questions asked? am I alone to be followed about by jealous prying eyes, which take note whether I go in at a back door or at the front, and who the men are who happen to call on me in the afternoon? Cowards! if I advanced one step, you would run away; it is not you that I fear: "Di me terrent, et Jupiter hostis."* It is because the Bishops still go on charging against me, though I have quite given up: it is that secret misgiving of heart which tells me that they do well, for I have neither lot nor part with them: this it is which weighs me down. I cannot walk into or out of my house, but curious eyes are upon me. Why will you not let me die in peace? Wounded brutes creep into some hole to die in, and no one grudges it them. Let me alone, I shall not trouble you long. This was the keen feeling which pierced me, and, I think, these are the very words in which I expressed it to myself. I asked in the words of a great motto, "Ubi lapsus? quid feci?"† One day when I entered my house, I found a flight of Under-graduates inside. Heads of Houses, as mounted patrols, walked their horses round those poor cottages. Doctors of Divinity dived into the hidden recesses of that private tenement uninvited, and drew domestic conclusions from what they saw there. I had thought that an Englishman's house was his castle; but the newspapers thought otherwise. . . .

I was in a humour, certainly, to bite off their ears. I will freely confess, indeed I said it some pages back, that I was angry with the Anglican divines. I thought they had taken me in; I had read the Fathers with their eyes; I had sometimes trusted their quotations or their reasonings; and from reliance on them, I had used words or made statements, which by right I ought rigidly to have examined myself. I had thought myself

---

* 'It is the gods that frighten me, and the emnity of Jupiter.' Virgil, *Aeneid*, 12.895. [Editor's note]

† 'Where is the fault? What have I done?' [Editor's note]

safe, while I had their warrant for what I said. I had exercised more faith than criticism in the matter. This did not imply any broad misstatements on my part, arising from reliance on their authority, but it implied carelessness in matters of detail. And this of course was a fault.

. . . The most oppressive thought, in the whole process of my change of opinion, was the clear anticipation, verified by the event, that it would issue in the triumph of Liberalism. Against the Anti-dogmatic principle I had thrown my whole mind; yet now I was doing more than any one else could do, to promote it. I was one of those who had kept it at bay in Oxford for so many years; and thus my very retirement was its triumph. The men who had driven me from Oxford were distinctly the Liberals; it was they who had opened the attack upon Tract 90, and it was they who would gain a second benefit, if I went on to abandon the Anglican Church. But this was not all. As I have already said, there are but two alternatives, the way to Rome, and the way to Atheism: Anglicanism is the halfway house on the one side, and Liberalism is the halfway house on the other. How many men were there, as I knew full well, who would not follow me now in my advance from Anglicanism to Rome, but would at once leave Anglicanism and me for the Liberal camp. It is not at all easy (humanly speaking) to wind up an Englishman to a dogmatic level. I had done so in good measure, in the case both of young men and of laymen, the Anglican *Via Media* being the representative of dogma. The dogmatic and the Anglican principle were one, as I had taught them; but I was breaking the *Via Media* to pieces, and would not dogmatic faith altogether be broken up, in the minds of a great number, by the demolition of the *Via Media*? Oh! how unhappy this made me! I heard once from an eye-witness the account of a poor sailor whose legs were shattered by a ball, in the action off Algiers in 1816, and who was taken below for an operation. The surgeon and the chaplain persuaded him to have a leg off; it was done and the torniquet applied to the wound. Then, they broke it to him that he must have the other off too. The poor fellow said, "You should have told me that, gentlemen," and deliberately unscrewed the instrument and bled to death. Would not that be the case with many friends of my own? How could I ever hope to make them believe in a second theology, when I had cheated them in the first? with what face could I publish a new edition of a dogmatic creed, and ask them to receive it as gospel? Would it not be plain to them that no certainty was to be found any where? Well, in my defence I could but make a lame apology; however, it was the true one, viz. that I had not read the Fathers cautiously enough; that in such nice points, as those which determine the angle of divergence between the two Churches, I had made considerable miscalculations. . . .

I left Oxford for good on Monday, February 23, 1846. On the

Saturday and Sunday before, I was in my house at Littlemore simply by myself, as I had been for the first day or two when I had originally taken possession of it. I slept on Sunday night at my dear friend's, Mr. Johnson's, at the Observatory. Various friends came to see the last of me; Mr. Copeland, Mr. Church, Mr. Buckle, Mr. Pattison, and Mr. Lewis. Dr. Pusey too came up to take leave of me; and I called on Dr. Ogle, one of my very oldest friends, for he was my private Tutor, when I was an Undergraduate. In him I took leave of my first College, Trinity, which was so dear to me, and which held on its foundations so many who had been kind to me both when I was a boy, and all through my Oxford life. Trinity had never been unkind to me. There used to be much snap-dragon growing on the walls opposite my freshman's room there, and I had for years taken it as the emblem of my own perpetual residence even unto death in my University.

On the morning of the 23rd I left the Observatory. I have never seen Oxford since, excepting its spires, as they are seen from the railway.

[ch. 4]

LETTERS

TO MRS NEWMAN

Dartington—July 7. 1831

To my Mother

. . . What strikes me most is the strange richness of every thing. The rocks blush into every variety of colour—the trees and fields are emeralds, and the cottages are rubies. A beetle I picked up at Torquay was as green and gold as the stone it lay on, and a squirrel which ran up a tree here just now was not a pale reddish brown, to which I am accustomed, but a bright brown red. Nay, my very hands and fingers look rosy, like Homer's Aurora, and I have been gazing on them with astonishment. All this wonder I know is simple, and therefore of course do not you repeat it. The exuberance of the grass and the foliage is oppressive, as if one had not room to breathe, though this is a fancy—the depth of the valleys and the steepness of the slopes increase the illusion—and the Duke of Wellington would be in a fidget to get some commanding point to see the country from. The scents are extremely fine, so very delicate, yet so powerful, and the colours of the flowers as if they were all shot with white. The sweet peas especially have the complexion of a beautiful face—they trail up the wall, mixed with myrtles, as creepers. As to the sunset, the Dartmoor heights look purple, and the sky close upon them a clear orange. When I turn back to think of Southampton water and the Isle of Wight, they seem by contrast to be drawn in india-ink or pencil. Now I cannot make out that this is fancy for why should I fancy? I am not especially in a poetic mood. I have heard of the brilliancy of Cintra, and still more of the East, and suppose that this region would fade beside them; yet I am content to marvel at what I see, and think of Virgil's description of the purple meads of Elysium—Let me enjoy what I feel, even though I may unconsciously exaggerate

TO MRS NEWMAN

On board the Hermes. Dec 11. 1832

My dear Mother

My Sisters will perhaps quarrell with me for writing three letters running to you—but I wish that you should receive the first letter I write home from foreign parts, and therefore they must wait the next time—though you will all receive from me letters at the same time. Today has been the most pleasurable day, as far as external causes go, I have ever had, that I can recollect, and now in the evening, I am sleepy and tired with the excitement. We are now off Cape Finisterre, but the moon is not yet risen, and we can see nothing of it, tho' some lights were just now visible from farm houses on shore, which is maybe 15 miles off. This morning early we saw the high mountains of Spain, (the first foreign land I ever saw,) having finished most prosperously our passage across the formidable Bay of Biscay. The land first discovered was Cape Ortegal and its neighbourhood—magnificent in its outline, and, as we neared it, marked with three lines of mountains, and in some places very precipitous and hoving over the sea. At first we were about 70 miles off them—then 25 perhaps. At the same time the day cleared, all clouds vanished, and the sea, which had ever hitherto been very fine, now became of a rich indigo colour, and the wind freshening was tipped with the most striking white edges, which breaking in foam turned into momentary rainbows. The sea gulls, quite at home, were sailing about, and the ship rocked to and fro with a motion, which unpleasant as it might have been in the bay or had the wind been from the southwest, yet was delightful as being from shore. I cannot describe the exquisite colour of the sea—which, tho' in no respects strange or novel, is yet unlike anything I have seen, unless this be a bull. In the sense I should call it a most gentlemanlike colour—i.e. so subdued, so destitute of all display, so sober. And then so deep and solemn, so resistlessly strong, if a colour may be so called—and the contrast between the white and indigo so striking—and in the wake of the vessel, it changed into all colours, transparent green, white, white-green, etc.—As evening came on, we had every appearance of being in a warmer latitude—The sea brighened to a glowing purple inclined to lilac—the sun set in a car of gold and was succeeded by a sky first pale orange, then gradually heightening to a dusky red—while Venus came out as the evening star, with a peculiar intense whiteness. . . .

We are altogether about 60 on board. We started from Falmouth about one on Saturday the 8th, having been kept in uncomfortable suspense about the arrival of the Vessel till the night before—She had suffered much hard weather in the downs—and we embarked (readily enough) at a short notice. It was most amusing to see the stores arrive. Fowls, Ducks, Turkeys, all alive and squatted down under legs of beef, hampers, and vegetables. One unfortunate Duck got away, and a chase ensued—I should have liked to have let him off, but the poor fool did not know how to use his fortune—and instead of making for the shore, kept quacking with absurd vehemence close to us—he was not caught for a considerable time, as he ducked and fluttered away, whenever the men got near him.—Then the decks had to be cleaned—for the vessel had been ordered off from Woolwich in a great hurry, and coals had since been taken in. This is a curious scene, if you never saw it, what with the spouting of the pipes and the scrubbing brooms of a dozen men. Cape Lizard was by this time past—and we were bidding adieu to old England—I got hungry (it was between 3 and 4) and what at home would have been lassitude and craving began now to discover itself in quamishness. When dinner came, I eat because I ought, but with such absurdly uncomfortable feelings, that I could not help laughing out. I will not here enter into the details of seasickness, but shall reserve the interesting subject for another letter, if I can by any reminiscences do justice to it. . . .

TO MRS NEWMAN

On board the Hermes, Dec 19. 1832

My dear Mother,

. . . I was very often and violently sick, but whenever it was over I was quite well, and my qualmishness was short; but the worst of seasickness is the sympathy which all things on board have with the illness, as if they were seasick too. First all the chairs, tables, much more the things on them are moving, moving, up and down, up and down—swing, swing—a tumbler turns over, knife and fork run down, wine is spilt—swing, swing. In this condition you go on talking and eating, as fast as you can, hiding your misery, which is provoking[ly] thrust upon you by every motion of the furniture which surrounds you. At length you are seized with sickness, up you get, swing, swing, you cannot move a step—you knock yourself against the table—run smack against the side of the cabin—you *cannot* make the door, the only point you want—you get into your berth at last, but the door will not shut—bang, bang you slam your fingers. At last things go right with you and down you lie. You are much better, but now a new misery begins, the noise of the bulkheads (i.e. the wooden partitions thro' the vessel). This is not heard on deck—in the cabin it is considerable—but when you lie down, you are in a perfect millhouse. All sorts of noises tenfold increased by the gale; creaking, clattering, shivering, and dashing. And then your bed is seasick too—up and down—swinging without exaggeration as high and as fast (to your feelings) as a swing in a fair—and incessantly—you cannot say, 'Let me out, I have had enough now—you are in for it'—This neverending motion is the great evil, and requires strong nerves to bear. . . .

TO MISS MUNRO

Oy Bm Feb 11/50

My dear Miss Munro,

. . . I return you Miss Moore's letter. You must undeceive her about me, though I suppose she uses words in a general sense. I have nothing of a Saint about me as every one knows, and it is a severe (and salutary) mortification to be thought next door to one. I may have a high view of many things, but it is the consequence of education and of a peculiar cast of intellect—but this is very different from *being* what I admire. I have no tendency to be a saint—it is a sad thing to say. Saints are not literary men, they do not love the classics, they do not write Tales. I may be well enough in my way, but it is not the 'high line.' People ought to feel this, most people do. But those who are at a distance have fee-fa-fum notions about one. It is enough for me to black the saints' shoes—if St Philip uses blacking, in heaven.

Ever Yours affecty John H Newman   Cong. Orat.

TO AUSTIN MILLS

Cork. February 22. 1854
(I shall not put this into the post for a day or two)

My dear Austin,

Though you are not Secretary, yet as Fr Edwards is a new hand, perhaps you will inform him how best to bring the following before the Congregatio Deputata. I submit part of a sketch of a *new work*, which must be submitted to *two Fathers*; I proposed to call it 'The doleful disasters and curious catastrophes of a traveller in the wilds of the west.' I have sketched five chapters as below.

1. The first will contain a series of varied and brilliant illustrations of the old proverb, 'more know Tom Fool than Tom Fool knows.'

2. The second will relate how at Carlow a large party of priests was asked to meet the author at dinner, after which the said author, being fatigued with the day, went to sleep—and was awakened from a refreshing repose by his next neighbour on the right shouting in his ear, 'Gentleman, Dr. N is about to explain to you the plan he proposes for establishing the new University,' an announcement, which the said Dr N. does aver most solemnly took him utterly by surprise, and he cannot think what he could have said in his sleep which could have been understood to mean something so altogether foreign to his intentions and his habits. However, upon this announcement, how the author was obliged to speak and answer questions, in which process he made mistakes and contradicted himself, to the clear consciousness and extreme disgust of the said author.

3. Chapter third will detail the merry conceit of the Paddy who drove him from the Kilkenny station, and who, instead of taking him to the Catholic Bishop's, took him to the Protestant Superintendent's palace, a certain O Brien, who now for 15 years past has been writing against the author and calling him hard names—and how the said carman deposited him at the door of the Protestant palace, and drove away, and how he kept ringing and no one came— and how at last he ventured to attempt and open the hall door without leave, and found himself inside the house, and made a noise in vain—and how, when his patience was exhausted, he advanced further in, and went up some steps and looked about him, and still found no one at all—all along thinking it the house of the true Bishop, and a very fine one too. And how at last he ventured to knock at a room door, and how at length out came a scullery maid, and assured him that the Bishop was in London—whereupon gradually the true state of the case unfolded itself to his mind, and he began to think that, had the Protestant Superintendant been at home, a servant

would have answered the bell, and he should have sent in his card or cartel with his own name upon it, for the inspection of the said Superintendant.

4. And the fourth chapter of the work will go on to relate how the Bishop of Ossory pleasantly suggested when he heard of the above, that the carman's mistake was caused by a certain shepherd's plaid which the author had upon his shoulders, by reason of which he, the author might be mistaken for a Protestant parson. And this remark will introduce the history of the said plaid, and how the author went to F Stanislas Flanagan's friend, Mr Geoghegan, in Sackville Street, and asked for a clerical wrapper, on which the said plaid was shown him. And he objecting to it as not clerical, the shop man on the contrary assured him it was. Whereupon in his simplicity he bought the said plaid, and took it with him on his travels, and left behind him his good Propaganda cloke; and how now he does not know what to do, for he is wandering over the wide world, in a fantastic dress, like a merry andrew, yet with a Roman collar on.

5. And the fifth chapter will narrate his misadventure at Waterford—how he went to the Ursuline Convent there, and the acting Superior determined he should see all the young ladies of the school to the number of 70, all dressed in blue, with medals on, some blue, some green, some red; and how he found he had to make them a speech, and how he puzzled and fussed himself what on earth he should say impromptu to a parcel of school girls—and how in his distress he *did* make what he considered his best speech—and how, when it was ended, the Mother Schoolmistress did not know he had made it, or even begun it, and still asked for his speech. And how he would not, because he could not make a second speech; and how, to make it up, he asked for a holiday for the girls, and how the Mother Schoolmistress flatly refused him, by reason (as he verily believes) because she would not recognise and accept his speech, and wanted another, and thought she had dressed up her girls for nothing—and how he nevertheless drank her rasberry's vinegar, which much resembles a nun's anger, being a sweet acid, and how he thought to himself, it being his birthday, that he was full old to be forgiven if he would not at a moment act the spiritual jack pudding to a girl's school.

This is as much as I have to send you—Would you kindly add your own criticisms on those of the two Fathers? Love to all

Ever Yrs affly J H N

TO GEORGE TALBOT

July 25. 1864

Dear Monsignor Talbot

I have received your letter, inviting me to preach next Lent in your Church at Rome, to 'an audience of Protestants more educated than could ever be the case in England.'

However, Birmingham people have souls; and I have neither taste nor talent for the sort of work, which you cut out for me: and I beg to decline your offer

I am &c J H N

TO AMBROSE ST JOHN

Buckland Grange Ryde Septr 13. 1865

My dear A

. . . Scarcely had I left Birmingham, when it struck me that, since Pusey was to be at Keble's that evening, there was no manner of doubt that he would get into my train at Oxford and journey down with me. I was sure of this. When he did not get into my carriage at Oxford, I felt sure we should recognise each other when we were thrown off the train at Reading—but no, he did not turn up—as it happened, he went by an earlier train. However, this expectation put me upon thinking on the subject—and I made up my mind to go to Keble's next morning and see him—and I did. I slept at the Railway Hotel at Southampton Dock—a very reasonable house, and good too—my bed only 2/6—(they are building close by a grand Imperial Hotel) and then yesterday morning (Tuesday) I retraced my steps to Bishopstoke, left my portmanteau there, and went over to Hursley. I had forgotten the country and was not prepared for such beauty, in the shape of Woods. Keble was at the door, he did not know me, nor I him. How mysterious that first sight of friends is! for when I came to contemplate him, it was the old face and manner, but the first effect and impression was different. His wife had been taken ill again in the night, and at the first moment he, I *think*, and *certainly* I, wished myself away. Then he said, Have you missed my letters? meaning Pusey is here, and I wrote to stop your coming. He then said I must go and prepare Pusey. He did so, and then took me into the room where Pusey was. I went in rapidly, and it is strange how action overcomes pain. Pusey, as being passive, was evidently shrinking back into the corner of the room—as I should have done if he had rushed in upon me. He could not help contemplating the look of me narrowly and long—Ah, I thought, you are thinking how old I am grown, and I see myself in you—though you, I do think, are more altered than I am. Indeed, the alteration in him shocked me (I would not say this to every one)—it pained and grieved me. I should have known him any where— his face is not changed, but it is as if you looked at him through a prodigious magnifier. I recollect him short and small—with a round head—smallish features—flaxen curly hair—huddled up together from his shoulders downward—and walking fast. This was as a young man— but comparing him even when last I saw him in 1846, when he was slow in his motions and staid in his figure, still there is a wonderful change. His head and his features are half as large again—his chest is very broad (*don't say all this*)—and he has, I think, a paunch—His voice is the same—were my eyes shut, I should not have been sensible of any lapse of time. As we three sat together at one table, I had as painful thoughts as I ever recollect, though it was a pain, not acute, but heavy. There were three old men, who had worked together vigorously in their prime. This is what they have come to—poor human nature—after 20 years they meet together round a table, but without a common cause, or free outspoken thoughts—but, though kind yet subdued, and antagonistic in their mode of speaking, and all of them with broken prospects. Pusey is full of his book which is all but out—against Manning; and full of his speech on the relations between physical science and the Bible, which he is to deliver at the Church Congress at Norwich. He is full of polemics and of hope. Keble is as different as possible; he is as delightful as ever— and, it *seemed* to me as if he felt a sympathy and intimacy with me which he did not find with Pusey. At least he spoke to me of him—and I don't think in the same tone he would have spoken to him of me. I took an early dinner with them, and when the bell chimed for evensong at 4 o'clock, I got into my gig . . .

Ever Yrs affly John H Newman

TO FRANCIS WILLIAM NEWMAN

End of October 1867

As to what you tell me of Archbishop Manning, I have heard that some also of our Irish bishops think that too many drink-shops are licensed. As for me, I do not know whether we have too many or too few.

TO ARCHBISHOP MANNING

Nov. 3 1869

My dear Archbishop
Thank you for your kind letter—I can only repeat what I said when you
last heard from me. I do not know whether I am on my head or my heels,
when I have active relations with you. In spite of my friendly feelings,
this is the judgment of my intellect

Yours affectionately in Christ, John H Newman
The Most Revd Dr Manning

'THE ANGLO-AMERICAN CHURCH'

. . . To tell the truth, we think one special enemy to which the American
Church, as well as our own, at present lies open is the influence of a
refined and covert Socinianism. Not that we fear any invasion of that
heresy within her pale now, any more than fifty years ago, but it is
difficult to be in the neighbourhood of icebergs without being chilled,
and the United States is, morally speaking, just in the latitude of ice and
snow. Here again, as our remarks will directly show, we mean nothing
disrespectful towards our Transatlantic relatives. We allude, not to their
national character, nor to their form of government, but to their
employments, which in truth we share with them. A trading country is the
habitat of Socinianism. . . . There is no accounting for tastes; and there is
a moral condition of mind to which this dismal creed *is* alluring. . . . Not
to the poor, the forlorn, the dejected, the afflicted, can the Unitarian
doctrine be alluring, but to those who are rich and have need of nothing,
and know not that they are "miserable and blind and naked;"—to such
men Unitarianism so-called is just fitted, suited to their need, fulfilling
their anticipations of religion, counterpart to their inward temper and
their modes of viewing things. Those who have nothing of this world to
rely upon need a firm hold of the next, they need a deep religion; they
are as if stripped of the body while here,—as if in the unseen state
between death and judgment; and as they are even now in one sense
what they then shall be, so they need to view God such as they then will
view Him; they endure, or rather eagerly desire, the bare vision of Him
stripped of disguise, as they are stripped of disguises too; they desire to
know that He is eternal, since they feel that they are mortal.

Such is the benefit of poverty; as to wealth, its providential corrective is the relative duties which it involves, as in the case of a landlord; but these do not fall upon the trader. He has rank without tangible responsibilities; he has made himself what he is, and becomes self-dependent; he has laboured hard or gone through anxieties, and indulgence is his reward. In many cases he has had little leisure for cultivation of mind, accordingly luxury and splendour will be his *beau ideal* of refinement. If he thinks of religion at all, he will not like from being a great man to become a little one; he bargains for some or other compensation to his self-importance, some little power of judging or managing, some small permission to have his own way. Commerce is free as air; it knows no distinctions; mutual intercourse is its medium of operation. Exclusiveness, separations, rules of life, observance of days, nice scruples of conscience, are odious to it. We are speaking of the general character of a trading community, not of individuals; and, so speaking, we shall hardly be contradicted. A religion which neither irritates their reason nor interferes with their comfort, will be all in all in such a society. Severity whether of creed or precept, high mysteries, corrective practices, subjection of whatever kind, whether to a doctrine or to a priest, will be offensive to them. They need nothing to fill the heart, to feed upon, or to live in; they despise enthusiasm, they abhor fanaticism, they persecute bigotry. They want only so much religion as will satisfy their natural perception of the propriety of being religious. Reason teaches them that utter disregard of their Maker is unbecoming, and they determine to be religious, not from love and fear, but from good sense.

Now it would be a miserable slander on the American Church to say that she was suited to such a form of mind as this; how can she, with her deep doctrines of the Apostolic Commission and the Eucharistic Sacrifice? but this is the very point; here we see around her the external influences which have a tendency to stifle her true development, and to make her inconsistent and unreal. If in the English Church the deep sea dried up more or less in the last century, why should it not in the American also. Let the latter dread her extension among the opulent merchants and traders in towns, where her success has principally been. Many undesirable persons will begin to see in the Church what they can find nowhere else; the Sectarian doctrines are more or less enthusiastic; the Roman Catholic despotic; in our Church there is (or may be) moderation, rationality, decency, and order, which are just the cardinal excellences, the highest "idea" of truth, the first and only fair, to which their minds attain. If this view of things is allowed a footing, a sleek gentlemanlike religion will grow up within the sacred pale, with well-warmed chapels, softly cushioned pews, and eloquent preachers. The poor and needy, the jewels of the Church, will dwindle away; the clergy will sink in honour, and rich laymen will culminate. . . .

We are aware it is a bold thing to speak of a Church which is a hemisphere off us: we are speaking from books, not from practical knowledge; but we think we may say without fear of mistake, that pews, carpets, cushions, and fine speaking are not developments of the Apostolical Succession. Fathers and brethren, we would say, if we might venture a word, dispense with this world when you enter the presence of another. Throw aside your pillows; set wide your closets; break down your partitions; tear away your carpets. Open a space whereon to worship freely, as those to whom worship was the first thing; who come to repent, not to repose; to give thanks, not to reason; to praise, not to enjoy yourselves. Dispense with your props and kneelers; learn to go down on the floor. What has possessed you and us to choose square boxes to pray in, while we despise Simeon upon his pillar? Why squeeze and huddle together as you neither do, nor would dream of doing, at a dinner-table or in a drawing-room? Let the visible be a type of the invisible. You have dispensed with the clerk, you are spared the royal arms; but still who would ever recognize in a large double cube, with bare walls, wide windows, high pulpit, capacious reading-desk, galleries projecting, and altar obscured, an outward emblem of the heavenly Jerusalem, the fount of grace, the resort of Angels? . . .

[Essays Critical and Historical, vol. i]

'PRIVATE JUDGMENT'

. . . this great people is not such a conscientious supporter of the sacred right of Private Judgment as a good Protestant would desire. Why should we go out of our way, one and all of us, to impute personal motives in explanation of the conversion of every individual convert, as he comes before us, if there were in us, the public, an adhesion to that absolute, and universal, and unalienable principle, as its titles are set forth in heraldic style, high and broad, sacred and awful, the right, and the duty, and the possibility of Private Judgment? . . . Is it not sheer wantonness and cruelty in Baptist, Independent, Irvingite, Wesleyan, Establishment-man, Jumper, and Mormonite, to delight in trampling on and crushing these manifestations of their own pure and precious charter, instead of dutifully and reverently exalting, at Bethel, or at Dan, each instance of it, as it occurs, to the gaze of its professing votaries? If a staunch Protestant's daughter turns Roman, and betakes herself to a convent, why does he not exult in the occurrence? Why does he not give a public breakfast, or hold a meeting, or erect a memorial, or write a pamphlet in honour of her, and of the great undying principle she has so gloriously vindicated? Why is he in this base, disloyal style muttering about priests, and Jesuits, and the horrors of nunneries, in solution of the phenomenon, when he has the fair and ample form of Private Judgment rising before his eyes, and pleading with him, and bidding him impute good motives, not bad, and in very charity ascribe to the influence of a high and holy principle, to a right and a duty of every member of the family of man, what his poor human instincts are fain to set down as a folly or a sin. All this would lead us to suspect that the doctrine of private judgment, in its simplicity, purity, and integrity,—private judgment, all private judgment, and nothing but private judgment,—is held by very few persons indeed; and that the great mass of the population are either stark unbelievers in it, or deplorably dark about it; and that even the minority who are in a manner faithful to it, have glossed and corrupted the true sense of it by a miserably faulty reading, and hold, not the right of private judgment, but the private right of judgment; in other words, their own private right, and no one's else. . . .

[*Essays Critical and Historical*, vol. ii]

THE TAMWORTH READING ROOM

Secular Knowledge not the Principle of Moral Improvement

A distinguished Conservative statesman tells us from the town-hall of Tamworth* that "in becoming wiser a man will become better;" meaning by wiser more conversant with the facts and theories of physical science; and that such a man will 'rise *at once* in the scale of intellectual and *moral* existence." "That," he adds, "is my belief." . . .

The first question which obviously suggests itself is *how* these wonderful moral effects are to be wrought under the instrumentality of the physical sciences. Can the process be analyzed and drawn out, or does it act like a dose or a charm which comes into general use empirically? Does Sir Robert Peel mean to say, that whatever be the occult reasons for the result, so it is; you have but to drench the popular mind with physics, and moral and religious advancement follows on the whole, in spite of individual failures? Yet where has the experiment been tried on so large a scale as to justify such anticipations? Or rather, does

* Sir Robert Peel (1788–1850) on 19 Jan. 1841 at the opening of a new library and reading-room at Tamworth. [Editor's note]

he mean, that, from the nature of the case, he who is imbued with science and literature, unless adverse influences interfere, cannot but be a better man? It is natural and becoming to seek for some clear idea of the meaning of so dark an oracle. To know is one thing, to do is another; the two things are altogether distinct. A man knows he should get up in the morning,—he lies a-bed; he knows he should not lose his temper, yet he cannot keep it. A labouring man knows he should not go to the ale-house, and his wife knows she should not filch when she goes out charing; but, nevertheless, in these cases, the consciousness of a duty is not all one with the performance of it. There are, then, large families of instances, to say the least, in which men may become wiser, without becoming better; what, then, is the meaning of this great maxim in the mouth of its promulgators?

Mr. Bentham* would answer, that the knowledge which carries virtue along with it, is the knowledge how to take care of number one—a clear appreciation of what is pleasurable, what painful, and what promotes the one and prevents the other. An uneducated man is ever mistaking his own interest, and standing in the way of his own true enjoyments. Useful Knowledge is that which tends to make us more useful to ourselves;—a most definite and intelligible account of the matter, and needing no explanation. But it would be a great injustice, both to Lord Brougham† and to Sir Robert, to suppose, when they talk of Knowledge being Virtue, that they are Benthamizing. Bentham had not a spark of poetry in him; on the contrary, there is much of high aspiration, generous sentiment, and impassioned feeling in the tone of Lord Brougham and Sir Robert. They speak of knowledge as something "pulchrum," fair and glorious, exalted above the range of ordinary humanity, and so little connected with the personal interest of its votaries, that, though Sir Robert does *obiter* talk of improved modes of draining, and the chemical properties of manure, yet he must not be supposed to come short of the lofty enthusiasm of Lord Brougham, who expressly panegyrizes certain ancient philosophers who gave up riches, retired into solitude, or embraced a life of travel, smit with a sacred curiosity about physical or mathematical truth.

Here Mr. Bentham, did it fall to him to offer a criticism, doubtless would take leave to inquire whether such language was anything better than a fine set of words "signifying nothing,"—flowers of rhetoric, which bloom, smell sweet, and die. But it is impossible to suspect so grave and practical a man as Sir Robert Peal of using words literally without any

---

* Jeremy Bentham (1748–1832) the Utilitarian philosopher. [Editor's note]

† Lord Brougham (1778–1868) as Lord Chancellor played a prominent part in the passing of the great Reform Bill; he was also instrumental in the founding of the secular, non-denominational London University. [Editor's note]

meaning at all; and though I think at best they have not a very profound meaning, yet, such as it is, we ought to attempt to draw it out.

Now, without using exact theological language, we may surely take it for granted, from the experience of facts, that the human mind is at best in a very unformed or disordered state; passions and conscience, likings and reason, conflicting,—might rising against right, with the prospect of things getting worse. Under these circumstances, what is it that the School of philosophy in which Sir Robert has enrolled himself proposes to accomplish? Not a victory of the mind over itself—not the supremacy of the law—not the reduction of the rebels—not the unity of our complex nature—not an harmonizing of the chaos—but the mere lulling of the passions to rest by turning the course of thought; not a change of character, but a mere removal of temptation. This should be carefully observed. When a husband is gloomy, or an old woman peevish and fretful, those who are about them do all they can to keep dangerous topics and causes of offence out of the way, and think themselves lucky, if, by such skilful management, they get through the day without an outbreak. When a child cries, the nurserymaid dances it about, or points to the pretty black horses out of window, or shows how ashamed poll-parrot or poor puss must be of its tantarums. Such is the sort of prescription which Sir Robert Peel offers to the good people of Tamworth. He makes no pretence of subduing the giant nature, in which we were born, of smiting the loins of the domestic enemies of our peace, of overthrowing passion and fortifying reason; he does but offer to bribe the foe for the nonce with gifts which will avail for that purpose just so long as they *will* avail, and no longer.

This was mainly the philosophy of the great Tully, except when it pleased him to speak as a disciple of the Porch. Cicero handed the recipe to Brougham, and Brougham has passed it on to Peel. . . . If a man was in grief, he was to be amused; if disappointed, to be excited; if in a rage, to be soothed; if in love, to be roused to the pursuit of glory. No inward change was contemplated, but a change of external objects; as if we were all White Ladies or Undines, our moral life being one of impulse and emotion, not subjected to laws, not consisting in habits, not capable of growth. When Cicero was outwitted by Cæsar, he solaced himself with Plato; when he lost his daughter, he wrote a treatise on Consolation. Such, too, was the philosophy of that Lydian city, mentioned by the historian, who in a famine played at dice to stay their stomachs.

And such is the rule of life advocated by Lord Brougham; . . . his notions of vigour and elevation, when analyzed, will be found to resolve themselves into a mere preternatural excitement under the influence of some stimulating object, or the peace which is attained by there being nothing to quarrel with. . . .

Whether Sir Robert Peel meant all this, which others before him have meant, it is impossible to say; but I will be bound, if he did not mean this, he meant nothing else, and his words will certainly insinuate this meaning, wherever a reader is not content to go without any meaning at all. They will countenance, with his high authority, what in one form or other is a chief error of the day, in very distinct schools of opinion,—that our true excellence comes not from within, but from without; not wrought out through personal struggles and sufferings, but following upon a passive exposure to influences over which we have no control. They will countenance the theory that diversion is the instrument of improvement, and excitement the condition of right action; and whereas diversions cease to be diversions if they are constant, and excitements by their very nature have a crisis and run through a course, they will tend to make novelty ever in request, and will set the great teachers of morals upon the incessant search after stimulants and sedatives, by which unruly nature may, *pro re natâ*, be kept in order.

Hence, be it observed, Lord Brougham ... frankly offers us a philosophy of expedients: he shows us how to live by medicine. Digestive pills half an hour before dinner, and a posset at bedtime at the best; and at the worst, dram-drinking and opium,—the very remedy against broken hearts, or remorse of conscience, which is in request among the many, in gin-palaces *not* intellectual.

And if these remedies be but of temporary effect at the utmost, more commonly they will have no effect at all. Strong liquors, indeed, do for a time succeed in their object; but who was ever consoled in real trouble by the small beer of literature or science? ... Or who was made to do any secret act of self-denial, or was steeled against pain, or peril, by all the lore of the infidel La Place, or those other "mighty spirits" which Lord Brougham and Sir Robert eulogize? Or when was a choleric temperament ever brought under by a scientific King Canute planting his professor's chair before the rising waves? And as to the "keen" and "ecstatic" pleasures which Lord Brougham, not to say Sir Robert, ascribes to intellectual pursuit and conquest, I cannot help thinking that in that line they will find themselves outbid in the market by gratifications much closer at hand, and on a level with the meanest capacity. Sir Robert makes it a boast that women are to be members of his institution; it is hardly necessary to remind so accomplished a classic, that Aspasia and other learned ladies in Greece are no very encouraging precedents in favour of the purifying effects of science. But the strangest and most painful topic which he urges, is one which Lord Brougham has had the good taste altogether to avoid,—the power, not of religion, but of scientific knowledge, on a deathbed; a subject which Sir Robert treats in language which it is far better to believe is mere oratory than is said in earnest.

Such is this new art of living, offered to the labouring classes,—we will say, for instance, in a severe winter, snow on the ground, glass falling, bread rising, coal at 20d. the cwt., and no work.

It does not require many words, then, to determine that, taking human nature as it is actually found, and assuming that there is an Art of life, to say that it consists, or in any essential manner is placed, in the cultivation of Knowledge, that the mind is changed by a discovery, or saved by a diversion, and can thus be amused into immortality,—that grief, anger, cowardice, selfconceit, pride, or passion, can be subdued by an examination of shells or grasses, or inhaling of gases, or chipping of rocks, or calculating the longitude, is the veriest of pretences which sophist or mountebank ever professed to a gaping auditory. If virtue be a mastery over the mind, if its end be action, if its perfection be inward order, harmony, and peace, we must seek it in graver and holier places than in Libraries and Reading-rooms.

[Letter 2]

Secular Knowledge not a Principle of Action

People say to me, that it is but a dream to suppose that Christianity should regain the organic power in human society which once it possessed. I cannot help that; I never said it could. I am not a politician; I am proposing no measures, but exposing a fallacy, and resisting a pretence. Let Benthamism reign, if men have no aspirations; but do not tell them to be romantic, and then solace them with glory; do not attempt by philosophy what once was done by religion. The ascendency of Faith may be impracticable, but the reign of Knowledge is incomprehensible. The problem for statesmen of this age is how to educate the masses, and literature and science cannot give the solution.
. . .

Science gives us the grounds or premises from which religious truths are to be inferred; but it does not set about inferring them, much less does it reach the inference;—that is not its province. It brings before us phenomena, and it leaves us, if we will, to call them works of design, wisdom, or benevolence; and further still, if we will, to proceed to confess an Intelligent Creator. We have to take its facts, and to give them a meaning, and to draw our own conclusions from them. First comes Knowledge, then a view, then reasoning, and then belief. This is why Science has so little of a religious tendency; deductions have no power of persuasion. The heart is commonly reached, not through the reason, but through the imagination, by means of direct impressions, by the testimony of facts and events, by history, by description. Persons influence us, voices melt us, looks subdue us, deeds inflame us. Many a man will live and die upon a dogma: no man will be a martyr for a conclusion. A conclusion is but an opinion; it is not a thing which *is*, but which *we are "certain about;"* and it has often been observed, that we never say we are certain without implying that we doubt. To say that a thing *must* be, is to admit that it *may not* be. No one, I say, will die for his own calculations; he dies for realities. This is why a literary religion is so little to be depended upon; it looks well in fair weather, but its doctrines are opinions, and when called to suffer for them, it slips them between its folios, or burns them at its hearth. And this again is the secret of the distrust and raillery with which moralists have been so commonly visited. They say and do not. Why? Because they are contemplating the fitness of things, and they live by the square, when they should be realizing their high maxims in the concrete. . . .

I have no confidence, then, in philosophers who cannot help being religious, and are Christians by implication. They sit at home, and reach forward to distances which astonish us; but they hit without grasping, and are sometimes as confident about shadows as about realities. They have worked out by a calculation the lie of a country which they never saw, and mapped it by means of a gazetteer; and like blind men, though they can put a stranger on his way, they cannot walk straight themselves, and do not feel it quite their business to walk at all.

Logic makes but a sorry rhetoric with the multitude; first shoot round corners, and you may not despair of converting by a syllogism. Tell men to gain notions of a Creator from His works, and, if they were to set about it (which nobody does), they would be jaded and wearied by the labyrinth they were tracing. Their minds would be gorged and surfeited by the logical operation. Logicians are more set upon concluding rightly, than on right conclusions. They cannot see the end for the process. Few men have that power of mind which may hold fast and firmly a variety of thoughts. We ridicule "men of one idea;" but a great many of us are born to be such, and we should be happier if we knew it. To most men argument makes the point in hand only more doubtful, and considerably less impressive. After all, man is *not* a reasoning animal; he is a seeing, feeling, contemplating, acting animal. He is influenced by what is direct and precise. It is very well to freshen our impressions and convictions from physics, but to create them we must go elsewhere. . . .

Life is not long enough for a religion of inferences; we shall never have done beginning, if we determine to begin with proof. We shall ever be laying our foundations; we shall turn theology into evidences, and divines into textuaries. We shall never get at our first principles. Resolve to believe nothing, and you must prove your proofs and analyze your elements, sinking further and further, and finding "in the lowest depth a lower deep," till you come to the broad bosom of scepticism. I would rather be bound to defend the reasonableness of assuming that Christianity is true, than to demonstrate a moral governance from the physical world. Life is for action. If we insist on proofs for everything, we shall never come to action: to act you must assume, and that assumption is faith.

Let no one suppose that in saying this I am maintaining that all proofs are equally difficult, and all propositions equally debatable. Some assumptions are greater than others, and some doctrines involve postulates larger than others, and more numerous. I only say that impressions lead to action, and that reasonings lead from it. Knowledge of premises, and inferences upon them,—this is not to *live*. It is very well as a matter of liberal curiosity and of philosophy to analyze our modes of thought; but let this come second, and when there is leisure for it, and then our examinations will in many ways even be subservient to action. But if we commence with scientific knowledge and argumentative proof, or lay any great stress upon it as the basis of personal Christianity, or attempt to make man moral and religious by Libraries and Museums, let us in consistency take chemists for our cooks, and mineralogists for our masons.

Now I wish to state all this as matter of fact, to be judged by the candid testimony of any persons whatever. Why we are so constituted that Faith, not Knowledge or Argument, is our principle of action, is a question with which I have nothing to do; but I think it is a fact, and if it be such, we must resign ourselves to it as best we may, unless we take refuge in the intolerable paradox, that the mass of men are created for nothing, and are meant to leave life as they entered it. So well has this practically been understood in all ages of the world, that no Religion has yet been a Religion of physics or of philosophy. It has ever been synonymous with Revelation. It never has been a deduction from what we know: it has ever been an assertion of what we are to believe. It has never lived in a conclusion; it has ever been a message, or a history, or a vision.

[Letter 6]

LECTURES ON THE
DIFFICULTIES OF ANGLICANS

On the Relation of the National Church to the Nation

I have said, we must not indulge our imagination in the view we take of the National Establishment. If, indeed, we dress it up in an ideal form, as if it were something real, with an independent and a continuous existence, and a proper history, as if it were in deed and not only in name a Church, then indeed we may feel interest in it, and reverence towards it, and affection for it, as men have fallen in love with pictures, or knights in romance do battle for high dames whom they have never seen. Thus it is that students of the Fathers, antiquaries, and poets, begin by assuming that the body to which they belong is that of which they read in times past, and then proceed to decorate it with that majesty and beauty of which history tells, or which their genius creates. Nor is it by an easy process or a light effort that their minds are disabused of this error. It is an error for many reasons too dear to them to be readily relinquished. But at length, either the force of circumstances or some unexpected accident dissipates it; and, as in fairy tales, the magic castle vanishes when the spell is broken, and nothing is seen but the wild heath, the barren rock, and the forlorn sheep-walk, so is it with us as regards the Church of England, when we look in amazement on that we thought so unearthly, and find so commonplace or worthless. Then we perceive, that aforetime we have not been guided by reason, but biassed by education and swayed by affection. We see in the English Church, I will not merely say no descent from the first ages, and no relationship to the Church in other lands, but we see no body politic of any kind; we see nothing more or less than an Establishment, a department of Government, or a function or operation of the State,—without a substance,—a mere collection of officials, depending on and living in the supreme civil power. Its unity and personality are gone, and with them its power of exciting feelings of any kind. It is easier to love or hate an abstraction, than so commonplace a framework or mechanism. We regard it neither with anger, nor with aversion, nor with contempt, any more than with respect or interest. It is but one aspect of the State, or mode of civil governance; it is responsible for nothing; it can appropriate neither praise nor blame; but, whatever feeling it raises is to be referred on, by the nature of the case, to the Supreme Power whom it represents, and whose will is its breath. And hence it has no real identity of existence in distinct periods, unless the present Legislature or the present Court can affect to be the offspring and disciple of its predecessor. Nor can it in consequence be said to have any antecedents, or any future; or to live, except in the passing moment. As a thing without a soul, it does not contemplate itself, define its intrinsic constitution, or ascertain its position. It has no traditions; it cannot be said to think; it does not know what it holds, and what it does not; it is not even conscious of its own existence. It has no love for its members, or what are sometimes called its children, nor any instinct whatever, unless attachment to its master, or love of its place, may be so called. Its fruits, as far as they are good, are to be made much of, as long as they last, for they are transient, and without succession; its former champions of orthodoxy are no earnest of orthodoxy now; they died, and there was no reason why they should be reproduced. Bishop is not like bishop, more than king is like king, or ministry like ministry; its Prayer-Book is an Act of Parliament of two centuries ago, and its cathedrals and its chapter-houses are the spoils of Catholicism.

I have said all this, my brethren, not in declamation, but to bring out clearly to you, why I cannot feel interest of any kind in the National Church, nor put any trust in it at all from its past history, as if it were, in however narrow a sense, a guardian of orthodoxy. It is as little bound by what it said or did formerly, as this morning's newspaper by its former numbers, except as it is bound by the Law; and while it is upheld by the Law, it will not be weakened by the subtraction of individuals, nor fortified by their continuance. Its life is an Act of Parliament. . . .

I say, that a nation's laws are a nation's property, and have their life in the nation's life, and their interpretation in the nation's sentiment: and where that living intelligence does not shine through them, they become worthless and are put aside, whether formally or on an understanding. Now Protestantism is, as it has been for centuries, the Religion of England; and since the semi-patristical Church, which was set up for the nation at the Reformation, is the organ of that religion, it must live for the nation; it must hide its Catholic aspirations in folios, or in college cloisters; it must call itself Protestant, when it gets into the pulpit; it must abjure Antiquity; for woe to it, if it attempt to thrust the wording of its own documents in its master's path, if it rely on a passage in its Visitation for the Sick, or on an Article of the Creed, or on the tone of its Collects, or on a catena of its divines, when the age has determined on a theology more in keeping with the progress of knowledge! The antiquary, the reader of history, the theologian, the philosopher, the Biblical student may make his protest; he may quote St. Austin, or appeal to the canons, or argue from the nature of the case; but *la Reine le veut*; the English people is sufficient for itself; it wills to be Protestant and progressive; and Fathers, Councils, Schoolmen, Scriptures, Saints, Angels, and what is above them, must give way. What are they to it? It thinks, argues, and acts according to its own practical, intelligible, shallow religion; and of that religion its Bishops and divines, will they or will they not, must be exponents.

In this way, I say, we are to explain, but in this way most naturally and satisfactorily, what otherwise would be startling, the late Royal decision to which I have several times referred. The great legal authorities, on whose report it was made, have not only pronounced, that, as a matter of fact, persons who have denied the grace of Baptism had held the highest preferments in the National Church, but they felt themselves authorised actually to interpret its ritual and its doctrine, and to report to her Majesty that the dogma of baptismal regeneration is not part and parcel of the national religion. . . . The question was, not what God had said, but what the English nation had willed and allowed; and, though it must be granted that they aimed at a critical examination of the letter of the documents, yet it must be granted on the other hand too, that their criticism was of a very national cast, and that the national sentiment was of great use to them in helping them to their conclusions. . . . What though the ritual categorically deposes to the regeneration of the infant baptized? The Evangelical party, who, in former years, had had the nerve to fix the charge of dishonesty on the explanations of the Thirty-nine Articles, put forth by their opponents, could all the while be cherishing in their own breasts an interpretation of the Baptismal Service, simply contradictory to its most luminous declarations. Inexplicable proceeding, if they were professing to handle the document in its letter; but not dishonourable, not dishonest, not hypocritical, but natural and obvious, on the condition or understanding that the Nation, which imposes the document, imposes its sense,— that by the breath of its mouth it had, as a god, made Establishment, Articles, Prayer-Book, and all that is therein, and could by the breath of its mouth as easily and absolutely unmake them again, whenever it was disposed.

Counsel, then, and pamphleteers may put forth unanswerable arguments in behalf of the Catholic interpretation of the Baptismal service; a long succession of Bishops, an unbroken tradition of writers, may have faithfully and anxiously guarded it. In vain has the Caroline school honoured it by ritual observance; in vain has the Restoration illustrated it by varied learning; in vain did the Revolution retain it as the price for other concessions; in vain did the eighteenth century use it as a sort of watchword against Wesley; in vain has it been persuasively developed and fearlessly proclaimed by the movement of 1833; all this is foreign to the matter before us. We have not to enquiry what is the dogma of a collegiate, antiquarian religion, but what, in the words of the Prime Minister, will give "general satisfaction;" what is the religion of Britons. May not the free-born, self-dependent, animal mind of the Englishman, choose his religion for himself? and have lawyers any more to do than to state, as a matter of fact and history, what that religion is, and for three centuries has been? are we to obtrude the mysteries of an objective, of a dogmatic, of a revealed system, upon a nation which intimately feels and has established, that each individual is to be his own judge in truth and falsehood in matters of the unseen world? How is it possible that the National Church, forsooth, should be allowed to

dogmatize on a point which so immediately affects the Nation itself?
Why, half the country is unbaptized; it is difficult to say for certain who
are baptized; shall the country unchristianize itself? it has not yet
advanced to indifference on such a matter. Shall it, by a suicidal act, use
its own Church against itself, as its instrument whereby to cut itself off
from the hope of another life? Shall it confine the Christian promise
within limits, and put restrictions upon grace, when it has thrown open
trade, removed disabilities, abolished monopolies, taken off agricultural
protection, and enlarged the franchise?—Such is the thought, such the
language of the England of to-day. What a day for the defenders of the
dogma in bygone times, if those times had anything to do with the
present! The giant ocean has suddenly swelled and heaved, and
majestically yet masterfully snaps the cables of the smaller craft which lie
upon its bosom, and strands them upon the beach. Hooker, Taylor, Bull,
Pearson, Barrow, Tillotson, Warburton, and Horne, names mighty in
their generation, are broken and wrecked before the power of a nation's
will. One vessel alone can ride those waves; it is the boat of Peter, the ark
of God.

[Lecture 1]

The Movement of 1833 Foreign to the National Church

If then "life" means strength, activity, energy, and well-being of any
kind whatever, in that case doubtless the national religion is alive. It is a
great power in the midst of us; it wields an enormous influence; it
represses a hundred foes; it conducts a hundred undertakings. It attracts
men to it, uses them, rewards them; it has thousands of beautiful homes
up and down the country, where quiet men may do its work and benefit
its people; it collects vast sums in the shape of voluntary offerings, and
with them it builds churches, prints and distributes innumerable Bibles,
books, and tracts and sustains missionaries in all parts of the earth. In all
parts of the earth it opposes the Catholic Church, denounces her as
antichristian, bribes the world against her, obstructs her influence, apes
her authority, and confuses her evidence. In all parts of the world it is the
religion of gentlemen, of scholars, of men of substance, and men of no
personal faith at all. If this be life,—if it be life to impart a tone to the
court and houses of parliament, to ministers of state, to law and
literature, to universities and schools, and to society,—if it be life to be a
principle of order in the population, and an organ of benevolence and
almsgiving towards the poor,—if it be life to make men decent,
respectable, and sensible, to embellish and refine the family circle, to
deprive vice of its grossness, and to shed a gloss over avarice and
ambition,—if indeed it is the life of religion to be the first jewel in the

Queen's crown, and the highest step of her throne, then doubtless the national Church is replete, it overflows with life; but the question has still to be answered, Life of what kind? . . .

. . . The very moment that Catholicism ventures out of books, and cloisters, and studies, towards the national house of prayer, when it lifts its hand or its very eyebrow towards this people so tolerant of heresy, at once the dull and earthly mass is on fire. It would be little or nothing though the minister baptized without water, though he chucked away the consecrated wine, though he denounced fasting, though he laughed at virginity, though he interchanged pulpits with a Wesleyan or a Baptist, though he defied his Bishop; he might be blamed, he might be disliked, he might be remonstrated with; but he would not touch the feelings of men; he would not inflame their minds;—but, bring home to them the very thought of Catholicism, hold up a surplice, and the religious building is as full of excitement and tumult as St. Victor's at Milan in the cause of orthodoxy, or St. Giles', Edinburgh, for the Kirk.

[Lecture 2]

The Movement not in the Direction of a Party in the National Church

The idea, then, of the divines of the movement was simply and absolutely submission to an external authority; to such an authority they appealed, to it they betook themselves; there they found a haven of rest; thence they looked out upon the troubled surge of human opinion and upon the crazy vessels which were labouring, without chart or compass, upon it. Judge then of their dismay, when, according to the Arabian tale, on their striking their anchors into the supposed soil, lighting their fires on it, and fixing in it the poles of their tents, suddenly their island began to move, to heave, to splash, to frisk to and fro, to dive, and at last to swim away, spouting out inhospitable jets of water upon the credulous mariners who had made it their home. And such, I suppose, was the undeniable fact: I mean, the time at length came, when first of all turning their minds (some of them, at least) more carefully to the doctrinal controversies of the early Church, they saw distinctly that in the reasonings of the Fathers, elicited by means of them, and in the decisions of authority, in which they issued, were contained at least the rudiments, the anticipation, the justification of what they had been accustomed to consider the corruptions of Rome. And if only one, or a few of them, were visited with this conviction, still even one was sufficient, of course, to destroy that cardinal point of their whole system, the objective perspicuity and distinctness of the teaching of the Fathers. But time went on, and there was no mistaking or denying the misfortune which was impending over them. They had reared a goodly house, but their foundations were falling in. The soil and the masonry both were bad. The Fathers *would* protect "Romanists" as well as extinguish Dissenters. The Anglican divines *would* misquote the Fathers, and shrink from the very doctors to whom they appealed. The Bishops of the seventeenth century were shy of the Bishops of the fourth; and the Bishops of the nineteenth were shy of the Bishops of the seventeenth. The ecclesiastical courts upheld the sixteenth century against the seventeenth, and, regardless of the flagrant irregularities of Protestant clergymen, chastised the mild misdemeanours of Anglo-Catholic. Soon the living rulers of the Establishment began to move. There are those who, reversing the Roman's maxim, are wont to shrink from the contumacious, and to be valiant towards the submissive; and the authorities in question gladly availed themselves of the power conferred on them by the movement against the movement itself. They fearlessly handselled their Apostolic weapons upon the Apostolical party. One after another, in long succession, they took up their song and their parable against it. It was a solemn war-dance, which they executed round victims, who by their very principles were bound hand and foot, and could only eye with disgust and perplexity this most unaccountable movement, on the part of their "holy Fathers, the representatives of the Apostles, and the Angels of the Churches." It was the beginning of the end.

My brethren, when it was at length plain that primitive Christianity ignored the National Church, and that the National Church cared little for primitive Christianity, or for those who appealed to it as her foundation; when Bishops spoke against them, and Bishops' courts sentenced them, and Universities degraded them, and the people rose against them, from that day their "occupation was gone." Their initial principle, their basis, external authority, was cut from under them; they had "set their fortunes on a cast;" they had lost; henceforward they had nothing left for them but to shut up their school, and retire into the country. Nothing else was left for them, unless, indeed, they took up some other theory, unless they changed their ground, unless they ceased to be what they were, and became what they were not; unless they belied their own principles, and strangely forgot their own luminous and most keen convictions; unless they vindicated the right of private judgment, took up some fancy-religion, retailed the Fathers, and jobbed theology. They had but a choice between doing nothing at all, and looking out for truth and peace elsewhere. . . .

. . . Free thinkers and broad thinkers, Laudians and Prayer-Book Christians, high and dry and Establishment-men, all these he [a Roman Catholic] would understand; but what he would feel so prodigious is this,—that such as you, my brethren, should consider Christianity given from heaven once for all, should protest against private judgment, should profess to transmit what you have received, and yet from diligent study of the Fathers, from your thorough knowledge of St. Basil and St. Chrysostom, from living, as you say, in the atmosphere of Antiquity, that you should come forth into open day with your new edition of the Catholic faith, different from that held in any existing body of Christians anywhere, which not half-a-dozen men all over the world would honour with their *imprimatur*; and then, withal, should be as positive about its truth in every part, as if the voice of mankind were with you instead of being against you.

You are a body of yesterday; you are a drop in the ocean of professing Christians; yet you would give the law to priest and prophet; and you fancy it an humble office, forsooth, suited to humble men, to testify the very truth of Revelation to a fallen generation, or rather to almost a long bi-millenary, which has been in unalleviated traditionary error. You have a mission to teach the National Church, which is to teach the British empire, which is to teach the world; you are more learned than Greece; you are purer than Rome; you know more than St. Bernard; you judge how far St. Thomas was right, and where he is to be read with caution, or held up to blame. You can bring to light juster views of grace, or of penance, or of invocation of saints, than St. Gregory or St. Augustine . . .

This is what you can do; yes, and when you have done all, to what have you attained? to do just what heretics have done before you, and, as doing, have incurred the anathema of Holy Church. . . .

And now, my brethren, will it not be so, as I have said, of simple necessity, if you attempt at this time to perpetuate in the National Church a form of opinion which the National Church disowns? You do not follow its Bishops; you disown its existing traditions; you are discontented with its divines; you protest against its law courts; you shrink from its laity; you outstrip its Prayer Book. You have in all respects an eclectic or an original religion of our own. . . . There is a *consensus* of divines, stronger than there is for Baptismal Regeneration or the Apostolical Succession, that Rome is, strictly and literally, an anti-Christian power:—Liberals and High Churchmen in your Communion in this agree with Evangelicals; you put it aside. There is a *consensus* against Transubstantiation, besides the declaration of the Article; yet many of you hold it notwithstanding. Nearly all your divines, if not all, call themselves Protestants, and you anathematize the name. Who makes the concessions to Catholics which you do, yet remains separate from them? Who, among Anglican authorities, would speak of Penance as a Sacrament, as you do? Who of them encourages, much less insists upon, auricular confession, as you? or makes fasting an obligation? or uses the crucifix and the rosary? or reserves the consecrated bread? or believes in miracles as existing in your communion? or administers, as I believe you do, Extreme Unction? In some points you prefer Rome, in others Greece, in others England, in others Scotland; and of that preference your own private judgment is the ultimate sanction.

What am I to say in answer to conduct so preposterous? Say you go by any authority whatever, and I shall know where to find you, and I shall respect you. Swear by any school of Religion, old or modern, by Ronge's Church, or the Evangelical Alliance, nay, by yourselves, and I shall know what you mean, and will listen to you. But do not come to me with the latest fashion of opinion which the world has seen, and protest to me that it is the oldest. Do not come to me at this time of day with views palpably new, isolated, original, *sui generis*, warranted old neither by Christian nor unbeliever, and challenge me to answer what I really have not the patience to read. Life is not long enough for such trifles. Go elsewhere, not to me, if you wish to make a proselyte. Your inconsistency, my dear brethren, is on your very front. Nor pretend that you are but executing the sacred duty of defending your own Communion: your Church does not thank you for a defence, which she has no dream of appropriating. You innovate on her professions of doctrine, and then you bid us love her for your innovations. You cling to her for what she denounces; and you almost anathematise us for taking a step which you would please her best by taking also. You call it restless, impatient, undutiful in us, to do what she would have us do; and you think it a loving and confiding course in her children to believe, not her, but you. She is to teach, and we are to hear, only according to your own private researches into St. Chrysostom and St. Augustine. "I began myself with doubting and inquiring," you seem to say; "I departed from the teaching I received; I was educated in some older type of Anglicanism; in the school of Newton, Cecil, and Scott, or in the Bartlett's-Building School, or in the Liberal Whig School. I was a Dissenter, or a Wesleyan, and by study and thought I became an Anglo-Catholic. And then I read the Fathers, and I have determined what works are genuine, and what are not; which of them apply to all times, which are occasional; which historical, and which doctrinal; what opinions are private, what authoritative; what they only seem to hold, what they ought to hold; what are fundamental, what ornamental. Having thus measured and cut and put together my creed by my own proper intellect, by my own lucubrations, and differing from the whole world in my results, I distinctly bid you, I solemnly warn you, not to do as I have done, but to accept what I have found, to revere that, to use that, to believe that, for it is the teaching of the old Fathers, and of your Mother the Church of England. Take my word for it, that this is the very truth of Christ; deny your own reason, for I know better than you, and it is as clear as day that some moral fault in you is the cause of your differing from me. It is pride, or vanity, or self-reliance, or fulness of bread. You require some medicine for your soul; you must fast; you must make a general confession; and look very sharp to yourself, for you are already next door to a rationalist or an infidel."

[Lecture 5]

The Movement not in the Direction of a Branch Church

. . . You wish to know whether the Establishment is what you began by assuming it to be—the grace-giving Church of God. If it be, you and your principles will surely find your position there and your home. When you proclaim it to be Apostolical, it will smile on you; when you kneel down and ask its blessing, it will stretch its hands over you; when you would strike at heresy, it will arm you for the fight; when you wind your dangerous way with steady tread between Sabellius, Nestorius, and Eutyches, between Pelagius and Calvin, it will follow you with anxious eyes and a beating heart; when you proclaim its relationship to Rome and Greece, it will in transport embrace you as its own dear children; you will sink happily into its arms, you will repose upon its breast, you will recognise your mother, and be at peace. If, however, on the contrary, you find that the more those great principles which you have imbibed from St. Athanasius and St. Augustine, and which have become the life and the form of your moral and intellectual being, vegetate and expand within you, the more awkward and unnatural you find your position in the Establishment, and the more difficult its explanation; if there is no lying, or standing, or sitting, or kneeling, or stooping there, in any possible attitude; if, as in the tyrant's cage, when you would rest your head, your legs are forced out between the Articles, and when you would relieve your back, your head strikes against the Prayer Book; when, place yourself as you will, on the right side or the left, and try to keep as still as you can, your flesh is ever being punctured and probed by the stings of Bishops, laity, and nine-tenths of the Clergy buzzing about you; is it not as plain as day that the Establishment is not your place, since it is no place for your principles? . . .

Here then, when you are investigating whither you shall go for your new succession and your new priesthood, I am going to offer you a suggestion which, if it approves itself to you, will do away with the opportunity, or the possibility, of choice altogether. It will reduce the claimants to one. Before entering, then, upon the inquiry, whither you shall betake yourselves, and what you shall be, bear with me while I give you one piece of advice; it is this:—While you are looking about for a new Communion, have nothing to do with a "Branch Church." You have had enough experience of branch churches already, and you know very well what they are. Depend upon it, such as is one, such is another. They may differ in accidents certainly; but, after all, a branch is a branch, and no branch is a tree. Depend on it, my brethren, it is not worth while leaving one branch for another. While you are doing so great a work, do it thoroughly; do it once for all; change for the better. Rather than go to another branch, remain where you are; do not put yourselves to trouble for nothing; do not sacrifice this world without gaining the next.

[Lecture 6]

Social State of Catholic Countries no Prejudice to the Sanctity of the Church

The world believes in the world's ends as the greatest of goods; it wishes society to be governed simply and entirely for the sake of this world. Provided it could gain one little islet in the ocean, one foot upon the coast, if it could cheapen tea by sixpence a pound, or make its flag respected among the Esquimaux or Otaheitans, at the cost of a hundred lives and a hundred souls, it would think it a very good bargain. What does it know of hell? it disbelieves it; it spits upon, it abominates, it curses its very name and notion. Next, as to the devil, it does not believe in him either. We next come to the flesh, and it is "free to confess" that it does not think there is any great harm in following the instincts of that nature which, perhaps it goes on to say, God has given. How could it be otherwise? who ever heard of the world fighting against the flesh and the devil? Well, then, what is its notion of evil? Evil, says the world, is whatever is an offence to me, whatever obscures my majesty, whatever disturbs my peace. . . .

. . . The Church aims, not at making a show, but at doing a work. She regards this world, and all that is in it, as a mere shadow, as dust and ashes, compared with the value of one single soul. She holds that, unless she can, in her own way, do good to souls, it is no use her doing anything; she holds that it were better for sun and moon to drop from heaven, for the earth to fail, and for all the many millions who are upon it to die of starvation in extremest agony, so far as temporal affliction goes, than that one soul, I will not say, should be lost, but should commit one single venial sin, should tell one wilful untruth, though it harmed no one, or steal one poor farthing without excuse. She considers the action of this world and the action of the soul simply incommensurate, viewed in their respective spheres; she would rather save the soul of one single wild bandit of Calabria, or whining beggar of Palermo, than draw a hundred lines of railroad through the breadth of Italy, or carry out a sanitary reform, in its fullest details, in every city of Sicily, except so far as these great national works tended to some spiritual good beyond them.

Such is the Church, O ye men of the world, and now you know her. Such she is, such she will be; and, though she aims at your good, it is in her own way,—and if you oppose her, she defies you. She has her mission, and do it she will, whether she be in rags, or in fine linen; whether with awkward or with refined carriage; whether by means of uncultivated intellects, or with the grace of accomplishments. Not that, in fact, she is not the source of numberless temporal and moral blessings to you also; the history of ages testifies it; but she makes no promises; she is sent to seek the lost; that is her first object, and she will fulfil it, whatever comes of it.

And now, in saying this, I think I have gone a great way towards suggesting one main solution of the difficulty which I proposed to consider. The question was this:—How is it, that at this time Catholic countries happen to be behind Protestants in civilization? In answer, I do not at all determine how far the fact is so, or what explanation there may be of the appearance of it; but anyhow the fact, granting it exists, is surely no objection to Catholicism, unless Catholicism has professed, or ought to have professed, directly to promote mere civilization; on the other hand, it has a work of its own, and this work is, first, *different* from that of the world; next, *difficult of attainment*, compared with that of the world; and lastly, *secret* from the world in its details and consequences. If, then, Spain or Italy be deficient in secular progress, if the national mind in those countries be but partially formed, if it be unable to develope into civil institutions, if it have no moral instinct of deference to a policeman, if the national finances be in disorder, if the people be excitable, and open to deception from political pretenders, if it know little or nothing of arts, sciences, and literature; I repeat, of course, I do not admit all this, except hypothetically, because it is difficult to draw the line between what is true in it and what is not:—then all I can say is, that it is not wonderful that civil governments, which profess certain objects, should succeed in them better than the Church, which does not. Not till the State is blamed for not making saints, may it fairly be laid to the fault of the Church that she cannot invent a steam-engine or construct a tariff. It is, in truth, merely because she has often done so much more than she professes, it is really in consequence of her very exuberance of benefit to the world, that the world is disappointed that she does not display that exuberance always,—like some hangers-on of the great, who come at length to think they have a claim on their bounty. . . .

Such being the extreme difference between the Church and the world, both as to the measure and the scale of moral good and evil, we may be prepared for those vast differences in matters of detail, which I hardly like to mention, lest they should be out of keeping with the gravity of the subject, as contemplated in its broad principle. For instance, the Church pronounces the momentary wish, if conscious and deliberate, that another should be struck down dead, or suffer any other grievous misfortune, as a blacker sin than a passionate, unpremeditated attempt on the life of the Sovereign. She considers direct unequivocal consent, though as quick as thought, to a single unchaste desire as indefinitely more heinous than any lie which can possibly be fancied, that is, when that lie is viewed, of course, in itself, and apart from its causes, motives, and consequences. Take a mere beggar-woman, lazy, ragged, and filthy, and not over-scrupulous of truth—(I do not say she had arrived at perfection)—but if she is chaste, and sober, and cheerful, and goes to her religious duties (and I am supposing not at all an impossible case), she will, in the eyes of the Church, have a prospect of heaven, which is quite closed and refused to the State's pattern-man, the just, the upright, the generous, the honourable, the conscientious, if he be all this, not from a supernatural power—(I do not determine whether this is likely to be the fact, but I am contrasting views and principles)— not from a supernatural power, but from mere natural virtue. Polished, delicate-minded ladies, with little of temptation around them, and no self-denial to practise, in spite of their refinement and taste, if they be nothing more, are objects of less interest to her, than many a poor outcase who sins, repents, and is with difficulty kept just within the territory of grace. Again, excess in drinking is one of the world's most disgraceful offences; odious it ever is in the eyes of the Church, but if it does not proceed to the loss of reason, she thinks it a far less sin than one deliberate act of detraction, though the matter of it be truth.

[Lecture 8]

LECTURES ON THE PRESENT POSITION OF CATHOLICS

Tradition the Sustaining Power of the Protestant View

. . . It is familiar to an Englishman to wonder at and to pity the recluse and the devotee who surround themselves with a high enclosure, and shut out what is on the other side of it; but was there ever such an instance of self-sufficient, dense, and ridiculous bigotry, as that which rises up and walls in the minds of our fellow-countrymen from all knowledge of one of the most remarkable phenomena which the history of the world has seen. This broad fact of Catholicism—as real as the continent of America, or the Milky Way—which Englishmen cannot deny, they will not entertain; they shut their eyes, they thrust their heads into the sand, and try to get rid of a great vision, a great reality, under the name of Popery. They drop a thousand years from the world's chronicle, and having steeped them thoroughly in sin and idolatry would fain drown them in oblivion. Whether for philosophic remark or for historical research, they will not recognise what infidels recognise as well as Catholics—the vastness, the grandeur, the splendour, the loveliness of the manifestations of this time-honoured ecclesiastical confederation. Catholicism is for fifteen hundred years as much a fact, and as great a one (to put it on the lowest ground) as is the imperial sway of Great Britain for a hundred; how can it then be actually imbecile or extravagant to believe in it and to join it, even granting it were an error? But this island, as far as religion is concerned, really must be called, one large convent, or rather workhouse; the old pictures hang on the walls; the world-wide Church is chalked up on every side as a wivern or a griffin; no pure gleam of light finds its way in from without; the thick atmosphere refracts and distorts such straggling rays as gain admittance. Why, it is not even a *camera obscura*; cut off from Christendom though it be, at least it might have a true picture of that Christendom cast in miniature upon its floor; but in this inquisitive age, when the Alps are crested, and seas fathomed, and mines ransacked, and sands sifted, and rocks cracked into specimens, and beasts caught and catalogued, as little is known by Englishmen of the religious sentiments, the religious usages, the religious motives, the religious ideas of two hundred millions of Christians poured to and fro, among them and around them, as if, I will not say, they were Tartars or Patagonians, but as if they inhabited the moon. Verily, were the Catholic Church in the moon, England would gaze on her with more patience, and delineate her with more accuracy, than England does now. . . .

Now, if I must give the main and proximate cause of this remarkable state of mind, I must simply say that Englishmen go by that very mode of information in its worst shape, which they are so fond of imputing against Catholics; they go by *tradition*, immemorial, unauthenticated *tradition*. I have no wish to make a rhetorical point, or to dress up a polemical argument. . . . I say, then, Englishmen entertain their present monstrous notions of us, mainly because those notions are received on information not authenticated, but immemorial. This it is that makes them entertain those notions; they talk much of free inquiry; but towards us they do not dream of practising it. . . .

Protestantism is also the Tradition of the Anglican Clergy . . . its especial duty as a religious body, is not to inculcate any particular theological system, but to watch over the anti-Catholic Tradition, to preserve it from rust and decay, to keep it bright and keen, and ready for action on any emergency or peril. It is the way with human nature to start with vigour, and then to flag; years tell upon the toughest frames; time introduces changes; prejudices are worn away; asperities are softened; views opened; errors are corrected; opponents are better understood; the mind wearies of warfare. The Protestant Tradition, left to itself, would in the course of time languish and decline; laws would become obsolete, the etiquette and usages of society would alter, literature would be enlivened with new views, and the old Truth might return with the freshness of novelty. It is almost the mission of the established clergy, by word and writing, to guard against this tendency of the public mind. In this specially consists its teaching; I repeat, not in the shreds of Catholic doctrine which it professes, not in proofs of the divinity of any creed whatever, not in separating opinion from faith, not in instructing in the details of morals, but mainly in furbishing up the old-fashioned weapons of centuries back; in cataloguing and classing the texts which are to batter us, and the objections which are to explode among us, and the insinuations and the slanders which are to mow us down. The Establishment is the keeper in ordinary of those national types and blocks from which Popery is ever to be printed off,—of the traditional view of every Catholic doctrine, the traditional account of every ecclesiastical event, the traditional lives of popes and bishops, abbots and monks, saints and confessors,—the traditional fictions, sophisms, calumnies, mockeries, sarcasms, and invectives with which Catholics are to be assailed.

This, I say, is the special charge laid upon the Establishment. Unitarians, Sabellians, Utilitarians, Wesleyans, Calvinists, Swedenborgians, Irvingites, Freethinkers, all these it can tolerate in its very bosom; no form of opinion comes amiss; but Rome it cannot abide. It agrees to differ with its own children on a thousand points, one is sacred—that her Majesty the Queen is "The Mother and Mistress of all Churches;" on one dogma it is infallible, on one it may securely insist without fear of being unseasonable or excessive—that "the Bishop of Rome hath no jurisdiction in this realm." Here is sunshine amid the darkness, sense amid confusion, an intelligible strain amid a Babel of sounds; whatever befalls, here is sure footing; it is, "No peace with Rome," "Down with the Pope," and "The Church in danger." Never has the Establishment failed in the use of these important and effective watchwords; many are its shortcomings, but it is without reproach in the execution of this its special charge. Heresy, and scepticism, and infidelity, and fanaticism, may challenge it in vain; but fling upon the gale the faintest whisper of Catholicism, and it recognises by instinct the presence of its connatural foe. . . .

[Lecture 2]

Fable the Basis of the Protestant View

. . . Coaches, omnibuses, carriages, and cars, day after day drive up and down the Hagley Road; passengers lounge to and fro on the foot-path; and close alongside of it are discovered one day the nascent foundations and rudiments of a considerable building. On inquiring, it is found to be intended for a Catholic, nay, even for a monastic establishment. This leads to a good deal of talk, especially when the bricks begin to show above the surface. Meantime the unsuspecting architect is taking his measurements, and ascertains that the ground is far from lying level; and then, since there is a prejudice among Catholics in favour of horizontal floors, he comes to the conclusion that the bricks of the basement must rise above the surface higher at one end of the building than at the other; in fact, that whether he will or no, there must be some construction of the nature of a vault or cellar at the extremity in question, a circumstance not at all inconvenient, considering it also happens to be the kitchen end of the building. Accordingly, he turns his necessity into a gain, and by the excavation of a few feet of earth, he forms a number of chambers convenient for various purposes, partly beneath, partly above the line of the ground. While he is thus intent on his work, loungers, gossipers, alarmists are busy at theirs too. They go round the building, they peep into the underground brickwork, and are curious about the drains; they moralise about Popery and its spread; at length they trespass upon the enclosure, they dive into the half-finished shell, and they take their fill of seeing what is to be seen, and imagining what is not. Every house is built on an idea; you do not build a mansion like a public office, or a palace like a prison, or a factory like a shooting box, or a church like a barn. Religious houses, in like manner, have their own idea; they have certain indispensable peculiarities of form and internal arrangement. Doubtless, there was much in the very idea of an Oratory perplexing to the Protestant intellect, and inconsistent with Protestant notions of comfort and utility. Why should so large a room be here? why so small a room there? why a passage so long and wide? and why so long a wall without a window? the very size of the house needs explanation. Judgments which had employed themselves on the high subject of a Catholic hierarchy and its need, found no difficulty in dogmatising on bedrooms and closets. There was much to suggest matter of suspicion, and to predispose the trespasser to doubt whether he had yet got to the bottom of the subject. At length one question flashed upon his mind: what can such a house have to do with cellars? cellars and monks, what can be their mutual relation? monks—to what possible use can they put pits and holes, and corners, and outhouses, and sheds? A sensation was created; it brought other visitors; it spread; it became an impression, a belief; the truth lay bare; a tradition was born; a fact was elicited which henceforth had many witnesses. *Those cellars were cells.* How obvious when once stated! and every one who entered the building, every one who passed by, became, I say, in some sort, ocular vouchers for what had often been read of in books, but for many generations had happily been unknown to England, for the incarcerations, the torturings, the starvings, the immurings, the murderings proper to a monastic establishment.

Now I am tempted to stop for a while in order to *improve* (as the evangelical pulpits call it) this most memorable discovery. . . .

. . . The nascent fable has indeed failed, as the tale about the Belgian sin-table has failed, but it might have thriven: it has been lost by bad nursing; it ought to have been cherished awhile in those underground receptacles where first it drew breath, till it could comfortably bear the light; till its limbs were grown, and its voice was strong, and we on whom it bore had run our course, and gone to our account; and then it might have raised its head without fear and without reproach, and might have magisterially asserted what there was none to deny. But men are all the creatures of circumstances; they are hurried on to a ruin which they may see, but cannot evade: so has it been with the Edgbaston Tradition. It was spoken on the house-tops when it should have been whispered in closets, and it expired in the effort. Yet it might have been allotted, let us never forget, a happier destiny. It might have smouldered and spread through a portion of our Birmingham population; it might have rested obscurely on their memories, and now and then risen upon their tongues; there might have been flitting notions, misgivings, rumours, voices, that the horrors of the Inquisition were from time to time renewed in our subterranean chambers; and fifty years hence, if some sudden frenzy of the hour roused the Anti-Catholic jealousy still lingering in the town, a mob might have swarmed about our innocent dwelling, to rescue certain legs of mutton and pats of butter from imprisonment, and to hold an inquest over a dozen packing-cases, some old hampers, a knife-board, and a range of empty blacking bottles.

[Lecture 3]

Logical Inconsistency of the Protestant View

Indeed, if the truth must be told, so one-sided is this Protestantism, that its supporters have not yet admitted the notion into their minds, that the Catholic Church has as much right to make converts in England, as any other denomination. It is a new idea to them; they had thought she ought to be content with vegetating, as a sickly plant, in some back-yard or garret window; but to attempt to spread her faith abroad—this is the real insidiousness, and the veritable insult. . . .

. . . At present, it is a matter of surprise to them that we dare to speak a word in our defence, and that we are not content with the liberty of breathing, eating, moving about, and dying in a Protestant soil. That we should have an opinion, that we should take a line of our own, that we should dare to convince people, that we should move on the offensive, is intolerable presumption, and takes away their breath. They think themselves martyrs of patience if they can keep quiet in our presence, and condescending in the heroic degree, if they offer us lofty civility. So was it the other day, when the late agitation began; the hangers-on of Government said to us, "Cling tight to our coat-tails; we are your best friends; we shall let you off easy; we shall only spit upon you; but beware of those rabid Conservatives;" and they marvelled that we did not feel it to be the highest preferment for the Catholic Church to wait in the ante-chambers of a political party. So it is with your Protestant controversialist, even when he shows to best advantage; his great principle of disputation is that he is up, and the Catholic is down; and his great duty is to show it. He is intensely conscious that he is in a very eligible situation, and his opponent in the gutter; and he lectures down upon him, as if out of a drawing-room window. It is against his nature to be courteous to those for whom he feels so cordial a disdain, and he cannot forgive himself for stooping to annihilate them. He mistakes sharpness for keenness, and haughtiness for strength; and never shows so high and mighty in manner as when he means to be unutterably conclusive. It is a standing rule with him to accuse his opponent of evasion and misstatement; and, when in fault of an argument, he always can impugn his motives, or question the honesty of his professions.

[Lecture 5]

Prejudice the Life of the Protestant View

And, as the huge giant, had he first been hit, not in the brain, but in the foot or the shoulder, would have yelled, not with pain, but with fury at the insult, and would not have been frightened at all or put upon the defensive, so our Prejudiced Man is but enraged so much the more, and almost put beside himself, by the presumption of those who, with their doubts or their objections, interfere with the great Protestant Tradition about the Catholic Church. To bring proof against us is, he thinks, but a matter of time; and we know in affairs of everyday, how annoyed and impatient we are likely to become, when obstacles are put in our way in any such case. We are angered at delays when they are but accidental, and the issue is certain; we are not angered, but we are sobered, we become careful and attentive to impediments, when there is a doubt about the issue. The very same difficulties put us on our mettle in the one case, and do but irritate us in the other. If, for instance, a person cannot open a door, or get a key into a lock, which he has done a hundred times before, you know how apt he is to shake, and to rattle, and to force it, as if some great insult were offered him by its resistance: you know how surprised a wasp, or other large insect is, that he cannot get through a window-pane; such is the feeling of the Prejudiced Man, when we urge our objections—not softened by them at all, but exasperated the more; for what is the use of even incontrovertible arguments against a conclusion which he already considers to be infallible? . . .

One word here as to the growth of Catholicism, of conversions and converts;—the Prejudiced Man has his own view of it all. First, he denies that there are any conversions or converts at all. This is a bold game, and will not succeed in England, though I have been told that in Ireland it has been strenuously maintained. However, let him grant the fact, that converts there are, and he has a second ground to fall back upon: the converts are weak and foolish persons,—notoriously so; all their friends think so; there is not a man of any strength of character or force of intellect among them. They have either been dreaming over their folios, or have been caught with the tinsel embellishments of Popish worship. They are lack-a-daisical women, or conceited young parsons, or silly squires, or the very dregs of our large towns, who have nothing to lose, and no means of knowing one thing from another. Thirdly, in corroboration:—they went over, he says, on such exceedingly wrong motives; not any one of them but you may trace his conversion to something distinctly wrong; it was love of notoriety, it was restlessness, it was resentment, it was lightness of mind, it was self-will. There was trickery in his mode of taking the step or inconsiderateness towards the feelings of others. They went too soon, or they ought to have gone sooner. They ought to have told every one their doubts as soon as ever they felt them, and before they knew whether or not they should overcome them or no: if they had clerical charges in the Protestant Church, they ought to have flung them up at once, even at the risk of afterwards finding they had made a commotion for nothing. Or, on the other hand, what, forsooth, must these men do when a doubt came on their mind, but at once abandon all their clerical duty and go to Rome, as if it were possible anywhere to be absolutely certain? In short, they did not become Catholics at the right moment; so that, however numerous they may be, no weight whatever attaches to their conversion. As for him, it does not affect him at all; he means to die just where he is; indeed these conversions are a positive argument in favour of Protestantism; he thinks still worse of Popery, in consequence of these men going over, than he did before. His fourth remark is of this sort: they are sure to come back. He prophesies that by this time next year, not one of them will be a Catholic. His fifth is as bold as the first;—they *have* come back. This argument, however, of the Prejudiced Man admits at times of being shown to great advantage, should it so happen that the subjects of his remarks have, for some reason or other, gone abroad, for then there is nothing to restrain his imagination. Hence, directly a new Catholic is safely lodged two or three thousand miles away, out comes the confident news that he has returned to Protestantism; when no friend has the means to refute it. When this argument fails, as fail it must, by the time a letter can be answered, our Prejudiced Man falls back on his sixth common-place, which is to the effect that the converts are very unhappy. He knows this on the first authority; he has seen letters declaring or showing it. They are quite altered men, very much disappointed with Catholicism, restless, and desirous to come back except from false shame. Seventhly, they are altogether deteriorated in character; they have become harsh, or overbearing, or conceited, or vulgar. They speak with extreme bitterness against Protestantism, have cast off their late friends, or seem to forget that they ever were Protestants themselves. Eighthly, they have become infidels;—alas! heedless of false witness, the Prejudiced Man spreads the news about, right and left, in a tone of great concern and distress; he considers it very awful.

Lastly, when every resource has failed, and in spite of all that can be said, and surmised, and expressed, and hoped, about the persons in question, Catholics they have become, and Catholics they remain, the Prejudiced Man has a last resource, he simply forgets that Protestants they ever were. They cease to have antecedents; they cease to have any character, any history to which they may appeal: they merge in the great fog, in which to his eyes everything Catholic is enveloped: they are dwellers in the land of romance and fable; and, if he dimly contemplates them plunging and floundering amid the gloom, it is as griffins, wiverns, salamanders, the spawn of Popery, such as are said to sport in the depths of the sea, or to range amid the central sands of Africa. He forgets he ever heard of them; he has no duties to their names, he is released from all anxiety about them; they die to him. . . .

. . . Our Prejudiced Man of course sees Catholics and Jesuits in everything, in every failure of the potato crop, every strike of the operatives, and every mercantile stoppage. His one idea of the Catholic Church haunts him incessantly, and he sees whole Popery, living and embodied, in every one of its professors, nay, in every word, gesture and motion of each. A Catholic Priest cannot be grave or gay, silent or talkative, without giving matter of offence or suspicion. There is peril in his frown, there is greater peril in his smile. His half sentences are filled up; his isolated acts are misdirected; nay, whether he eats or sleeps, in every mouthful and every nod he ever has in view one and one only object, the aggrandizement of the unwearied, relentless foe of freedom and of progress, the Catholic Church. The Prejudiced Man applauds himself for his sagacity, in seeing evidences of a plot at every turn; he groans to think that so many sensible men should doubt its extension all through Europe, though he begins to entertain the hope that the fact is breaking on the apprehension of the Government.

The Prejudiced Man travels, and then everything he sees in Catholic countries only serves to make him more thankful that his notions are so true; and the more he sees of Popery, the more abominable he feels it to be. If there is any sin, any evil in a foreign population, though it be found among Protestants also, still Popery is clearly the cause of it. If great cities are the schools of vice, it is owing to Popery. If Sunday is profaned, if there is a Carnival, it is the fault of the Catholic Church. Then, there are no private homes, as in England, families live on staircases; see what it is to belong to a Popish country. Why do the Roman labourers wheel their barrows so slow on the Forum? why do the Lazzaroni of Naples lie so listlessly on the beach? why, but because they are under the *malaria* of a false religion. Rage, as is well-known, is in the Roman like a falling sickness, almost as if his will had no part in it, and he had no responsibility; see what it is to be a Papist. Bloodletting is as frequent and as much a matter of course in the South, as hair-cutting in England; it is a trick borrowed from the convents, when they wish to tame down refractory spirits.

The Prejudiced man gets up at an English hour, has his breakfast at his leisure, and then saunters into some of the churches of the place; he is scandalized to have proof of what he has so often heard, the infrequency of communions among Catholics. Again and again, in the course of his tour, has he entered them, and never by any chance did he see a solitary communicant:—hundreds, perhaps, having communicated in those very churches, according to their custom, before he was out of his bedroom. But what scandalizes him most, is that even bishops and priests, nay, the Pope himself, does not communicate at the great festivals of the Church. He was at a great ceremonial, a High Mass, on Lady Day, at the Minerva; not one Cardinal communicated; Pope and Cardinals, and every Priest present but the celebrant, having communicated, of course, each in his own Mass, and in his own chapel or church early in the morning. Then the churches are so dirty; faded splendour, tawdriness, squalidness are the fashion of the day;—thanks to the Protestants and Infidels, who, in almost every country where Catholicism is found, have stolen the revenues by which they were kept decent.

He walks about and looks at the monuments, what is this? the figure of a woman: who can it be? His Protestant cicerone at his elbow, who perhaps has been chosen by his good father or guardian to protect him on his travels from a Catholic taint, whispers that it is Pope Joan, and he notes it down in his pocket-book accordingly. . . .

. . . He carries England with him abroad; and though he has ascended mountains and traversed cities, knows scarcely more of Europe than when he set out.

But perhaps he does not leave England at all; he never has been abroad; it is all the same; he can scrape together quite as good evidence against Catholicism at home. One day he pays a visit to some Catholic chapel, or he casually finds the door open, and walks in. He enters and gazes about him, with a mixed feeling of wonder, expectation and disgust; and according to circumstances, this or that feeling predominates, and shows itself in his bearing and his countenance. In one man it is curiosity; in another, scorn; in another, conscious superiority; in another, abhorrence; over all of their faces, however, there is a sort of uncomfortable feeling, as if they were in the cave of Trophonius or in a Mesmerist's lecture-room. One and all seem to believe that something strange and dreadful may happen any moment; and they crowd up together, if some great ceremony is going on, tiptoeing and staring, and making strange faces, like the gargoyles or screen ornaments of the church itself. Every sound of the bell, every movement of the candles, every change in the grouping of the sacred ministers and the assistants, puts their hands and limbs in motion, to see what is coming next. . . .

[Lecture 6]

Assumed Principles the Intellectual Ground of the Protestant View

. . . Why may not my First Principles contest the prize with yours? they have been longer in the world; they have lasted longer, they have done harder work, they have seen rougher service. You sit in your easy-chairs, you dogmatize in your lecture-rooms, you wield your pens: it all looks well on paper: you write exceedingly well: there never was an age in which there was better writing; logical, nervous, eloquent, and pure,— go and carry it all out in the world. Take your First Principles, of which you are so proud, into the crowded streets of our cities, into the formidable classes which make up the bulk of our population; try to work society by them. You think you can; I say you cannot—at least you have not as yet; it is yet to be seen if you can. "Let not him that putteth on his armour boast as he who taketh it off." Do not take it for granted that that is certain which is waiting the test of reason and experiment. Be modest until you are victorious. My principles, which I believe to be

eternal, have at least lasted eighteen hundred years; let yours live as
many months. That man can sin, that he has duties, that the Divine
Being hears prayer, that He gives His favours through visible
ordinances, that He is really present in the midst of them, these
principles have been the life of nations; they have shown they could be
carried out; let any single nation carry out yours, and you will have
better claim to speak contemptuously of Catholic rites, of Catholic
devotions, of Catholic belief.

What is all this but the very state of mind which we ridicule, and call
narrowness, in the case of those who have never travelled? We call them,
and rightly, men of contracted ideas, who cannot fancy things going on
differently from what they have themselves witnessed at home, and
laugh at everything because it is strange. They themselves are the
pattern men; their height, their dress, their manners, their food, their
language, are all founded in the nature of things; and everything else is
good or bad, just in that very degree in which it partakes, or does not
partake, of them. All men ought to get up at half-past eight, breakfast
between nine and ten, read the newspapers, lunch, take a ride or drive,
dine. Here is the great principle of the day—dine; no one is a man who
does not dine; yes, dine, and at the right hour; and it must *be* a dinner,
with a certain time after dinner, and then, in due time, to bed. Tea and
toast, port wine, roast beef, mince-pies at Christmas, lamb at Easter,
goose at Michaelmas, these are their great principles. They suspect any
one who does otherwise. Figs and maccaroni for the day's fare, or
Burgundy and grapes for breakfast!—they are aghast at the atrocity of
the notion. And hence you read of some good country gentleman, who,
on undertaking a Continental tour, was warned of the privations and
mortifications that lay before him from the difference between foreign
habits and his own, stretching his imagination to a point of enlargement
answerable to the occasion, and making reply that he knew it, that he
had dwelt upon the idea, that he had made up his mind to it, and
thought himself prepared for anything abroad, provided he could but
bargain for a clean table-cloth and a good beef-steak every day.

[Lecture 7]

Ignorance Concerning Catholics the Protection of the Protestant View

They like to think as they please; and as they would by no means
welcome St. Paul, did he come from heaven to instruct them in the
actual meaning of his "texts" in Romans iii. or Galatians ii., so they
would think it a hardship to be told that they must not go on
maintaining and proving, that we were really what their eyes then
would testify we were not. And then, too, dear scandal and romancing

put in their claim; how would the world go on, and whence would come its staple food and its cheap luxuries, if Catholicism were taken from the market? Why it would be like the cotton crop failing, or a new tax put upon tea. And then, too, comes prejudice . . . how is prejudice to exist without Catholic iniquities and enormities? prejudice, which could not fast for a day, which would be in torment inexpressible, and call it Popish persecution, to be kept on this sort of meagre for a Lent, and would shake down Queen and Parliament with the violence of its convulsions, rather than it should never suck a Catholic's sweet bones and drink his blood any more. . . .

. . . And another consideration weighs with such Protestants as are in a responsible situation in their own communion, or are its ministers and functionaries. These persons feel that while they hold office in a body which is at war with Catholics, they are as little at liberty to hold friendly intercourse with them, even with the open avowal of their differing from them in serious matters, as an English officer or a member of Parliament may lawfully correspond with the French Government during a time of hostilities. These various motives, and others besides, better and worse, are, I repeat, almost an insuperable barrier in the way of any real and familiar intercourse between Protestants and ourselves: and they act, in consequence, as the means of perpetuating what may be considered the chief negative cause, and the simplest explanation of the absurdities so commonly entertained about us by all classes of society. Personal intercourse, then, being practically just as much out of the question with us, as with the Apostles themselves or the Jewish prophets, Protestantism has nothing left for it, when it would argue about us, but to have recourse, as in the case of Scripture, to its "texts," its chips, shavings, brickbats, potsherds, and other odds and ends of the Heavenly City, which form the authenticated and ticketed specimens of what the Catholic Religion is in its great national Museum.

[Lecture 8]

Duties of Catholics Towards the Protestant View

Protestantism is established in the widest sense of the word; its doctrine, religious, political, ecclesiastical, moral, is placed in exclusive possession of all the high places of the land. It is forced upon all persons in station and office, or almost all, under sanction of an oath; it is endowed with the amplest estates, and with revenues supplied by Government and by chartered and other bodies. It has innumerable fine churches, planted up and down in every town, and village, and hamlet in the land. In consequence, everyone speaks Protestantism, even those who do not in their hearts love it; it is the current coin of the realm. As English is the natural tongue, so Protestantism is the intellectual and moral language of the body politic. The Queen *ex officio* speaks Protestantism; so does the court, so do her ministers. All but a small portion of the two Houses of Parliament; and those who do not are forced to apologize for not speaking it, and to speak as much of it as they conscientiously can. The Law speaks Protestantism, and the Lawyers; and the State Bishops and clergy of course. All the great authors of the nation, the multitudinous literature of the day, the public press, speak Protestantism. Protestantism the Universities; Protestantism the schools, high, and low, and middle. Thus there is an incessant, unwearied circulation of Protestantism all over the whole country, for 365 days in the year from morning till night; and this, for nearly three centuries, has been almost one of the functions of national life. As the pulse, the lungs, the absorbents, the nerves, the pores of the animal body, are ever at their work, as that motion is its life, so in the political structure of the country there is an action of the life of Protestantism, constant and regular. It is a vocal life; and in this consists its perpetuation, its reproduction. . . .

. . . You see, the Protestant Tradition had it all its own way; Elizabeth, and her great men, and her preachers, killed and drove away all the Catholics they could; knocked down the remainder, and then at their leisure proved unanswerably and triumphantly the absurdity of Popery, and the heavenly beauty and perfection of Protestantism. Never did we undergo so utter and complete a refutation; we had not one word to utter in our defence. When she had thus beaten the breath out of us, and made us simply ridiculous, she put us on our feet against, thrust us into a chair, hoisted us up aloft, and carried us about as a sort of Guy Faux, to show to all the boys and riff-raff of the towns what a Papist was like. Then, as if this were not enough, lest anyone should come and ask us anything about our religion, she and her preachers put it about that we had the plague, so that, for fear of a moral infection, scarce a soul had the courage to look at us, or breathe the same air with us.

This was a fair beginning for the Protestantizing of the people, and everything else that was needed followed in due time, as a matter of course. Protestantism being taught everywhere, Protestant principles were taught with it, which are necessarily the very reverse of Catholic principles. The consequence was plain—viz., that even before the people heard a Catholic open his mouth, they were forearmed against what he would say, for they had been taught this or that as if a precious truth, belief in which was *ipso facto* the disbelief and condemnation of some Catholic doctrine or other. When a person goes to a fever ward, he takes some essence with him to prevent his catching the disorder; and of this kind are the anti-Catholic principles in which Protestants are instructed from the cradle. For instance, they are taught to get by heart without any sort of proof, as a kind of alphabet or spelling lesson, such propositions as these:—"miracles have ceased long ago;" "all truth is in the Bible;" "any one can understand the Bible;" "all penance is absurd;" "a priesthood is pagan, not Christian," and a multitude of others. These are universally taught and accepted, as if equally true and equally important, just as are the principles "it is wrong to murder or thieve," or "there is a judgment to come." When then a person sets out in life with these maxims as a sort of stock in trade in all religious speculations, and encounters Catholics, whose opinions hitherto he had known nothing at all about, you see he has been made quite proof against them, and unsusceptible of their doctrines, their worship, and their reasoning, by the preparation to which he has been subjected. He feels an instinctive repugnance to everything Catholic, by reason of these arbitrary principles, which he has been taught to hold, and which he thinks identical with reason. "What? you have priests in your religion," he says; "but do you not know, are you so behind the world as not to know, that priests are pagan, not Christian?" And sometimes he thinks that, directly he has uttered some such great maxim, the Catholic will turn Protestant at once, or, at least, ought to do so, and if he does not, is either dull or hypocritical. And so again, "You hold saints are to be invoked, but the practice is not in the Bible, and nothing is true that is not there." And again, "They say that in Ireland and elsewhere the priests impose heavy penances; but this is against common sense, for all penances are absurd." Thus the Protestant takes the whole question for granted on starting;—and this was the subject of my seventh Lecture.

This fault of mind I called Assumption or Theorizing; and another quite as great, and far more odious, is Prejudice; and this came into discussion in the sixth Lecture. The perpetual talk against Catholicism, which goes on everywhere, in the higher classes, in literary circles, in the public press, and in the Protestant Church and its various dependencies, makes an impression, or fixes a stain, which it is continually deepening, on the minds which are exposed to its influence; and thus, quite independent of any distinct reasons and facts for thinking so, the multitude of men are quite certain that something very horrible is going on among Catholics. They are convinced that we are all but fiends, so that there is no doubt at all, even before going into the matter, that all that is said against us is true, and all that is said for us is false.

These, then, are the two special daughters, as they may be called, of the Protestant Tradition, Theory or Assumption on the one hand, and Prejudice on the other,—Theory which scorns us, and Prejudice which hates us; yet, though coming of one stock, they are very different in their constitution, for Theory is of so ethereal a nature, that it needs nothing to feed upon; it lives on its own thoughts, and in a world of its own, whereas Prejudice is ever craving for food, victuals are in constant request for its consumption every day; and accordingly they are served up in unceasing succession, Titus Oates, Maria Monk, and Jeffreys, being the purveyors, and platform and pulpit speakers being the cooks. And this formed the subject of the third, fourth, and fifth Lectures.

Such, then, is Popular Protestantism, considered in its opposition to Catholics. Its truth is Establishment by law; its philosophy is Theory; its faith is Prejudice; its facts are Fictions; its reasonings Fallacies; and its security is Ignorance about those whom it is opposing. The Law says that white is black; Ignorance says, why not? Theory says it ought to be, Fallacy says it must be, Fiction says it is, and Prejudice says it shall be.

And now, what are our duties at this moment towards this enemy of ours? How are we to bear ourselves towards it? what are we to do with it? what is to come of the survey we have taken of it? with what practical remark and seasonable advice am I to conclude this attempt to determine our relation with it? The lesson we gain is obvious and simple, but as difficult, you will say, as it is simple; for the means and the end are almost identical, and in executing the one we have already reached the other. Protestantism is fierce, because it does not know you; ignorance is its strength; error is its life. Therefore bring yourselves before it, press yourselves upon it, force yourselves into notice against its will. Oblige men to know you; persuade them, importune them, shame them into knowing you. Make it so clear what you are, that they cannot affect not to see you, nor refuse to justify you. Do not even let them off with silence, but give them no escape from confessing that you are not what they have thought you were. They will look down, they will look aside, they will look in the air, they will shut their eyes, they will keep them shut. They will do all in their power not to see you; the nearer you come, they will close their eyelids all the tighter; they will be very angry and frightened, and give the alarm as if you were going to murder them.

[Lecture 9]

SELECT BIBLIOGRAPHY

NEWMAN'S WRITINGS

Newman collected most of his works in a uniform edition (36 vols.) between 1868 and 1881, which was published by Longmans, Green and Co. of London from 1886 until the destruction of the stock in World War II. Most of the volumes of this edition have been photographically reprinted in recent years by Christian Classics, Westminster, Maryland. Apart from the uniform edition, a dozen or so further volumes of writings have since been published, nearly all posthumously. See the list given in the bibliography to Charles Stephen Dessain's *John Henry Newman* (1966; 3rd edn. 1980). The Catholic correspondence has been edited by Charles Stephen Dessain *et al.* in *The Letters and Diaries of John Henry Newman*, vols. xi–xxii (1961–72), xxiii–xxxi (1973–7); for the Anglican period, vols. i–vi (1978–84) have so far been edited by Ian Ker, Thomas Gornall, SJ, and Gerard Tracy.

Editions

There are Oxford critical editions of the *Apologia pro Vita Sua*, ed. Martin J. Svaglic (1967); *An Essay in Aid of A Grammar of Assent*, ed. I. T. Ker (1985); *The Idea of a University*, ed. I. T. Ker (1976). There are various reprints and paperback editions of the *Apologia*, *Callista*, *Discourses to Mixed Congregations*, *Essay on the Development of Christian Doctrine*, *Grammar of Assent*, *Loss and Gain*, *A Letter to the Duke of Norfolk*, *Meditations and Devotions*, *Oxford University Sermons*, and *Parochial and Plain Sermons*.

Bibliographies

There are excellent comprehensive bibliographical surveys of secondary materials by Martin J. Svaglic and Charles Stephen Dessain in David J. DeLaura (ed.), *Victorian Prose: A Guide to Research* (1973), as well as the select annotated bibliography in Dessain's *John Henry Newman*. See also John R. Griffin, *Newman: A Bibliography of Secondary Studies* (1980). For a bibliography of Newman's writings, see Vincent Ferrer Blehl, SJ, *John Henry Newman: A Bibliographical Catalogue of his Writings* (1978).

SECONDARY WORKS

Only general, introductory, and standard secondary works relevant to the selections in this volume are mentioned below; for more specialized studies, see the bibliographies cited.

Biographical

The best short introduction to Newman's life is Meriol Trevor, *Newman's Journey* (1974), an abridged version of her two-volume biography, *Newman: The Pillar of the Cloud* and *Newman: Light in Winter* (1962). Dessain's excellent short life *John Henry Newman* is the best introduction to his thought and writings. Ian Ker's *John Henry Newman: A Biography* (1988) is a full-length life as well as an intellectual and literary biography.

Educational

The introduction to Ker's edition of *The Idea of a University* contains a critique of Newman's educational thought which it sets in its historical context. The first part of A. Dwight Culler's *The Imperial Intellect: A Study of Newman's Educational Ideal* (1955) is an illuminating educational biography of Newman as student and teacher, but the last part, which is a discussion of his educational ideas, is much less reliable.

Literary

A useful survey is C. F. Harrold, *John Henry Newman: An Expository and Critical Study of His Mind, Thought and Art* (1945). Individual works are discussed in Ker's *John Henry Newman: A Biography*. Geoffrey Tillotson's 'Newman the Writer', in Geoffrey and Kathleen Tillotson, *Mid-Victorian Studies* (1965) (a revised version of the introduction to his Reynard Library selection *Newman: Prose and Poetry* (1957)) is perhaps the best single example of general literary criticism. For a discussion of Newman's rhetoric, see John Holloway, *The Victorian Sage: Studies in Argument* (1953), and also Ker's *John Henry Newman: A Biography*. For the *Apologia* the introduction to Svaglic's edition provides the essential background information; for criticism see Walter E. Houghton, *The Art of Newman's* Apologia (1945), and the essays in David J. DeLaura (ed.), *Apologia pro Vita Sua* (1968). Ker's *John Henry Newman: A Biography*, which is closely based on the letters, affords access to the riches of the vast correspondence; Joyce Sugg's *A Packet of Letters* (1983) is a useful selection with a good introduction. The satirical writings, which have been largely neglected by critics, are highlighted in Ker's *John Henry Newman: A Biography*.

Philosophical

Newman's philosophy of religion may be approached through the introduction to Ker's edition of the *Grammar of Assent*, which provides a concise account of the development of Newman's thought on the justification of religious belief, particularly in the *Oxford University Sermons*, as well as an elucidation of the central ideas of the *Grammar*. The first volume of Edward Sillem's edition of *The Philosophical Notebook of John Henry Newman* (1969–70) contains the fullest and most detailed study of Newman's philosophy, including a survey of the background influences. Thomas Vargish's *Newman: The Contemplation of Mind* (1970) is a readable introduction to Newman's general epistemology.

Preaching

The Scriptural and Patristic theology of *Parochial and Plain Sermons* is elucidated in Dessain's *John Henry Newman*, ch. 2. For their spirituality, see Hilda Graef, *God and Myself: The Spirituality of John Henry Newman* (1967), and Ker's *John Henry Newman: A Biography*, ch. 2. For a general introduction to Newman as a preacher, see W. D. White (ed.), *The Preaching of John Henry Newman* (1969).

Theology

The best concise introduction to Newman's theology is Dessain's *John Henry Newman*. Jan H. Walgrave, *Newman the Theologian: The Nature of Belief and Doctrine as Exemplified in his Life and Works* (1960) is the nearest thing to a systematic study. Ker's *John Henry Newman: A Biography* provides a comprehensive survey of the individual theological writings, complemented by material from the letters, including a full study of Newman's Catholic ecclesiology.